MORE THAN JUST NAMES

Remembering Townsville's WW1 Fallen

ISBN: 978-1-7638240-1-0 eBook
ISBN: 978-1-7638240-0-3 Paperback

Cover Design and Layout by Author Services Australia.

"We bought your freedom with our lives. So take this freedom. Guard it as we have guarded it, use it as we can no longer use it, and with it as a foundation, build. Build a world in which meanness and poverty, tyranny and hate, have no existence."

John Gorton: Kerang RSL 1946

CONTENTS

PART 4 – 1918 357

For the 172 men from Townsville and North Queensland who paid the ultimate price in the Great War

Lest We Forget

AUTHORS NOTE

World War I, also known as the Great War, reshaped the world in ways that still resonate today. From 1914 to 1918, the world witnessed an unprecedented scale of human sacrifice, the collapse of old empires, and the birth of new political and social orders. Yet amid the sweeping global changes, individual lives—each unique, precious, and deeply mourned—were caught up in the relentless machinery of conflict.

This book is a tribute to 172 men from Townsville and the greater North Queensland region—men who left their homes and families to serve in distant lands and never returned. In Townsville, the cenotaph stands as a silent guardian of their legacy. Each name etched in stone is a reminder that history is more than grand events chronicled in textbooks; it is also woven from the personal stories of young men with dreams and futures, who answered the call of duty in a time of war. These names represent sons, brothers, fathers, and friends who were once part of the community before joining the Australian Imperial Force (AIF) and embarking for battlefields in Gallipoli, Europe, and the Middle East.

For many who pass by the cenotaph, these names may seem like just that—names. Yet behind each one lies a story of courage, duty, and loss. The purpose of this book is to bring these names to life, to restore the individuality that time and history have gradually faded. Each short biography within these pages provides a glimpse into the lives of these 172 men: where they were born, the lives they led before enlisting, and the battles where they fought and fell. It is a humble attempt to honour their memory, ensuring their sacrifice is neither forgotten nor reduced to mere dates and places in history books.

The book is organized by year, reflecting the order in which these sons of Townsville and the wider North Queensland region gave their lives in service to their country. Through these pages, it is hoped that readers will gain a deeper understanding of the true human cost of war. The stories of these men are as varied as the lives they left behind, yet they are united by a shared thread of bravery, sacrifice, and commitment to something greater than themselves.

As you read these biographies, remember that each name represents a family forever changed, a future forever altered, and a community forever touched by their absence. May this book stand as a reminder that the freedoms and peace we enjoy today were dearly won, and that these names etched in stone represent lives lost too soon.

INTRODUCTION

The Call to War

In 1914, when the First World War broke out, Australia was still a young nation, federated just 13 years earlier. Yet, despite its geographical isolation from Europe and its small population, Australia committed to the war effort with remarkable zeal. Over 416,000 Australians enlisted to serve in World War I—an extraordinary number for a country with fewer than five million people at the time. Among these volunteers were 1,786 volunteers who enlisted in or had close links with Townsville. They were young men whose lives were shaped by the rhythms of commerce, farming, trade, and the sea.

Some enlisted out of a sense of adventure, lured by promises of seeing the world beyond the shores of Australia. Others felt a deep sense of duty to the British Empire, of which Australia was a part, and a conviction that they were defending freedom and justice. Many more were motivated by a mixture of both—a desire to be part of something bigger than themselves, to do their part in a war that was supposed to end all wars. Whatever their reasons, they left behind the familiar to face the unknown.

The Kennedy Regiment

Townsville's military contribution began even before the war's outbreak. Established in 1909, the Kennedy Regiment, a citizen force battalion based in North Queensland, was one of the first to mobilize upon the declaration of war. This unit was tasked with defending Thursday Island, a strategic point off the coast of Australia. When war broke out in August 1914, the Kennedy Regiment requisitioned the steamer *Kanowna* , setting off to reinforce the garrison at Thursday Island. Hearing of a proposed mission in German New Guinea, around 500 of the regiment's men volunteered for service beyond Australia's borders.

This journey, however, was fraught with difficulties. Lacking adequate supplies, proper equipment, and even suitable living conditions aboard the *Kanowna* , the men of the Kennedy Regiment soon found themselves in challenging circumstances. Ultimately, the unit was recalled to Townsville due to issues with age, fitness, inadequate preparation for tropical deployment and problems with the ship's crew. Despite their recall, this experience underscored the regiment's dedication to the war effort. It also highlighted the logistical challenges Australia faced in mobilizing a military force from such a remote area. The Kennedy Regiment's story became emblematic of the resilience and enthusiasm North Queensland brought to the war effort, even in the face of adversity.

From the Trenches to the Desert: Australian Contributions

The young men of Townsville played a significant role in Australia's contribution to World War I, particularly during the Gallipoli campaign in 1915. Like thousands of Australians, many Townsville soldiers landed at Anzac Cove on April 25, facing harsh conditions and fierce Ottoman resistance. The brutal campaign, aimed at securing the Dardanelles Strait, resulted in a stalemate that lasted months. Of the 8,700 Australian lives lost, 19 were from Townsville.

However, Townsville's involvement extended beyond Gallipoli. Its soldiers were scattered across the major theatres of war, including the Western Front, where the horror of trench warfare claimed thousands more lives. In battles like Fromelles, Pozières, and Passchendaele, these men faced relentless artillery bombardments, mud-filled trenches, and relentless enemy fire.

Townsville men served in the Middle East, enduring the gruelling conditions of desert warfare in Egypt and Palestine. The Australian Light Horse units, in which Townsville soldiers served, became famous for courage on Gallipoli and their service in Palestine, especially during the decisive Battle of Beersheba in 1917.

The Cost of War

The toll of World War I in Australia was staggering. Of the 416,000 who enlisted, over 60,000 never returned. Among the dead were the 171 soldiers and one merchant marine sailor whose names are etched into the Townsville cenotaph. For each of these men, there were families who received that dreaded telegram or letter informing them that their loved one had been killed in action, was missing, or had died from wounds or illness. The grief was profound, and it reverberated through the community, casting a long shadow over Townsville in the years that followed.

In a country so far removed from the battlefields of Europe, the impact of these losses was keenly felt. The cenotaph in Townsville became a focal point for communal mourning and remembrance. Every year on Anzac Day, as the sun rises over the cenotaph, the people of Townsville gather to remember the sacrifice of these men. But remembrance goes beyond a single day. The very existence of this monument speaks to a collective effort to ensure that future generations never forget the cost of war.

Gary Ewart
2024

PART 1 - 1915

The Gallipoli Year

In 1915, the town of Townsville suffered the heartbreaking loss of 20 local soldiers, all of whom had volunteered to serve in World War I. While most of these men died in the brutal Gallipoli campaign, two lost their lives during training — one in Australia and one in Egypt. The first soldier from Townsville to die in service was 90 Private Francis Joseph Gilvarry who died of pneumonia in Egypt on the 1st of March 1915. The first soldier to be Killed in Action was 900 Private Edward Augustine Cahill, who fell on the very first day of the Gallipoli landings, the 25th of April 1915. His death marked the beginning of a series of tragedies for the community, as more families received the devastating news of fallen sons, brothers, and friends.

The death of Townsville's men began to shift the community's perception of the Great War. What may have initially been seen as a noble cause, with young men proudly answering the call of duty, quickly turned into a sobering reality of sacrifice and loss. The once-distant conflict had come home to Townsville, leaving a lasting mark on the town as more soldiers succumbed to the violence of Gallipoli.

The loss of the 20 soldiers, whether in battle or training, deeply impacted the community. The town began to grasp the high human cost of the war, realizing that the conflict was not only shaping distant lands but also tearing apart their own. These early losses foreshadowed the long and painful years ahead, as the Great War continued to claim the lives of countless Australians. The individual stories of Townsville's losses in 1915 are told below.

90 PRIVATE FRANCIS JOSEPH GILVARRY (1890 – 1915)

9th Infantry Battalion

Francis Joseph Gilvarry (Joe) was born on the 27th of December 1890 in Hughenden, Queensland. He was the fourth of seven children born to Michael and Annie Gilvarry (née Rutledge). He had three brothers and three sisters. Joe's father worked as a labourer, and by 1903, the family was living on Morris St in Townsville. Tragically, Michael Gilvarry died in Townsville in February 1909. Following her husband's death, Annie, a midwife, moved to Cairns and by 1913 was living at a property on Sheridan Street.

Joe Gilvarry had also relocated and was living on Martin St, Fortitude Valley, Brisbane when Australia declared war on the 4th of August 1914. He was among the first to enlist, applying to join the AIF on the 23rd of August 1914, less than three weeks after the war declaration.

At enlistment, Joe Gilvarry declared he was single, employed as a Postal Assistant, and 23 years and eight months old. He also declared previous service in the Australian Garrison Artillery. His medical examination noted that he was 5'9" (175 cm) tall, weighed 140 lbs (63.5 kg), had a fresh complexion, grey eyes, and brown hair. He identified as Roman Catholic and listed his mother as his next of kin.

Accepted into the AIF, Joe Gilvarry was given service number 90, assigned to the Infantry, and posted as an original member of "A" Company of the 9th Battalion, the first battalion recruited in Queensland. Raised within weeks of the war's declaration in August 1914, the 9th Battalion embarked for Egypt just two months later. They sailed from Brisbane with Convoy No.1 aboard Transport A5 *Omrah* on the 24th of September 1914. The *Omrah* carried 1,146 troops and 15 horses, while the entire convoy transported 21,529 troops and 7,882 horses. The 9th Battalion arrived in Egypt on the 2nd of December 1914 and underwent rigorous training for their deployment to Gallipoli. As part of the 3rd Brigade, they were the covering force during the ANZAC landing on the 25th of April 1915, being the first to land at approximately 4.30 am. Lieutenant Duncan Chapman held the distinction of being the first soldier to set foot on Gallipoli soil. Once ashore, the 9th Battalion was heavily involved in establishing and defending the ANZAC beachhead.

However, Private Joe Gilvarry was not part of the Gallipoli landing. He had remained in Egypt, ill with pneumonia. Tragically, he succumbed to the disease and died on the 1st of March 1915. He was 24 years and two months old and had served with the AIF for six months at the time of his death. Private Joe Gilvarry was laid to rest at Cairo War Cemetery (Row M, Grave No.20). He was the first soldier from Townsville to die. His death was reported in *The Argus* (Melbourne) and other newspapers on the 3rd of March 1915.

Francis Joseph Gilvarry's name is inscribed on The Australian War Memorial Roll of Honour (panel 56), the Postmaster-General's Dept Qld WW1 Honour Roll, and The Townsville Cenotaph (incorrectly engraved as "J.P." Gilvarry).

Private Francis Joseph Gilvarry served in WW1 in Egypt and was awarded the 1914/15 Star (no: 2134), the British War Medal (no: 400), and the Victory Medal (no: 357). A memorial scroll and a memorial plaque were given to Annie Gilvarry in 1921 and 1922, respectively. She also received a pension of £52 per annum for the loss of her son, half of the wage paid to AIF soldiers in 1915.

Joe Gilvarry's death was announced in *The Argus* (Melbourne) and other newspapers on the 3rd of March 1915:

DEATH FROM PNEUMONIA.

The Defence department has announced the death of Private Francis Joseph Gilvarry, of "A" Company, 9th Battalion, Queensland. Death occurred at Mena, Egypt, on March 1, as the result of pneumonia. Next of kin is given as Mrs. Annie Eliza Gilvarry (mother), Sheridan Street, Cairns, Queensland.

The following notice was published in the *Brisbane Courier* on the 1st of March 1916:

GILVARRY — In sad but loving memory of Private Francis Joseph Gilvarry, who died of pneumonia, at Mena Camp, Egypt, on March 1, 1915.

Ah, pale-cold now those rosy lips we aft have kissed sae fondly,
And closed for aye the sparkling glance that dwelt on us sae kindly,
And mouldering now in silent dust that heart that loved us dearly,
For dear to us as light and life, was our dear, darling Joe.
Inserted by his loving mother, brothers, and sisters.

900 PRIVATE EDWARD AUGUSTINE CAHILL (1883 – 1915)

9th Infantry Battalion

PRIVATE E. A. CAHILL.

Edward (Eddie) A. Cahill was born on the 15th of August 1883 in Cardwell, Queensland, to Patrick and Bridget Cahill (née Kenney). He had four brothers and two sisters. By the early 1900s, the Cahill family had moved to Townsville and lived on Victoria Street. Patrick Cahill worked as a carpenter. Eddie attended both Catholic and state schools during his early years. After completing his education, Eddie qualified and gained employment as an accountant. By 1913, he was living in premises on Melton Street, Townsville.

On the 19th of August 1914, two weeks after Australia entered WW1, Eddie Cahill applied in Townsville to enlist in the Australian Imperial Force (AIF). He declared that he was single and employed as an insurance manager with Thomas Brown & Sons in Townsville. He also declared that he had been a member of the Townsville Rifle Club for two years and stated that he was 29 years old on his enlistment application, though his birth records show he was 31 years old. Eddie Cahill's medical examination noted that he was 5'7½" (171 cm) tall, weighed 146 lbs (66 kg), had a sallow complexion, grey eyes, and black hair. He identified as Roman Catholic and with his parents deceased, Eddie nominated his brother, J.A. Cahill, as his next of kin giving his address as Albert St, Auckland, New Zealand.

Accepted into the AIF, Eddie Cahill was allocated service number 900, assigned to the Infantry, and posted as an original member of the newly raised 9th Battalion, the first battalion recruited in Queensland. It was later joined by the 10th, 11th, and 12th Battalions to form the 3rd Brigade. After a short introduction to the AIF of 25 days, Private Eddie Cahill embarked with the 9th Battalion from Brisbane, Queensland, on board Transport A5 *Omrah* on the 24th of September 1914, bound for Egypt. The *Omrah* was part of Convoy No.1 and carried 1,142 troops and 15 horses on the voyage. The entire convoy of 29 Australian ships carried 21,431 troops, 70 naval personnel, 28 civilians, 7,882 horses, and towed the submarine AE2 behind HMAS *Berrima*.

The 9th Battalion arrived in Egypt in early December 1914 and moved to Mena Camp from Alexandria on the 1st of January 1915. The Battalion continued training for the upcoming operations until the 27th of February 1915, when orders were received to support naval operations in the Dardanelles.

On the 4th of March 1915, the 9th Battalion reached Lemnos, where they began preparations for their task of occupying Turkish forts following naval action. These preparations included training exercises in small boat disembarkation, hill assaults, and route marches. The 3rd Brigade was the covering force for the ANZAC landing on the 25th of April 1915 and was the first ashore at around 4.30 am. The 9th Battalion was heavily involved in establishing and defending the front line of the ANZAC beachhead.

During this phase of operations, Private Eddie Cahill was Killed in Action on the 25th of April 1915. He has no known grave. He was 31 years and eight months old and had served with the AIF for eight months at the time of his death. His death was notified in the 46th list of Australian casualties released on the 30th of June 1915.

Edward Cahill's name is inscribed on The Australian War Memorial Roll of Honour (panel 55), The Lone Pine Memorial, Gallipoli, The City of Townsville WW1 Honour Roll, and The Townsville Cenotaph. Cahill Street, Aitkenvale, Townsville, is named in his memory.

Private Edward Augustine Cahill served in WW1 in Egypt and Gallipoli. He was awarded the 1914/15 Star (no: 1945), the British War Medal (no: 340), and the Victory Medal (no: 296). A memorial scroll and a memorial plaque were given to his brother in 1922.

The following article was published in *The Telegraph* (Brisbane) on the 2nd of July 1915:

NOTES ON CASUALTIES

Pte. E. A. Cahill, reported killed in action, was the second son of the late Mrs. B. M. Cahill, of North Ward, Townsville. He was one of the first to answer his country's call and left Queensland with the first expeditionary force. Prior to enlisting, he was with Messrs. Thomas Brown and Sons, as manager of their insurance department. Deceased was 28 years of age and was a brother of Mrs. A. McDonald, Agnes Street, Valley, and a nephew of Mr. A. Cahill, of Spring Street.

The following tributes were published in the *Townsville Daily Bulletin* on the: 25th April 1918:

CAHILL. —In loving memory of Pte Eddie Cahill, who was Killed in Action, at Gallipoli, April 25th, 1915.

> In a soldier's grave he lies sleeping,
> One of earth's bravest and best.
> In our hearts we shall miss him forever,
> Though we know he is only at rest.
> Inserted by his loving sister, Alys.

CAHILL. —In loving memory of Pte E. A. (Eddie) Cahill, who was Killed in Action, at Gallipoli, April 25th, 1916. (sic)

> "They shall not grow old as we that are left grow old.
> Age shall not weary them, nor years condemn,
> At the going down of the sun and in the morning,
> We will remember them."

CAHILL. —In fond and loving memory of Edward (Eddie) Cahill, who was killed on Gallipoli, on April 25th, 1915.

Inserted by his sincere friends, A. and L. Kinging. (and M. B.)

759 PRIVATE PATRICK ALLEN YEATMAN (1878 - 1915)

9th Infantry Battalion

PRIVATE A. YEATMAN.

Patrick Allen Yeatman was born on the 25th of October 1878 in Chislehurst, Kent, England. He was the third son and fourth of six children born to his father, John Yeatman, a barrister-at-law, and his mother, Charlotte (née McDonogh). He had three brothers and two sisters. Patrick grew up in Birmingham and attended St. Philip's Grammar School. Later in life, he pursued a career in journalism. After 1899, he joined Bethune's Horse Regiment and served in the Boer War, earning the South African War Medal.

Following the Boer War, Patrick moved to Australia and worked as a journalist for the *Townsville Daily Bulletin* , residing at the Exchange Hotel in Townsville. On the 31st of August 1914, shortly after the outbreak of WW1, he enlisted in the AIF in Townsville. He stated he was born in England, was 36 years and ten months old, single, and working as a journalist, also noting his prior Boer War service.

Patrick's medical examination recorded his height as 5'8½" (174 cm) and weight as 163 lbs (74 kg), with a fresh complexion, blue eyes, and fair hair. He identified as Protestant and listed his family solicitors, Messrs Griffith and Gardiner, 13 St Swithin Lane, London, as his next of kin.

Accepted into the AIF, Patrick was allocated service number 759, assigned to the Infantry, and posted as an original member of "G" Company of the newly formed 9th Battalion. The 9th Battalion, the first recruited in Queensland, joined the 10th, 11th, and 12th Battalions to form the 3rd Brigade. After just 25 days with the AIF, Private Patrick Yeatman embarked with the 9th Battalion from Brisbane on HMAT A5 *Omrah* on the 24th of September 1914, bound for Egypt. The *Omrah,* part of Convoy No.1, carried 1,142 troops and 15 horses. The convoy, comprising 29 Australian ships, transported 21,431 troops, 70 naval personnel, 28 civilians, 7,882 horses, and towed the submarine AE2 behind HMAS *Berrima.*

The 9th Battalion arrived in Egypt in early December 1914 and moved to Mena Camp from Alexandria on the 1st of January 1915. After training for operations, they received orders on the 27th of February 1915 to support naval actions in the Dardanelles.

On the 4th of March 1915, the 9th Battalion reached Lemnos, preparing for their role in occupying Turkish forts post-naval action. Preparations included training in small boat disembarkation, hill assaults, and route marches. The 3rd Brigade covered the ANZAC landing on the 25th of April 1915, with the 9th Battalion among the first ashore around 4:30 a.m. They were heavily involved in establishing and defending the ANZAC beachhead. After this fighting, Private Patrick Yeatman was declared Missing in Action on the 25th of April 1915. He was later declared Killed in Action on that same day.

Patrick Yeatman's body was recovered, and he was buried at Plugge's Plateau Cemetery (Row B, Grave No.10), Gallipoli, Turkey. Patrick was 36 years and six months old and had served with the AIF for eight months. His death was reported in the 46th list of Australian Casualties on the 30th of June 1915.

Patrick Allen Yeatman's name is inscribed on The Australian War Memorial Roll of Honour (panel 58), The City of Townsville WW1 Honour Roll, and The Townsville Cenotaph. Yeatman Street in Hyde Park, Townsville, is named in his memory.

759 Private Patrick Allen Yeatman served in WW1 in Egypt and Gallipoli and was awarded the 1914/15 Star (no:2343), the British War Medal (no:611), and the Victory Medal (no:568). In 1922, a memorial scroll, and plaque were sent to Griffith & Gardner in England.

Patrick's younger brother, 2nd Lieutenant Bernard Joseph Pym Yeatman D.S.O., joined the British Army in WW1 and was Killed in Action at Le Cateau on the 23rd of October 1918, aged 34.

The following letter to the Editor was published in the *Townsville Daily Bulletin* on the 14th of July 1915:

HOW PRIVATE YEATMAN DIED

(To the Editor)

Sir, — I very much regret to have to inform you of the death of Private A. Yeatman of the 9th Battalion (late of the "Townsville Bulletin staff). A corporal of his company was wounded in the leg and arm and could not move. Yeatman at great risk rescued and carried him to a place he thought was safe and put him down to have a rest, when a shrapnel shell burst and killed him instantly, the corporal that he had been carrying escaping. I would be pleased if you would print a copy of this, so that Yeatman's many friends in Townsville will hear of the gallant way in which he died. I am very well myself and will endeavor to write you a full account of our experiences as soon as the censorship is lifted.

Yours, etc
ALEX LANG
3rd Field Artillery Brigade.,
First Australian Expeditionary Force.

An article in The Northern Miner (Charters Towers) on the 1st of July 1915 reported:

Pte. P.A. Yeatman, who was reported missing, is now included among those killed in action. He was a member of the '*Townsville Daily Bulletin* ' staff for some years, volunteering at the outbreak of the war. He was an Englishman with South African war experience. He paid one or two visits to Charters Towers and will be remembered by several for his pleasant and sociable disposition, particularly by Mr. T. Webb, who once, in a spirit of fun, paced him at top speed up a long rise in the Deeps.

1308 CORPORAL NEIL BAXTER (1893 – 1915)

15th Infantry Battalion

Neil Baxter was born on the 1st of June 1893 in Glasgow, Scotland, to Duncan and Annie Baxter (née McDonald). He was the firstborn and had three brothers and three sisters. Duncan Baxter and his brother Neil moved to Townsville from Scotland in 1903, bringing their families a year later. They operated the Baxter Furniture Warehouse until Neil's death in 1908, after which Duncan became the proprietor. The Baxter family lived on Paxton Street in Townsville.

Neil was 11 years old when he arrived in Australia. He grew up in Townsville, attending Central State School and Gatton College in the Lockyer Valley, Queensland. After college, he worked on the Baxter farm at Alligator Creek. He was an active member of the City Football Club, Oddfellows, and the Rifle Club.

On the 16th of November 1914, Neil Baxter applied in Townsville to join the AIF. He declared that he was single, a farmer, and 21 years and five months old, also noting his three-month membership in the Rifle Club. His enlistment medical examination recorded his height as 5'4" (162 cm) and weight as 117 lbs (53 kg), with a sallow complexion, blue eyes, and fair hair. He identified as Presbyterian and listed his father, Duncan Baxter, of Flinders Street, Townsville, as his next of kin.

Neil was accepted into the AIF, allocated service number 1308, assigned to the Infantry, and posted to the 15th Battalion/2nd reinforcements. The 15th Battalion was raised from late September 1914, six weeks after the First World War began. Three-quarters of the battalion were Queensland volunteers, the rest from Tasmania. The 15th, along with the 13th, 14th, and 16th Battalions, formed the 4th Brigade, commanded by Colonel John Monash. When the 4th Brigade set out for Egypt, Neil Baxter was undergoing basic training in Australia.

After three months, Private Neil Baxter and his fellow reinforcements embarked from Brisbane on the 13th of February 1915, aboard HMAT *Seang Bee* as part of Convoy No.3, which transported 1,027 troops. The entire convoy of 21 ships carried 8,829 troops and 6,027 horses to Egypt.

The 4th Brigade arrived in Egypt in February 1915 and joined the New Zealand and Australian Division (ANZAC). From the 3rd of February to the 11th of April 1915, the 15th Battalion trained at Aerodrome Camp, Heliopolis. Private Neil Baxter was promoted to Acting Corporal before the Battalion moved to Lemnos from Alexandria on the 14th of April 1915. The 15th Battalion then deployed to the Gallipoli Peninsula, landing at ANZAC Cove late in the afternoon of the 25th of April 1915. Corporal Neil Baxter was Killed in Action on the 26th of April 1915, aged 21 years and 10 months. He served with the AIF for five months and one week and was a member of the 15th Battalion for 10 weeks. He has no known grave, and his death was reported in the 35th list of Australian Casualties on the 10th of June 1915.

Following the report of Neil's death, his parents refused to believe he was dead. Neil's letters were returned marked 'Missing', and his parents hoped he was recovering in a hospital or a prisoner of war. It was not until April 1916 that the AIF officially declared Corporal Neil Baxter Killed in Action.

Neil Baxter's name is inscribed on The Australian War Memorial Roll of Honour (Panel 74), The Lone Pine Memorial Gallipoli (Panel 45), The City of Townsville WW1 Honour Roll, and The Townsville Cenotaph. Baxter Street, West End, Townsville, is named in memory of both Duncan and Neil Baxter.

1308 Corporal Neil Baxter served in Egypt and Gallipoli in WW1 and was awarded the 1914/15 Star (no:15570), the British War Medal (no:8017), and the Victory Medal (no:7977). A memorial scroll and a memorial plaque were sent to his father in 1922.

A tribute published in the *Townsville Daily Bulletin* on the 26th of April 1917, reads:

BAXTER. —In loving memory of our dear Son and Brother, Neil, who was Killed in Action at Gallipoli, 26th April 1915, first-born to Mrs. and late C.Q.M.S. Duncan Baxter, aged 21 years and 10 months.

"Lead kindly light."

135 LANCE CORPORAL HENRY DAINTREE FOOT (1883 – 1915)

15th Infantry Battalion

Henry Daintree Foot, known as Hal, was born on the 10th of September 1883 in Townsville. He was the second of 10 children and the eldest son of Alfred and Mary Foot (née Madden). In 1899, the Foot family settled on Cardington Station, located 42 miles (67 km) from Townsville on the Haughton River. During his youth, Hal attended Townsville Grammar School as a boarder. He was an accomplished rifleman and a member of the Townsville Rifle Club. After finishing school, Hal returned to the family property and worked as a grazier.

On the 14th of August 1914, 10 days after Australia entered WW1, Henry Foot enlisted in the 2nd Battalion, Australian Naval and Military Expeditionary Force (AN&MEF). The AN&MEF's mission was to occupy the Pacific colony of German New Guinea between September and November 1914. Henry left Cairns on the 8th of August 1914, heading for Thursday Island to join the Queensland contingent bound for Port Moresby. The contingent boarded the SS *Kanowna* for the journey toward capturing German New Guinea, but they did not participate due to *"trouble caused by the firemen on the troopship."* Hal Foot returned to Townsville on the 18th of September 1914 and was discharged from the AN&MEF. Three weeks later, on the 9th of October, he applied to enlist in the Australian Imperial Force (AIF) at Enoggera in Brisbane. He declared that he was 31 years old, single, and a grazier. His enlistment medical examination noted that he was 5'6½" (169 cm) tall, weighed 124 lbs (56 kg), had a medium complexion, greyish-brown eyes, and black hair. Hal Foot identified as Church of England and listed his mother as his next of kin, giving her address as Cardington G.N. Rly Townsville.

Accepted into the AIF, Hal Foot was allocated service number 135, assigned to the Infantry, and posted to the 15th Battalion, which was raised in late September 1914, six weeks after the outbreak of the war. Three-quarters of the 15th Battalion were recruited from Queensland, with the rest from Tasmania.

Alongside the 13th, 14th, and 16th Battalions, it formed the 4th Brigade, commanded by Colonel John Monash. The Queensland companies concentrated at Enoggera in September 1914 and began initial training, while the Tasmanians formed at Claremont in October. In late November 1914, the battalion concentrated at Broadmeadows, Victoria. Private Hal Foot underwent training in Enoggera in Brisbane and was promoted to Lance Corporal on the 1st of October 1914.

The 15th Battalion embarked for Egypt from Melbourne on the 22nd of December 1914 as part of Convoy No.2. The Battalion sailed aboard A40 *Ceramic*, which carried 2,708 troops and 21 horses. The entire convoy transported 11,305 troops and 4,643 horses to Egypt. After arrival in February 1915, the Battalion underwent extensive training in the desert throughout February and March. On the 10th of April 1915, the 15th Battalion moved by train to Alexandria, where they embarked on the troopships *Seeang Bee* and *Australind,* bound for Gallipoli. The Battalion landed at Anzac Cove on the afternoon of the 25th of April 1915 and was rushed into the line on the left flank of the beachhead. Later, they helped secure the line before occupying positions around Pope's Hill and Russell's Top, where they joined an attack on the 1st of May 1915 around "Quinn's Post". Lance Corporal Henry Foot was Killed in Action on the 1st of May 1915. He was 31 years and seven months old and had served with the AIF for seven months. He has no known grave.

Henry Daintree Foot's name is inscribed on The Australian War Memorial Roll of Honour (Panel 75), The Lone Pine Memorial Gallipoli (Panel 44), The City of Townsville WW1 Honour Roll, The Townsville Grammar School Roll of Honour, and The Townsville Cenotaph. Foot Street, West End, Townsville, was named in 1919 after Lance Corporal Henry Foot and his brother Trooper Alexander Madden Foot, who was also Killed in Action during WW1. The street has since been closed and is now part of Keyatta Park.

135 Lance Corporal Henry Daintree Foot served in WW1 at Gallipoli and was awarded the 1914/15 Star (no:12417), the British War Medal (no:6608), and the Victory Medal (no:6582). A memorial scroll and a memorial plaque were sent to his father in 1922.

Henry had three brothers who also served in WW1: 244 Trooper Alexander Madden Foot, 5th Light Horse Regiment, Killed in Action at Gallipoli on the 22nd of November 1915 and buried at Shell Green Cemetery on Gallipoli, 74 Private George Pierce Foot, 2nd Light Horse Regiment, returned to Australia on the 22nd of February 1917, 4064 Private Eric Douglas Foot, 4th Pioneer Battalion, returned to Australia on the 18th of July 1919.

366 PRIVATE MERVYN DANE COLLINGS (1891 – 1915)

15th Infantry Battalion

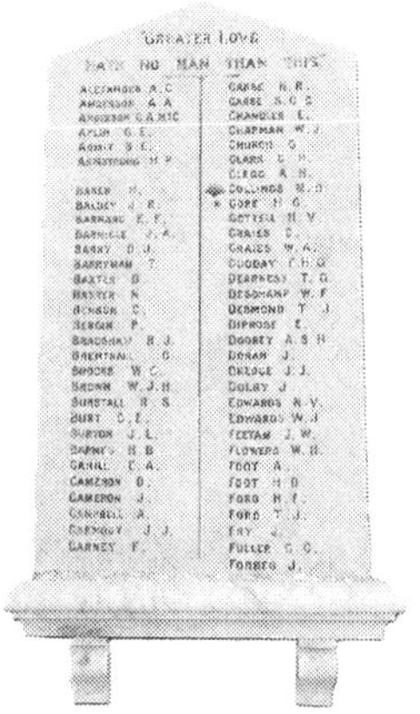

Mervyn Dane Collings (sometimes recorded as Mervyn Dave Collings) was born in Brisbane on the 17th of May 1891. He was the eldest of five boys born to Daniel and Amy Collings (née Leaver). Daniel Collings was a tea merchant, and the Collings family lived in Indooroopilly, Brisbane.

Mervyn attended high school in Southport, Queensland, after which he trained as an insurance clerk. Upon qualifying, he moved to Townsville and assumed the position of Accountant at New Zealand Insurance Co Ltd on Flinders Street. He lived at *"Valetta"* on Melton Hill.

On the 16th of November 1914, Mervyn applied in Townsville to join the AIF. He declared his age as 23 years and five months, his marital status as single, and his occupation as a clerk with no prior military experience. His medical examination noted that he was 5'9½" (176 cm) tall, weighed 150 lbs (68 kg), had a dark complexion, brown eyes, and brown hair. He identified as Church of England and nominated his father as his next of kin, giving his address as *"Indoocombe"* Woodbridge, Tasmania.

Accepted into the AIF, Mervyn was allocated service number 366, assigned to the Infantry, and posted to the 15th Battalion. The 15th Battalion's formation began in late September 1914, approximately six weeks after the outbreak of the First World War. About three-quarters of its personnel were volunteers from Queensland, while the remainder came from Tasmania. Under the command of Colonel John Monash, the 15th Battalion, along with the 13th, 14th, and 16th Battalions, constituted the 4th Brigade. The two elements of the battalion trained together in Victoria.

Private Mervyn Collings was with the 15th Battalion when the unit embarked for Egypt from Melbourne on the 22nd of December 1914. They sailed as part of Convoy No.2, aboard HMAT A40 *Ceramic*, which carried 2,708 troops and 21 horses. The entire convoy transported 11,305 troops and 4,643 horses to Egypt.

Upon arriving in Egypt in February 1915, the 15th Battalion became part of the New Zealand and Australian Division (ANZAC). From the 3rd of February to the 11th of April 1915, the 15th Battalion was stationed at Aerodrome Camp, Heliopolis, Egypt, where they completed Battalion, Brigade, and Divisional training.

On the 10th of April 1915, the 15th Battalion was moved by train to Alexandria, where it embarked on the troopships *Seeang Bee* and *Australind*, bound for Gallipoli. Private Mervyn Collings landed on the Gallipoli Peninsula at ANZAC late in the afternoon of the 25th of April 1915 with the rest of the 4th Brigade. With the arrival of Turkish reinforcements, the ANZAC position became more precarious as the assaulting force could not secure their initial objectives. They retreated to rudimentary and shallow entrenchments, steadfastly holding on during the critical first night. By that evening, 16,000 men had been deployed, with over 2,000 Australians either killed or wounded.

Between May and August 1915, the 15th Battalion played a significant role in creating and safeguarding the ANZAC beachhead's front line. It was during this phase of operations that Private Mervyn Collings was Killed in Action on the 3rd of May 1915. Unfortunately, there are no recorded details regarding the circumstances of his death. However, the Battalion war diary of the 3rd of May 1915 records that *"The enemy made an attack upon our position early in the morning but was repulsed."*

Private Mervyn Collings was 23 years and 11 months old and had served with the AIF for five months and two weeks at the time of his death. He has no known grave. His death was notified in the thirty-fifth list of Australian Casualties released on the 10th of June 1915.

Mervyn Dane Collings' name is inscribed on The Australian War Memorial Roll of Honour (Panel 75), The Lone Pine Memorial, Gallipoli (Panel 45), The City of Townsville WW1 Honour Roll, and The Townsville Cenotaph. Collings Street in Stuart, Townsville, is named in his memory.

366 Private Mervyn Dane Collings served in WW1 in Gallipoli and was awarded the 1914/15 Star (no:14137), the British War Medal (no:6576), and the Victory Medal (no:6550). A memorial scroll and a memorial plaque were sent to his father in 1921 and 1922, respectively.

The following notice was published in *The Brisbane Courier* on the 12th of June 1915:

PERSONAL NOTES.

Pte Mervyn D. Collings, reported killed in action, served for a number of years in the Brisbane office of the New Zealand Insurance Co. and later became an accountant in the Townsville office.

207/1057 LANCE CORPORAL KEITH WATSON (1894 - 1915)

15th Infantry Battalion

Keith Watson was born on the 2nd of October 1894 in Townsville, Queensland. He was the only son and youngest of two children born to Elijah Watson and his wife Emily (née Cheyne). Elijah Watson was a traveling salesman, and the Watson family lived in the suburb of Eagle Junction in Brisbane.

Keith Watson attended the local state school, where he excelled academically and won a state scholarship. He then attended Brisbane Grammar School, continuing his academic success, and winning the

Lilley Medals in his first and third years—Brisbane Grammar School's most prestigious academic awards.

Keith Watson passed the Sydney junior and senior exams, matriculating with honours. He was also a talented sportsman, playing cricket and football and winning prizes in running. After graduation, he studied surveying and served a cadetship in the central district near Winton. He temporarily gave up surveying to undertake commercial training with his brother Bruce in Townsville.

Keith Watson was a member of the Cadets and then the Citizen Forces. In August 1914, he was mobilized for war service and embarked at Cairns, Queensland, for Thursday Island (War Station) Garrison Duty. On the 14th of August 1914, Private Keith Watson enlisted for service outside Australia. He was taken on the strength of the Australian Naval and Military Expedition Force (A.N.&M.E.F.), allocated service number 1057, and re-embarked from Thursday Island on the 16th of August 1914 aboard the troopship *Kanowna* to take part in the capture of German New Guinea. Mid-voyage, however, the firemen aboard went on strike, and the *Kanowna* was required to return to Townsville, stoked by the troops. As a result, Private Keith Watson did not participate in the capture of German New Guinea. The *Kanowna* docked in Townsville on the 18th of September 1914, and Keith Watson was discharged from the A.N.&M.E.F.

Three days later, on the 21st of September 1914, Keith Watson applied to enlist in the Australian Imperial Force (AIF) at Townsville. He was one week short of his twentieth birthday, single, and qualified as a clerk. His enlistment medical documents noted that he was 5'7" (170 cm) tall, weighed 129 lbs (59 kg), had a sallow complexion, brown eyes, and black hair. He identified as Church of England and listed his father as his next of kin, giving his address as Cressington, Sydney Street, Eagle Junction, Brisbane.

Accepted into the AIF, Keith Watson was allocated service number 207, assigned to the Infantry, and posted as an original member of 'A' Company of the newly formed 15th Infantry Battalion, raised as part of the all-volunteer AIF in September 1914. Drawing personnel from Queensland and Tasmania volunteers, the 15th formed part of the 4th Brigade, along with the 13th,

14th, and 16th Battalions. On the 1st of October 1914, Private Keith Watson was promoted to Lance Corporal.

The 15th Battalion embarked for Egypt from Melbourne on the 22nd of December 1914 as part of Convoy No.2. The Battalion sailed aboard A40 *Ceramic*, which carried 2,708 troops and 21 horses. The entire convoy carried 11,305 troops and 4,643 horses to Egypt. Upon arrival in Egypt in February 1915, the 15th Battalion became part of the New Zealand and Australian Division (ANZAC).

From the 3rd of February to the 11th of April 1915, the 15th Battalion was stationed at Aerodrome Camp, Heliopolis, Egypt, where they completed Battalion, Brigade, and Divisional training. On the 10th of April 1915, the 15th Battalion was moved by train to Alexandria, where it embarked on the troopships *Seeang Bee* and *Australind*, bound for Gallipoli.

Lance Corporal Keith Watson landed on the Gallipoli Peninsula at ANZAC late in the afternoon of the 25th of April 1915 with the rest of the 4th Brigade. With the arrival of Turkish reinforcements, the ANZAC position became more precarious as the assaulting force could not secure their initial objectives. They retreated to rudimentary and shallow entrenchments, steadfastly holding on during the critical first night. By that evening, 16,000 men had been deployed, with over 2,000 Australians either killed or wounded.

Later, the 15th Battalion helped shore up the line before occupying positions around Pope's Hill and Russell's Top, where they joined an attack on the 1st of May 1915 around Quinn's Post. It was during this operation that Lance Corporal Keith Watson was wounded by a bullet in the neck. He was evacuated back to HMHS *Gloucester Castle* but succumbed to his wounds and died on the 4th of May 1915. He was laid to rest at sea on the 5th of May 1915. Lance Corporal Keith Watson was 20 years and seven months old at the time of his death. He had been a member of the AIF for seven months. His death was notified in the 22nd list of Australian Casualties released on the 25th of May 1915.

Keith Watson's name is inscribed on the Australian War Memorial Roll of Honour (Panel 78), The Lone Pine Memorial, Gallipoli, Turkey (Panel 44),

The Brisbane Grammar School Memorial Library WW1 Honour Board 2, The City of Townsville WW1 Honour Roll, and The Townsville Cenotaph.

207 Lance Corporal Keith Watson was awarded the 1914/15 Star (No.15239), the British War Medal (No.6794), and the Victory Medal (No.6768). A memorial scroll and a memorial plaque were passed to his father in 1921 and 1922, respectively.

A story detailing Keith Watson's life and death was published in The *Telegraph* (Brisbane, Qld. 1872 - 1947) on Wednesday the 2nd of June 1915.

LIEUTENANT HUTTON PERKINS ARMSTRONG (1891 – 1915)

15th Infantry Battalion

LIEUTENANT H. P. ARMSTRONG

Hutton Perkins Armstrong was born to George Armstrong and Elizabeth Armstrong (née Crocker) of Denham Street Townsville on the 24th of February 1891. He was the youngest of three surviving children, all boys. After schooling at Central State School & Technical College in Townsville, he qualified as a clerk and was employed by Howard Smith Ltd., a shipping company located in Flinders St Townsville.

Hutton Armstrong was a member of the Townsville Rifle Club and was an expert rifleman. He was so good at shooting that he was selected as a member of the Australian rifle team that competed at Bisley England in 1913 where he achieved the highest aggregate score.

On the outbreak of WW1, he enlisted and was appointed as a 2nd Lieutenant in the Australian Naval and Military Expeditionary Force. The AN&MEF was tasked with the Australian takeover and occupation of the Pacific colony of German New Guinea in September – November 1914. 2Lt Armstrong left Townsville aboard the *Kanowna* on the 16th of August 1914 bound for Thursday Island and Port Moresby with the Queensland contingent. As events turned out, Hutton Armstrong did not participate in the capture of German

New Guinea on account of *"trouble caused by the firemen on troopship."* He returned to Townsville on the 18th of September 1914 and was discharged from the AN&MEF.

Three months later, on the 22nd of December 1914, Hutton Armstrong applied to enlist in the Australian Imperial Force (AIF). He was 23 years, 10 months old. He declared that he was single, qualified as a clerk, and had previous military service with the AN&MEF. His medical assessment noted that he was 5'7" (170 cm) tall, weighed 140lbs (63.5kg), had a fair complexion, grey eyes, and fair hair. He identified as Church of England and nominated his parents as his next of kin giving their address as Denham St Townsville. This was later changed to Rosecliffe St Dutton Park, South Brisbane.

Accepted into the AIF, Hutton Armstrong was allocated service number 102, assigned to the Infantry, and posted as a Private to the newly raised 15th Battalion. The 15th was originally raised as part of the all-volunteer AIF in September 1914. Drawing personnel from volunteers from Queensland and Tasmania, it formed part of the 4th Brigade. Basic training followed enlistment, and Private Hutton Armstrong was promoted to Sergeant in November 1914.

The 15th Battalion embarked for Egypt from Melbourne on the 22nd of December 1914 as part of Convoy No.2. The Battalion sailed aboard A40 *Ceramic*, which carried 2,708 troops and 21 horses on the voyage. The entire convoy carried 11,305 troops and 4,643 horses to Egypt. Upon arrival in Egypt in February 1915, the 15th Battalion became part of the New Zealand and Australian Division (ANZAC).

From the 3rd of February to the 11th of April 1915, the 15th Battalion was stationed at Aerodrome Camp, Heliopolis, Egypt, during which they completed Battalion, Brigade, and Divisional training. Sergeant Hutton Armstrong was appointed as a Lieutenant on the 25th of March 1915.

On the 10th of April 1915, the 15th Battalion was moved by train to Alexandria, where it embarked on the troopships *Seeang Bee* and *Australind*, bound for Gallipoli. Assigned to the follow-up waves, the 15th Battalion landed at Anzac Cove on the afternoon of the 25th of April 1915. With the arrival of Turkish reinforcements, the ANZAC position became more precarious as the assaulting force could not secure their initial objectives. They retreated to rudimentary and shallow entrenchments, steadfastly holding on during the critical first night. By that evening, 16,000 men had been deployed, with over 2,000 Australians either killed or wounded.

Later, the 15th Battalion helped shore up the line before occupying positions around Pope's Hill and Russell's Top, where they joined an attack on the 1st of May 1915 around Quinn's Post. After that, they occupied Quinn's Post. It was here that Lieutenant Hutton Armstrong was Killed in Action while holding a trench, which he first charged and captured on the 10th of May 1915. He was 24 years and two months old when he died. Lieutenant Hutton Armstrong had served in the AIF for six months and two weeks and had been a member of the 15th Battalion for four months and two weeks. He was buried at Quinn's Post, Gallipoli but has no known grave.

Hutton Armstrong's name is inscribed on The Australian War Memorial Roll of Honour (Panel 74), The Lone Pine Memorial, Gallipoli (Panel 43), The City of Townsville WW1 Honour Roll, and The Townsville Cenotaph. Armstrong St, Hermit Park, Townsville is named in his memory.

Lieutenant Hutton Perkins Armstrong served in Egypt and Gallipoli in WW1. He was awarded the 1914/15 Star (no:15288), the British War Medal (no:6523), and the Victory Medal (no:6498). These medals, as well as a memorial scroll and a memorial plaque, were passed to his father.

Lieutenant Hutton Armstrong's death was notified in the 24th list of Australian Casualties released on the 28th.May 1915.

Elizabeth Armstrong was granted a war pension of 39 pounds (£39/-/-) per annum from the 17th of July 1915 for the loss of her son. This was approximately 35 percent of the wage paid to a private soldier in the AIF 1915.

One of Armstrong's brothers also enlisted in WW1. His details are:

3760 Private John Armstrong, 25th Battalion -returned to Australia, 5th January 1919.

The following tributes were published in the *Queensland Times* and the *Sydney Morning Herald* respectively on the dates indicated:

THE ROLL OF HONOR
LT HUTTON PERKINS ARMSTRONG

Lieutenant Hutton Perkins Armstrong, who is reported Killed in Action at the Dardanelles, was the third son of Mr. and Mrs. George Armstrong, old residents of Townsville and a brother of Mr George Armstrong, Union Bank, Kingaroy and cousin of Mr and Mrs Stahlschmidt of Kangaroo Point Brisbane. (abridged) – *Queensland Times* 29th May 1915

LIEUT. H.P. ARMSTRONG.

Lieut. Hutton Perkins Armstrong, a Queensland officer, who is reported killed in action, was 24 years of age, and only received his first appointment to the Commonwealth Military Forces in April of last year in the 2nd Infantry Regiment. He belonged to the Mount Morgan-Gladstone Company and went to the front with the 15th Battalion. – *Sydney Morning Herald* 27th May 1915.

1375 PRIVATE GEORGE SANDILANDS ROGERS (1896 – 1915)

15th Infantry Battalion

George Sandilands Rogers was born on the 11th of October 1896 in Rockhampton, Queensland. He was the only son and the youngest of four children born to Alfred and Jane Rogers (née Cable), with three older sisters. Alfred Rogers worked as a plumber.

George Rogers worked as a salesman at Smellie and Co., a major supplier of various hardware, engineering requisites, and farming equipment. He was also a Lieutenant in the cadets and was a member of the Cathedral choir.

On the 16th of November 1914, George Rogers applied at Rockhampton to join the AIF. He declared that he was single, 18 years, one month old, and had no previous military experience (despite the later newspaper reports that he was a lieutenant in the cadets). His enlistment medical examination noted that he was 5'9" (175 cm) tall, weighed 9st. (57 kg), had a fair complexion, brown eyes, and brown hair. He identified as Church of England and listed his father as his next of kin, giving his address as Walker St, Townsville.

Accepted into the AIF, George Rogers was allocated service number 1375, assigned to the Infantry, and posted to the 15th Battalion, 1st reinforcements. The 15th Battalion was established in late September 1914, six weeks after the start of the First World War. Approximately three-quarters of its personnel were volunteers from Queensland, while the rest came from Tasmania. Together with the 13th, 14th, and 16th Battalions, the 15th Battalion formed the 4th Brigade under Colonel John Monash (later Lieutenant General Sir John Monash).

The 15th Battalion embarked for Egypt from Melbourne on the 22nd of December 1914 as part of Convoy No.2. The Battalion sailed aboard A40 *Ceramic*, which carried 2,708 troops and 21 horses on the voyage. The entire convoy carried 11,305 troops and 4,643 horses to Egypt. HMAT *Berrima* also towed the submarine *AE2*.

Upon arrival in Egypt in February 1915, the 15th Battalion became part of the New Zealand and Australian Division (ANZAC). From the 3rd of February until the 11th of April, Private Rogers trained with the rest of the 15th Battalion at Heliopolis Camp near Cairo. The battalion completed Battalion, Brigade, and Divisional training in preparation for the coming battle in the Dardanelles.

On the afternoon of the 24th of April 1915, the 15th boarded HMAT *Australind* for a short overnight voyage. At 5:30 a.m. on the 25th of April, the *Australind* was at the entrance to the Dardanelles, and Private Rogers would have been able to observe the bombardment and landing. The 15th Battalion landed at Anzac Cove on the afternoon of the 25th of April 1915.

The Ottoman defenders halted the Allied advance inland, and the arrival of Turkish reinforcements made the ANZAC position increasingly precarious as the assaulting force struggled to secure initial objectives. The 15th Battalion was rushed to the left flank of the beachhead, relying on improvised, shallow entrenchments to hold on for a crucial first night. By that evening, 16,000 men had landed, with over 2,000 Australians killed or wounded.

The 15th Battalion positioned itself between the 1st and 2nd Brigades, remaining there until the 30th of April. Captain Hugh Quinn led part of the 15th to establish what became known as Quinn's Post.

During the consolidation operations at ANZAC, Private George Sandilands Rogers was Killed in Action on the 27th of April 1915. He was 18 years and six months old and had served with the AIF for five months and one week.

Initially buried at Popes Hill Cemetery, Private George Sandilands was later reinterred and laid to rest at Quinn's Post. Cemetery (Row D, Grave No. 9), Gallipoli, Turkey. His death was reported in the 27th list of Australian Casualties released on the 31st of May 1915.

An undated letter from 733 Sergeant Jack Craven (later Captain J. Craven D.C.M.) written to Duncan Baxter, about his son Neil Baxter, states that "*Neil went missing from the affair on the 27th of April, and that there is very little hope of him being alive. Neil was out with George Rogers on the outpost, and men were seen to drop everywhere, but it was impossible to see who they were.' George Rogers was one of the few to get back, only to die in my arms.*"

George Sandilands Rogers' name is inscribed on The Australian War Memorial Roll of Honour (Panel 77), The City of Townsville WW1 Honour Roll, and The Townsville Cenotaph. Rogers Street, Aitkenvale, Townsville, is named in his memory.

1375 Private George Sandilands Rogers served in Egypt and Gallipoli in WW1 and was awarded the 1914/15 Star (no: 12835), the British War Medal (no: 6730), and the Victory Medal (no: 6704). A memorial scroll and a memorial plaque were passed to his father in 1922 and 1923, respectively.

The following article was published in the *Townsville Daily Bulletin* on the 28th of May 1915:

THE WAR IN EUROPE.

Townsville boys are suffering heavily at the Dardanelles. Mr & Mrs A G Rogers, of Walker Street, were advised yesterday that their only son, George Sandilands Rogers, had been Killed in Action on May 23. Young Rogers was a lieutenant in the cadets and left with the second contingent, attached to the signallers. He was only 18 years of age but was an enthusiast at his work. Before his departure, the deceased was engaged at Smellie and Co's and was a member of the Cathedral choir. General sympathy will be extended to Mr and Mrs Rogers in the loss of their only son.

The following tribute was published in the *Townsville Daily Bulletin* on the 27th of April 1916:

Family Notices

ROGERS. —In loving memory of our dear and only son and brother, Private George Sandilands Rogers, 15th Battalion, who fell in action at Gallipoli, April 27th, 1915.

MAJOR HUGH QUINN (1888 – 1915)

15th Infantry Battalion

Hugh Quinn was born on the 6th of May 1888, in Charters Towers, Queensland. He was the eldest son of John Quinn, an Irish-born mounted police constable, and his Australian wife, Mary Jane (née Irwin). He attended Millchester State School and Dixon's School at Southport. John Quinn died in 1912 when Hugh Quinn was 24 years old. Mary Quinn became the manager of the Gladstone Hotel in Charters Towers but later moved to Uanda with her son Frank.

On leaving school, Hugh Quinn joined Cummins & Campbell, Wine and Spirit merchants of Townsville. Later, he set up his own business as a commission agent.

A talented boxer, Quinn became the North Queensland champion. He was also a member of the militia in the Kennedy Infantry Regiment. Rising through the ranks, he was commissioned as a Second Lieutenant in 1908, promoted to Lieutenant in 1911, and to Captain in 1912.

After the declaration of war in August 1914, the Kennedy Regiment briefly served garrison duty on Thursday Island. Captain Quinn and 500 others from the unit then volunteered for the Australian Naval and Military Expeditionary Force (AN&MEF), where he was appointed adjutant.

The AN&MEF was assigned to the Australian takeover of German New Guinea from September to November 1914. However, Captain Quinn did not participate due to a strike by the firemen aboard the troopship *Kanowna* . After the ship's departure from Port Moresby for Rabaul on the 7th of September 1914, the force returned to Townsville.

Hugh Quinn arrived back in Townsville on 18th September 1914 and, six days later, applied for a commission in the AIF on 24th September. He was 26 years and four months old. He declared himself single with eight years of previous military service in the Kennedy Regiment and AN&MEF. His medical assessment noted that he was 5'9¾" (177 cm) tall and weighed 12st 8lbs (80 kg). He identified as Roman Catholic and listed his mother, living at the Gladstone Hotel in Charters Towers, as his next of kin.

Accepted into the AIF, Hugh Quinn was appointed as a Captain in the newly formed 15th Infantry Battalion and posted as Officer Commanding 'C' Company.

The 15th Battalion was formed in late September 1914, six weeks after the start of the First World War. It was made up of volunteers, with three-quarters from Queensland and the remainder from Tasmania. Together with the 13th, 14th, and 16th Battalions, it became part of the 4th Brigade, commanded by Colonel John Monash (later Lieutenant General Sir John Monash).

The 15th Battalion embarked for Egypt from Melbourne on the 22nd of December 1914 as part of Convoy No.2. The Battalion sailed aboard A40 *Ceramic*, which carried 2,708 troops and 21 horses on the voyage. The entire convoy carried 11,305 troops and 4,643 horses to Egypt. HMAT *Berrima* also towed the submarine *AE2*.

Disembarking in Egypt in February 1915, the 15th Battalion became part of the New Zealand and Australian Division (ANZAC). From the 3rd of February until the 10th of April, the battalion completed Battalion, Brigade, and Divisional training at Heliopolis Camp near Cairo, in preparation for the coming battle in the Dardanelles.

On the 10th of April 1915, the 15th Battalion was moved by train to Alexandria, and on the afternoon of 24th April 1915, the 15th boarded HMAT *Australind* for a short overnight voyage. At 5:30 a.m. on the 25th of April, the *Australind* was at the entrance to the Dardanelles, and all aboard would have been able to observe the bombardment and landing.

The 15th Battalion landed at Anzac Cove on the afternoon of the 25th of April 1915 and was immediately sent to the left flank of the beachhead. Captain Quinn's company landed the next day, on the 26th of April.

On the 29th of April 1915, Captain Quinn was ordered to defend a crucial position at the head of Monash Valley, which protected the main supply route from the beach. The position consisted of shallow rifle-pits across a 150-yard (137m) front, exposed on two sides, with part of the rear open to enemy fire. In some areas, the front lines were only 10 yards (9m) apart. This position was initially called Quinn's Hill and Quinn's Corner, but later became known as Quinn's Post.

The Turks held a significant advantage in firepower. On the 1st of May 1915, they launched several attacks but were repelled by the Australians. That same day, Captain Quinn was promoted to Major. Despite setbacks, the Turks continued tunnelling and, on the 29th of May, detonated mines, and seized part of Quinn's Post.

Under Major Quinn's leadership, the Australians launched a successful counterattack, reclaiming much of the position. Ordered to lead a second attack, Major Quinn needed to reconnoitre the front lines. During this effort, he was Killed in Action by rifle fire.

The eventual second attack was successful. Major Hugh Quinn was 27 years and one month old and had served with the AIF for eight months and five days. He was laid to rest in Shrapnel Valley Cemetery (Plot III, Row C, Grave

No.21), Anzac, Gallipoli, Turkey and his death was notified in the 38th list of Australian Casualties released on the 14th of June 1915.

Hugh Quinn's name is inscribed on the Australian War Memorial Roll of Honour (Panel 77), The City of Townsville WW1 Honour Roll, the Boys Central School Great War Honor Board, and The Townsville Cenotaph. Quinn Street Rosslea, Townsville is named in his memory.

Major Hugh Quinn served in WW1 in Gallipoli. He was awarded the 1914/15 Star (no:15298), the British War Medal (no:6725) and the Victory Medal (no:6699). A memorial scroll and a memorial plaque was passed to his mother in 1921 and 1922, respectively. She was also granted a war pension of 104 pounds (£104) per annum, which was approximately three quarters of the national minimum award wage in 1915.

The following notice was published in *The Northern Miner* (Charters Towers, Qld.) on 29th May 1919:

ROLL OF HONOR.

QUINN. —In proud and loving memory of our dear beloved son and brother, Major Hugh Quinn, 15th Battalion, 4th Brigade, Killed in Action at Quinn's Post, Gallipoli.

332/617 TROOPER LOUIS GERALD GRIFFIN (1874 - 1915)

4th Queensland Bushmen & 8th Light Horse Regiment

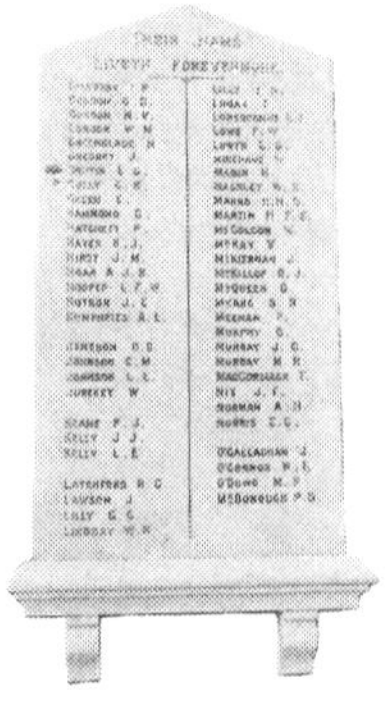

Louis Gerald Griffin was born in about February 1874 at Bundaberg, Queensland, to Gerald and Margaret Griffin (née Fogarty). His father was a skilled lithographer who later became the licensee of the Criterion Hotel at Bundaberg in 1877. Gerald Griffin passed away in 1881 from a lengthy illness when Louis was about seven years old.

After Gerald's death, Margaret Griffin managed the Criterion Hotel until 1884, when she married Lewis

O'Mara. Margaret Griffin died in November 1889, shortly after giving birth to her last child. Louis Griffin was approximately 15 years old at the time of her death.

Records of Louis Griffin's whereabouts after his mother's death are unclear. He had an uncle in Townsville, where he stayed with his aunt Gertrude and later Eleanor. His uncle, Daniel Ryan, was elected to the Queensland state parliament in 1915 and held the seat of Townsville until 1920.

In 1900, at around 26 years old, Louis Griffin enlisted in the Fourth (Queensland Imperial Bushmen) Contingent for service in the Boer War, receiving service number 332. The contingent required candidates to be skilled marksmen, riders, experienced bushmen, and in good health. They left on 18th May 1900, with 26 officers, 368 other ranks, and 512 horses. The contingent fought in several engagements and returned to Australia on the 5th of August 1901 aboard the *Britannic*. Whether Louis Griffin Louis returned to Australia or stayed in Africa is unclear, but he was in Africa in 1901, trying his hand at mining. In 1907, he applied for assisted passage back to Australia, which was approved, requiring him to reimburse the Australian Government eight pounds (£8). At that time, he stated he was 35 years old and born in 1872.

Upon his return to Australia, Louis Griffin worked as a bush labourer in the Cloncurry, Cairns, and Townsville regions for seven years. On the 31st of August 1914, he applied to join the Australian Imperial Force (AIF) in Townsville, claiming to be 36 years and seven months old, though he was older. His medical records noted he was 6' tall, weighed 11st 12lbs, had a fresh complexion, grey eyes, and black hair. He listed his uncle, Mr. D. Ryan, as his next of kin.

Accepted into the AIF, Louis Griffin was allocated service number 617, assigned to the Infantry, and posted to the 9th Battalion. He began training in Brisbane but later re-enlisted in the 8th Light Horse Regiment (LHR), 2nd reinforcements, in Melbourne on the 9th of October 1914, receiving service number 698. He listed his brother, Paul Griffin of Bundaberg, as his next of kin.

The 8th LHR, formed on the 3rd of October at the Broadmeadows camp, was a light mounted infantry unit. Trooper Louis Griffin and his unit embarked from Melbourne on the 12th of February 1915 aboard HMAT A26 *Armadale*, part of a convoy transporting troops and horses to Egypt.

In May 1915, reinforcements were urgently needed at Gallipoli, and the Light Horse Regiments were deployed without their horses. Trooper Louis Griffin joined the 8th LHR on Gallipoli on the 27th of July 1915. On the 7th of August 1915, the 8th LHR suffered devastating losses in the Battle of the Nek, with 372 out of 600 Australians either killed or wounded. Trooper Griffin was initially reported Missing in Action and later confirmed Killed in Action. He was approximately 41 years old and had served in the AIF for 11 months.

Louis Griffin's name is inscribed on The Australian War Memorial Roll of Honour, The Lone Pine Memorial Gallipoli, The City of Townsville WW1 Honour Roll, and The Townsville Cenotaph. Griffin Street in Townsville was named in his memory.

698 Trooper Louis Gerald Griffin served in WW1 at Gallipoli. He was awarded the 1914/15 Star (no:16038), the British War Medal (no:8794), and the Victory Medal. A memorial scroll and a memorial plaque were passed to his brother in 1921 and 1922, respectively.

The following notice was published in the *Queensland Times* on 1st September 1915:

Mr. D. Ryan, M.L.A., received advice on Monday that one of his nephews in the Dardanelles was missing since the 7th of August. Yesterday he was advised that another nephew, Louis Gerald Griffin, formerly of Bundaberg and Townsville, was killed on the 7th instant.

1837 CORPORAL EDWIN CHARLES NORRIS (1885 – 1915)

3rd Infantry Battalion

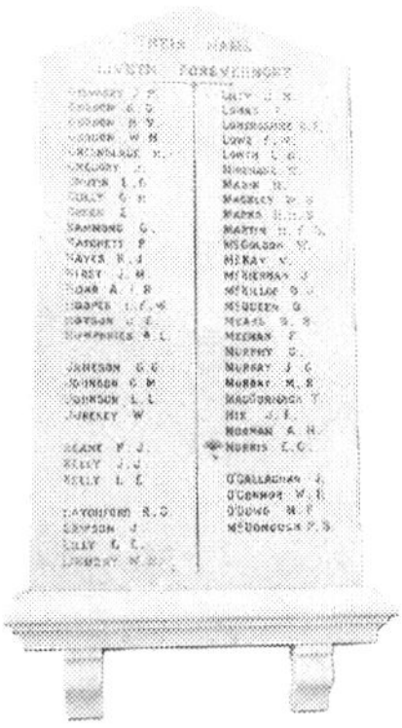

Edwin Charles Norris, also known as Jack Vincent, was born on the 16th of November 1885 in Brisbane to Edwin and Emily Norris (née Kennedy). He was the only surviving child of the Norris family. Edwin Snr. was a solicitor, and Edwin Jnr. (Jack) grew up in Townsville and attended Townsville West State School as a child. After finishing school, he qualified as an accountant.

"Jack" was also a gifted singer, and he both sang in and conducted the choir at St Peters' Church of England, West Townsville. He was working as an accountant with Burns Phillips Merchants in Townsville before he joined J. C. Williamson's opera, being described as *"[possessing] a capital baritone voice"*. Now a professional singer, he applied to enlist in the Australian Imperial Force (AIF) on the 4th of January 1915 at Liverpool, NSW, using his stage name "Jack Vincent". He declared that he was single, an accountant, and 29 years old. He also noted he had two years of service with the Infantry in Townsville. His medical examination recorded that he was 5'7½" (171 cm) tall and weighed 13st 6lbs (85kg). He had a dark complexion, grey eyes, and light dark hair. His religious denomination was Church of England, and he nominated his mother as his next of kin, giving her address as Flinders Street West, Townsville.

Accepted into the AIF, Jack Vincent was allocated service number 1837, assigned to the Infantry, and posted to the 3rd Battalion, 4th Reinforcements. The 3rd was one of the first Battalions raised for the AIF. When Jack Vincent joined the AIF, the Battalion was already in Egypt preparing for the coming operations on the Gallipoli Peninsula. For the next 10 weeks, Private Jack Vincent completed his basic training in Australia. He was promoted to Lance Corporal due to his previous military service with the CMF Infantry. The Reinforcements and Lance Corporal Vincent embarked from Sydney with Convoy No. 4 on the 17th of March 1915 (some records indicate the 7th of March 1915 as the embarkation date) bound for Egypt. They sailed aboard

HMAT *Shropshire*, which carried 560 troops and 424 horses. The five ships of the convoy transported a total of 3,299 troops and 2,124 horses on the voyage.

The 4th Reinforcements joined the 3rd Battalion on Gallipoli on the 31st of May 1915. Lance Corporal Vincent reverted to Private upon joining the unit but was promoted to Corporal three weeks later, on the 23rd of June 1915.

In August 1915, the ANZACs were tasked with a crucial operation to break the Gallipoli stalemate. The plan involved a diversionary attack at Lone Pine, while the main assault targeted the northern area. The 1st Australian Brigade, including the 1st, 2nd, and 3rd Infantry Battalions with about 3,000 soldiers, was selected for the Lone Pine mission. The ensuing battle was among the fiercest the Australian troops had faced in the campaign so far. The Australians incurred 2,277 casualties, including both deaths and injuries.

Corporal Jack Vincent was reported missing after the attack on Lone Pine in August 1915 and was later declared by a Court of Inquiry to have been Killed in Action sometime between the 7th and 12th of August 1915. He was 29 years, nine months old and had served in the AIF for seven months. His body was not recovered, and he has no known grave. His death was notified in the 240th list of Australian Casualties released on the 10th of November 1916.

Corporal Edwin Charles Norris' name is inscribed on The Australian War Memorial Roll of Honour (Panel 37), The Lone Pine Memorial Gallipoli, Turkey (Panel 6), The City of Townsville WW1 Honour Roll, and The Townsville Cenotaph. Norris St, Hermit Park, Townsville, is named in his memory.

1837 Corporal Edwin Charles Norris was awarded the 1914/1915 Star (no:11886), the British War Medal (no:8880), and the Victory Medal (no:8837). A memorial scroll and a memorial plaque were passed to his father in 1921.

2204 PRIVATE JOHN INGLIS SMITH (1895 – 1915)

15th Infantry Battalion

John Inglis (Jack) Smith was born on the 16th of May 1895 in Waverley, New South Wales. He was the sixth of seven children and the second son born to John and Florence Smith (née Cheesmond). He had one brother and four surviving sisters.

The Smith family moved to Townsville when John was a young boy. His father, John Smith Sr., was a draper who opened and owned a retail drapery in Flinders St, Townsville (Inglis Smith and Co.). John Smith Jr. grew up in Townsville and attended the Central and West End State Schools, followed by Townsville Grammar School and Gatton Agricultural College. He completed his education at Scotch College in Melbourne.

After finishing school, John Smith joined his father's drapery business in Townsville. He was also a member of the Cadets and then the Citizen Forces. In August 1914, John Smith was mobilized for war service and embarked at Cairns, Queensland, for Thursday Island (War Station) Garrison Duty. On the 14th of August 1914, Private John Smith enlisted for service outside Australia. He joined the Australian Naval and Military Expedition Force (A.N.&M.E.F.) and embarked from Thursday Island on the 16th of August 1914 aboard the Troopship *Kanowna* , with the intention of participating in the capture of German New Guinea.

However, during the voyage, a strike among the firemen on the *Kanowna* forced the ship to return to Townsville, stoked by the troops. Consequently, Private John Smith was unable to take part in the capture of German New Guinea. The *Kanowna* docked in Townsville on the 18th of September 1914, and on the same date, Private Smith was discharged.

Six months later, on the 15th of March 1915, John Smith re-enlisted in Townsville, this time applying to join the AIF. He was 19 years and 10 months old, and both of his parents provided written consent for him to enlist for active service. John Smith declared that he was single, working as a draper, and had served in the Cadets for six years and the Citizen Forces for two years. His enlistment medical noted that he was 5'9½" (177 cm) tall, weighed 9st (57 kg), had a fair complexion, blue eyes, and light brown hair. He identified as Presbyterian and listed his father as his next of kin, giving his address as Flinders St, Townsville.

Accepted into the AIF, John was allocated service number 2204, assigned to the Infantry, and posted to the 15th Battalion, 6th reinforcements. The 15th Battalion was established in late September 1914, six weeks following the commencement of the First World War. Approximately three-quarters of its personnel were volunteers from Queensland, while the remaining recruits came from Tasmania. Together with the 13th, 14th, and 16th Battalions, the 15th Battalion constituted the 4th Brigade under the leadership of Colonel John Monash (later Lieutenant General Sir John Monash). The 15th Battalion was training in Egypt when John Smith enlisted.

After initial training, Private John Smith and his fellow reinforcements embarked from Brisbane with Convoy No. 8 on the 12th of June 1915, bound for Egypt. They sailed aboard HMAT A63 *Karoola*, which transported 1,047 troops. The entire convoy transported 2,470 troops on the voyage. Private John Smith was promoted to Acting Corporal.

After disembarking in Egypt, a brief period of further training was undertaken in the desert conditions. Training completed, Corporal John Smith, along with the rest of the reinforcements, joined the 15th Battalion on the Gallipoli Peninsula on the 2nd of August 1915. Corporal John Smith reverted to Private when he joined the 15th at Gallipoli.

Private John Smith and his fellow reinforcements arrived at Gallipoli in time to participate in the August Offensive, also known as the Battle of Sari Bair. This was the final attempt by British and ANZAC forces to capture the Gallipoli peninsula from the Turks. The Allied strategy focused on securing the high ground of the Sari Bair ridge, linking the ANZAC front with a new front to the north at Suvla. The main objectives of the offensive were to capture two key peaks on the Sari Bair range: Chunuk Bair and Hill 971.

On the 7th and 8th of August 1915, the 15th Battalion participated in the assault on Hill 971 (Koja Chemen Tepe), the highest point of the Sari Bair Ridge at Gallipoli. Advancing as part of the 4th Brigade during the night of the 6th of August, they faced challenging terrain and by dawn on the 7th had fallen short of their objective. An attempt to capture the summit on the 8th of August resulted in a costly failure.

It was during this offensive that Private John Inglis Smith was Killed in Action on the 8th of August 1915. He was 20 years and three months old and had served with the AIF for six months. He had been a member of the 15th Battalion for six days. Private John Inglis Smith has no known grave, and his death was notified in the 79th list of Australian casualties released on the 17th of September 1915.

An Australian Red Cross enquiry was conducted to determine the details of Private John Inglis Smith's death. In that report, 2128 Private W.J. Kemp stated that Private Smith was killed about 8:00 am on the 8th of August and that he *"passed him lying dead shot through the head."* He goes on to state that they had to retire from where Smith lay.

John Inglis Smith's name is inscribed on The Australian War Memorial Roll of Honour (Panel 77), The Lone Pine Memorial Gallipoli, Turkey (Panel 44), The Townsville West State School Memorial Gate, The Townsville Grammar School Memorial, The Gatton Agricultural College WW1 Honour Roll, The City of Townsville WW1 Honour Roll, and The Townsville Cenotaph. He is listed on the Scotch College World War I Commemorative Website. Inglis-Smith Street, Rosslea, Townsville is named in his memory, although his surname is clearly listed as "Smith," not "Inglis-Smith," as written in his own hand on his attestation papers.

2204 Private John Inglis Smith was awarded the 1914/15 Star (No.14958), the British War Medal (No.14072), and the Victory Medal (No.14015). A memorial scroll and a memorial plaque were passed to his father in 1921 and 1922, respectively.

John Smith's older brother, Morrison Smith, also enlisted in the AIF. A member of the 11th Light Horse Regiment, he survived the war and returned to Australia on the 9th of July 1919. A sad circumstance is that whilst Corporal Inglis Smith was in Cairo, Trooper Morrison Smith arrived there. Each brother called at the other brother's camp but had the bad luck to miss seeing each other.

A story detailing John Inglis Smith's life and death was published in the *Townsville Daily Bulletin* (Qld) on Saturday, the 11th of September 1915.

2201 PRIVATE ARTHUR SNELHAM (1890 – 1915)

15th Infantry Battalion

Arthur Snelham was born in June 1890 in Preston, Lancashire, England. He was the youngest child of Thomas and Christiana Snelham (née Fawcett). Thomas Snelham worked as a bookkeeper at the biscuit works in Preston. Arthur grew up in Preston and attended the local Grammar School. After completing his schooling, Arthur Snelham trained as an electrical apprentice with Dick, Kerr and Co., a locomotive and tramcar manufacturer.

His two older brothers, Robert and Frederick, left England for Australia in the early 1900s, eventually settling in Townsville, and sharing a house in Morris St, West End, Townsville. Arthur Snelham followed his brothers to Australia in 1911, arriving in Brisbane on the 1st of January 1912. He also moved to Townsville and worked as a labourer.

On the 22nd of March 1915, he applied in Townsville to enlist in the Australian Imperial Force (AIF). He was 24 years and nine months old, single, and working as a labourer, with no previous military experience. Arthur's enlistment medical noted that he was 5'7½" (171 cm) tall, weighed 8st 12lbs (56 kg), had a fair complexion, hazel eyes, and light brown hair. He identified as Church of England and listed his father as his next of kin, giving his address as Ainslee Road, Preston, Lancashire, England.

Accepted into the AIF, Arthur Snelham was allocated service number 2201, assigned to the Infantry, and posted to the 15th Battalion, 6th reinforcements. The 15th Battalion was formed in late September 1914, six weeks after the outbreak of the First World War. About three-quarters of its members were volunteers from Queensland, with the rest recruited from Tasmania. Alongside the 13th, 14th, and 16th Battalions, the 15th Battalion made up the 4th Brigade, commanded by Colonel John Monash (later Lieutenant General Sir John Monash). The battalion was training in Egypt when Arthur Snelham enlisted.

Following initial training, Private Arthur Snelham and his fellow reinforcements embarked from Brisbane with Convoy No.8 on the 12th of June 1915, bound for Egypt. They sailed aboard HMAT A63 *Karoola*, which transported 1,047 troops. The entire convoy of four ships transported 2,470 troops on the voyage. After disembarking in Egypt, a brief period of further training was undertaken in the desert conditions.

Training completed, Private Snelham, along with the rest of the reinforcements, joined the 15th Battalion on the Gallipoli Peninsula on the 2nd of August 1915. The reinforcements arrived at ANZAC in time to participate in the August Offensive (also known as the Battle of Sari Bair), which marked the final attempt by British and ANZAC forces to capture the Gallipoli peninsula from the Turks. The Allies planned to secure the high ground along the Sari Bair ridge, linking the ANZAC front with a new front to the north at Suvla. Key objectives of the August Offensive included capturing two prominent peaks on the Sari Bair Range: Chunuk Bair and Hill 971.

On the 7th and 8th of August 1915, the 15th Battalion took part in the assault on Hill 971 (also known as Koja Chemen Tepe), the highest point on the Sari Bair Ridge at Gallipoli. The 4th Brigade, which included the 15th Battalion, advanced on the night of the 6th of August but made slow progress due to the difficult terrain, and by dawn on the 7th of August, they had not reached their objective. A further attempt to take the summit on the 8th of August ended in a costly failure.

It was during this offensive that Private Arthur Snelham was reported as wounded and missing. A later Court of Enquiry determined that he had been

Killed in Action on the 8th of August 1915. He was 25 years and two months old. He had served with the AIF for 22 weeks and had been a member of the 15th Battalion, in the field, for nine days. Arthur Snelham's body was not recovered, and he has no known grave. His death was notified in the 80th list of Australian casualties released on the 19th of September 1915.

Arthur Snelham's name is inscribed on The Australian War Memorial Roll of Honour (Panel 77), The Lone Pine Memorial, Gallipoli, Turkey (Panel 49), The City of Townsville WW1 Honour Roll, and The Townsville Cenotaph.

2201 Private Arthur Snelham was awarded the 1914/15 Star (No.17923), the British War Medal (No.14073), and the Victory Medal (No.14016). A memorial scroll and a memorial plaque were passed to his father in 1922.

2001 PRIVATE JAMES ALBERT SOONING (1892 – 1915)

15th Infantry Battalion

J. A. SOONING

James Albert Soon Hing was born on the 11th of October 1892 in Townsville, Queensland, the son of William and Mary Soon Hing (née Waddell). William Hoon Sing was born in China, and Mary in Scotland. James was the third of six children and had four brothers and one sister. William Hoon Sing died in 1900 when James was eight years old. The family was raised in Townsville by Mary Soon Hing, who remarried in 1916.

James Soon Hing attended the West End State School, and the family lived in Sturt Street in Townsville. After finishing school, he trained as a gasfitter apprentice for four years with the Townsville Gas Company. On the 18th of January 1915, James Soon Hing applied in Townsville to enlist in the Australian Imperial Force (AIF) under the name James Albert Sooning. He was 22 years and three months old, single, and had one year's previous military experience with the Senior Cadets. His enlistment medical noted that he was

5'6" (168 cm) tall, weighed 8st 10lbs (55 kg), had a dark complexion, brown eyes, and dark hair. He identified as Presbyterian and listed his mother as his next of kin, giving her address as Livingstone Street, West End, Townsville.

Accepted into the AIF, James Sooning was allocated service number 2001, assigned to the Infantry, and posted to the 15th Battalion, 5th reinforcements. The 15th Battalion was formed in late September 1914, six weeks after the outbreak of the First World War. About three-quarters of its members were volunteers from Queensland, with the rest recruited from Tasmania. Alongside the 13th, 14th, and 16th Battalions, the 15th Battalion made up the 4th Brigade, commanded by Colonel John Monash (later Lieutenant General Sir John Monash). The battalion was on its way to Egypt when James Sooning enlisted.

Following three months of initial training, Private James Sooning and his fellow reinforcements embarked from Brisbane, Queensland, on the 16th of April 1915 with Convoy No.5, bound for Egypt. They sailed aboard the HMAT A55 *Kyarra*, which carried 913 troops on the voyage. The seven ships of Convoy No.5 transported a combined total of 3,220 troops and 2,018 horses to Egypt. After further training in Egypt, Private James Sooning joined the 15th Battalion on the Gallipoli Peninsula on the 16th of June 1915, in the company of fifty other reinforcements.

The reinforcements arrived at ANZAC in time to participate in the August Offensive (also known as the Battle of Sari Bair), which marked the final attempt by British and ANZAC forces to capture the Gallipoli peninsula from the Turks. The Allies planned to secure the high ground along the Sari Bair ridge, linking the ANZAC front with a new front to the north at Suvla. Key objectives of the August Offensive included capturing two prominent peaks on the Sari Bair Range: Chunuk Bair and Hill 971.

On the 7th and 8th of August 1915, the 15th Battalion took part in the assault on Hill 971 (also known as Koja Chemen Tepe), the highest point on the Sari Bair Ridge at Gallipoli. The 4th Brigade, which included the 15th Battalion, advanced on the night of the 6th of August but made slow progress due to the difficult terrain, and by dawn on the 7th of August, they had not reached their objective. A further attempt to take the summit on the 8th of August ended in a costly failure.

It was during this offensive that Private James Sooning was Killed in Action on the 8th of August 1915. A note in Sooning's military file records that "*1955 Pte W. Humphrey, 15th Bn, C Coy (formerly B Coy), says that in the charge on the morning of 7.8.15 past No.3 Outpost, he passed Sooning lying on the ground shot through the head. He was dead and must have been killed outright.*" Private James Sooning was 22 years and nine months old at the time of his death. He had served with the AIF for seven months and had been a member of the 15th Battalion at Gallipoli for seven weeks. His body was not recovered, and he has no known grave. His death was notified in the 79th list of Australian Casualties released on the 17th of September 1915.

James Sooning's name is inscribed on The Australian War Memorial Roll of Honour (Panel 77), The Lone Pine Memorial, Gallipoli, Turkey (Panel 49), The City of Townsville WW1 Honour Roll, and The Townsville Cenotaph. Sooning Street, Hermit Park, and Sooning Street, Nelly Bay (originally part of Nelly Bay Esplanade), Townsville, are named in his memory.

2001 Private James Albert Sooning was awarded the 1914/15 Star (No.13852), the British War Medal (No.10670), and the Victory Medal (No.1064). A memorial scroll and a memorial plaque were passed to his mother in 1921 and 1922, respectively.

Mary Soon Hing was granted a pension of 26 pounds (£26/-/-) per annum from the 8th of October 1915, later increased to £2 per fortnight, for the loss of her son. This was half of the wage paid to AIF soldiers in 1915.

1943 PRIVATE JOHN GEORGE FRY (1896 – 1915)

15th Infantry Battalion

J. G. FRY.

John George Fry was born on the 24th of January 1896 in Devon, England. He was the second child of Thomas and Edith Fry (née Brown). He had two brothers and four sisters. Edith Fry arrived in Australia in 1901 with three of the children, including John, who was five years old. Thomas Fry was a sailmaker, and the family settled in Townsville, living in Bell St, South Townsville.

John Fry grew up in Townsville and attended South Townsville State School. After completing his schooling, he started a boilermaker apprenticeship. On the 18th of January 1915, John Fry applied in Townsville to enlist in the Australian Imperial Force (AIF). He declared that he was single, 18 years and 11 months old, and working as an apprentice boilermaker. He also declared that he had served in the Naval Cadets for three and a half years and had won three silver medals for rifle shooting. His enlistment medical noted that he was 5'5" (165 cm) tall, weighed 116lbs (52 kg), had a fair complexion, blue eyes, and light grey hair. He identified as Church of England and nominated his mother as his next of kin, giving her address as Allen St, South Townsville.

Accepted into the AIF, with his mother's permission, John Fry was allocated service number 1943, assigned to the Infantry, and posted to the 15th Battalion, 5th reinforcements. The 15th Battalion was formed in late September 1914, six weeks after the beginning of the First World War. Around three-quarters of its members were volunteers from Queensland, while the rest came from Tasmania. Alongside the 13th, 14th, and 16th Battalions, the 15th Battalion made up the 4th Brigade under the command of Colonel John Monash (later Lieutenant General Sir John Monash). The 15th Battalion was on its way to Egypt when John Fry enlisted.

Following three months of initial training, Private John Fry and his fellow reinforcements embarked from Brisbane, Queensland, on the 16th of April 1915, with Convoy No.5, bound for Egypt. They sailed aboard the HMAT A55 *Kyarra*, which carried 913 troops on the voyage. The seven ships of Convoy No.5 transported a combined total of 3,220 troops and 2,018 horses to Egypt.

After further training in Egypt, Private John Fry joined the 15th Battalion on the Gallipoli Peninsula on the 16th of June 1915 in the company of fifty other reinforcements. He arrived at ANZAC in time to join the August Offensive (also known as the Battle of Sari Bair), which was the final attempt made by the British and ANZAC forces to seize control of the Gallipoli Peninsula from the Turks. It was during this offensive that Private John Fry was wounded by a gunshot in the abdomen on the 21st of August 1915. He was evacuated to the 4th Field Ambulance at ANZAC and died of his wounds on the 22nd of August 1915.

Private John Fry was 19 years and seven months old at the time of his death. He had served with the AIF for seven months and three weeks and been a member of the 15th Battalion for one month. He has no known grave. His death was notified in the 94th list of Australian Casualties released on the 19th of October 1915.

John Fry's name is inscribed on The Australian War Memorial Roll of Honour (Panel 75), The Lone Pine Memorial Gallipoli, Turkey (Panel 46), The City of Townsville WW1 Honour Roll, and The Townsville Cenotaph. Fry Street, Belgian Gardens, in Townsville, is named in his memory.

1943 Private John George Fry was awarded the 1914/15 Star, the British War Medal, and the Victory Medal. These medals, as well as a memorial scroll and a memorial plaque, were passed to his father.

The following articles were published in the *Townsville Daily Bulletin* (Qld) on the 22nd of August 1918:

IN MEMORIAM.

FRY.—In loving memory of John Fry, who died at Anzac from wounds, August 22nd, 1915. In sure and certain hope of a glorious resurrection.

(Inserted by his loving father, mother, brothers, and sisters, Thos. Fry and family.

FRY.—In loving memory of our dear son and brother, Private John G. R. Fry, who Died of Wounds at Gallipoli, August 22nd, 1915.

Eternal rest grant to him, O Lord, and let light perpetual shine upon him.

(Inserted by his loving parents, brothers, and sisters.)

246 CORPORAL JOHN EDWARD HOTSON (1888 – 1915)

18th Infantry Battalion

John Edward Hotson was born on the 26th of January 1888, in Charters Towers, Queensland, to Richard and Elizabeth Hotson (née Guthrie). Richard Hotson died in August 1888 when John Hotson was only about seven months old. Elizabeth Hotson remarried in June 1900, and John Hotson was raised by his stepfather, Henry Haylock. Elizabeth and Henry had two sons together, making them John Hotson's younger half-brothers.

Henry Haylock and Elizabeth lived at the Commonwealth Hotel in Townsville, with Elizabeth listed as Licensed Victualler and Henry as Manager. John Hotson grew up in Townsville and attended the South Townsville State School. After completing his schooling, he started working in the footwear industry.

In 1915, John Hotson was living in Sydney, NSW, and employed as a boot salesman with T. Guthrie and Coy, located on Elizabeth Street. On the 3rd of February 1915, he applied to enlist in the Australian Imperial Force (AIF) at Liverpool, NSW. He declared that he was 27 years old and single. His medical examination noted that he was 5'7" (170 cm) tall, weighed 148 lbs (67 kg), had a dark complexion, brown eyes, and dark hair. He identified as Church of England and listed his mother as his next of kin, giving her address as the Commonwealth Hotel, Townsville.

Accepted into the AIF, John Hotson was allocated service number 246, assigned to the Infantry, and posted to "A" Company of the 18th Battalion. The 18th Battalion was raised at Liverpool in NSW in March 1915 as part of the 5th Brigade. Five months of basic training, equipment issues, and battalion organization followed enlistment for Private John Hotson. During this period, Private John Hotson was promoted to Lance Corporal on the 29th of March 1915.

Once training and organization were completed, the 18th embarked from Sydney with Convoy No. 6 on the 25th of June 1915, bound for Egypt. They sailed aboard Transport A40 *Ceramic*, which carried 2,776 troops. The fifteen ships that comprised Convoy No. 6 transported a total of 13,951 military personnel, 156 civilians, and 2,594 horses on the voyage.

After arriving in Egypt, the 18th Battalion trained from mid-June until mid-August 1915. Lance Corporal Hotson was promoted to Corporal on the 23rd of July 1915. On the 22nd of August, the 18th Battalion landed at ANZAC Cove. They had not been ashore for a single day before the battalion was committed to the costly August Offensive. The 18th was initially deployed to Damakjelik Bair and, from there, participated in the attack on Hill 60, which lasted until the 29th of August. Beginning the battle with about 1,000 men, by the end, the battalion could muster only 386.

It was during the attack on Hill 60 that Corporal John Hotson was Killed in Action on the 22nd of August 1915. He was 27 years and five months old and had been a member of the AIF for five and a half months. He has no known grave, and his death was notified in the 89th List of Australian Casualties released on the 11th of October 1915.

John Edward Hotson's name is inscribed on The Australian War Memorial Roll of Honour (Panel 85), The Lone Pine Memorial Gallipoli, Turkey (Panel 60), The City of Townsville WW1 Honour Roll, and The Townsville Cenotaph. Hotson Street, Railway Estate, Townsville, was named in his memory in 1926.

Corporal John Edward Hotson was awarded the 1914/15 Star (no: 16531), the British War Medal (no: 20423), and the Victory Medal (no: 20436). A memorial plaque and a memorial scroll were passed to his mother in 1922 and 1923, respectively.

The following tribute was published in *The Tamworth Daily Observer* (NSW) on the 13th of October 1915:

LATE CORPORAL HOTSON.

"Corporal T. E. Hotson Killed in Action at the Dardanelles was well known in Tamworth where he carried on a boot and shoe business, which however, he sold before enlisting for active service. He was very popular and took a great interest in sport, particularly Wednesday afternoon Football. It is understood that his relatives reside at Townsville, Queensland."

2016 PRIVATE THOMAS SMITH (1884 – 1915)

9th Infantry Battalion

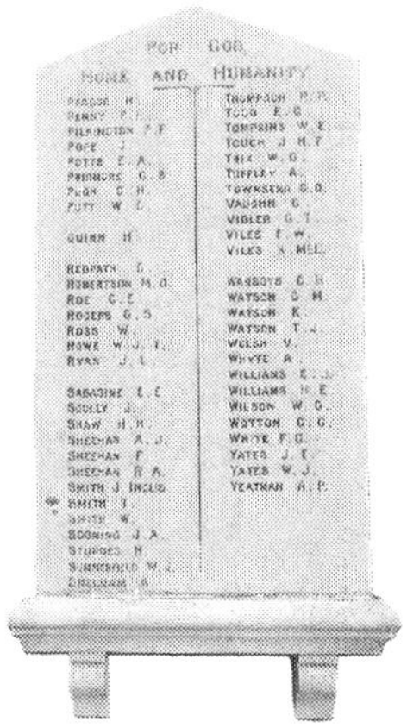

Thomas Smith was born on the 31st of March 1884 in Townsville, Queensland. He was the first of six children born to Thomas and Elizabeth Smith (née Hogan). He had three brothers and two sisters. Thomas Sr. worked as a plumber, and the family lived at Morris St and then Flinders Street West in Townsville. Thomas Smith Jr. grew up in Townsville and found work as a labourer once his school years were finished.

In 1915, Thomas Smith was in Brisbane when he applied to enlist in the Australian Imperial Force (AIF) on the 5th of January. He declared that he was single, had no previous military experience, and was 28 years and nine months old (his birth records show that he was two years older). His enlistment medical examination noted that he was 5'5½" (166 cm) tall, weighed 130 lbs (59 kg), had a fair complexion, blue eyes, and light brown hair. He identified as Roman Catholic and listed his father as his next of kin, giving his address as West End, Townsville.

Accepted into the AIF, Thomas Smith was allocated service number 2016, assigned to the Infantry, and posted to the 9th Battalion, 5th reinforcements. The 9th Battalion was one of the earliest infantry units raised for the Australian Imperial Force (AIF) during the First World War, being the first battalion formed in Queensland. Along with the 10th, 11th, and 12th Battalions, it constituted the 3rd Brigade. The battalion was raised shortly after the declaration of war in August 1914 and was ready for deployment within two months. By the time Thomas Smith enlisted, the 9th Battalion was stationed in Egypt, preparing for upcoming operations.

After three months of initial training, Private Thomas Smith and his fellow reinforcements embarked from Brisbane, Queensland, bound for Egypt on the 16th of April 1915. They sailed aboard the HMAT A55 *Kyarra,* which carried 913 troops on the voyage as part of Convoy No. 5. The seven ships of the convoy shipped a combined total of 3,215 military personnel, 42 civilians, and 2,018 horses to Egypt. After further training in Egypt, Private Thomas Smith, along with the rest of the reinforcements, joined the 9th Battalion on the Gallipoli Peninsula on the 10th of July 1915.

In September 1915, Private Smith was admitted to the 2nd Field Ambulance suffering from an intestinal problem. It was serious enough for him to be evacuated back to the medical facilities on Imbros. It was three weeks before Private Smith recovered and was transported back to Gallipoli. He re-joined the 9th Battalion on the 28th of September 1915.

On the 1st of October 1915, the 9th Battalion reorganized their position on ANZAC because of additional troops joining the battalion. It was during this reorganization that Private Thomas Smith was killed in action. He was 31 years and six months old and had served with the AIF for nine months.

An Australian Red Cross investigation into the circumstances of Private Thomas Smith's death was conducted. In the report of the investigation, a statement made by Private A. McDonald of the 9th Battalion was recorded. Private McDonald stated:

"On Oct. 1st, 1915, at the right flank of the ANZAC position, Pte. Smith was killed on an 'ordinary day' in the trenches. He was shot in the head with a bullet and died in the trench about an hour afterwards. He was buried in the battalion burying ground by 'Shrapnel Green' – near the sea, the same day, and the burial service was taken by a minister of religion."

Private Thomas Smith was laid to rest at Shell Green Cemetery (Plot I, Row D, Grave No. II), Gallipoli, Turkey. His death was notified in the 86th list of Australian casualties released on the 4th of October 1915.

Thomas Smith's name is inscribed on The Australian War Memorial Roll of Honour (Panel 57), The City of Townsville WW1 Honour Roll, and The Townsville Cenotaph.

2016 Private Thomas Smith was awarded the 1914/15 Star (no. 2231), the British War Medal (no. 19386), and the Victory Medal (no. 10342). A memorial scroll and a memorial plaque were passed to his father in 1921 and 1922, respectively.

PRIVATE EDWARD CHARLES TODD (1897 – 1915)

No.5 Depot Battalion

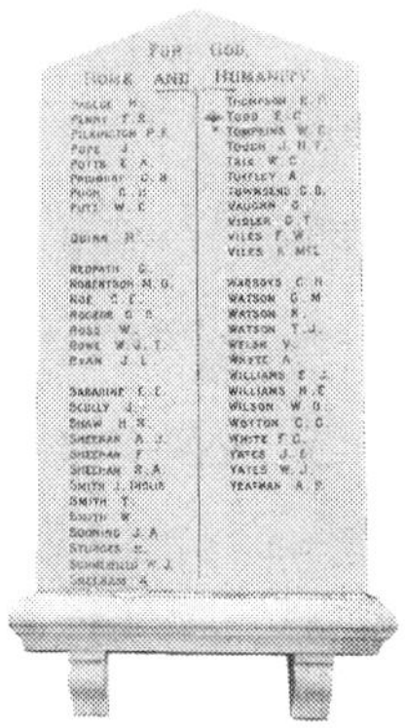

Edward Charles Todd was born on the 29th of March 1897 in Maryborough, Queensland, to Charles and Alvina Todd (née Dahlke). He was the second born and second son of six children, having two brothers and three sisters.

In 1903, the Todd family lived in Rockhampton, where Charles Todd worked as a Prison Warder. By 1913, the family had moved to Townsville and lived at Stewart's Creek. Charles Todd was employed as a Warder at Stewart's Creek Gaol.

On the 27th of September 1915, Edward Todd applied in Townsville to enlist in the Australian Imperial Force (AIF). He declared that he was single, employed as a labourer, had no previous military experience, and was 18 years and six months old. Both parents signed a consent form for him to enlist because of his age. Edward Todd's enlistment medical examination noted that he was 5'5¾" (167 cm) tall, weighed 8st 8lbs (54 kg), had a fair complexion, blue eyes, and brown hair. He identified as Church of England and nominated his mother, Alvina, as his next of kin.

Accepted into the AIF, Edward Todd was transferred to the 5th Depot Battalion at Enoggera, Queensland, for basic training and allocation to a unit. Six weeks after enlistment, he contracted cerebro-spinal meningitis. The disease is an infection affecting the fluid and membranes surrounding the brain and spinal cord and can swiftly propagate throughout the body. This condition has the potential to induce brain damage within hours and may result in fatality within a mere 24 hours. This is the worst outcome and tragically it is exactly what happened in Private Todd's case. He died at the Brisbane General Hospital on the 18th of November 1915. He was laid to rest at Toowong Cemetery, Brisbane, Queensland, Portion 20 - Section 1 - Plot 8 (L1/813).

Edward Charles Todd's name is inscribed on the Australian War Memorial Roll of Honour (Panel 187), the City of Townsville WW1 Honour Roll, and the Townsville Cenotaph. Todd Street, Railway Estate, Townsville is named in his memory.

A memorial scroll and a memorial plaque were passed to his father in 1922. One week before Edward's death, his older brother Fred joined the AIF and saw action on the Western Front. Fred was repatriated back to Australia on the 15th of April 1918, suffering from being "gassed".

The following notices were published in the *Townsville Daily Bulletin* on the 18th of November 1916:

THANKS.

Mr. and Mrs. Todd and Family, of Stewart's Creek, wish to THANK the residents of Stewart's Creek, Roseneath and Brookhill, for their kindness and sympathy in placing on the deceased son's grave at Toowong Cemetery, a nice wreath in glass case, with wire protection, also marble tablet suitably inscribed. Edward Todd enlisted for the front in September 1915 and was only seven weeks in the Training Camp when he took suddenly ill and died from meningitis at the age of 18½ years.

TODD.

In loving memory of our dear son and brother, Private Edward Charles Todd, who departed this life on the 18th day of November 1915.

> Though far from home and kindred,
> And Friends who wished him well,
> His memory is with us,
> And will forever dwell.
> And whilst the sad years vanish
> His name we will oft recall
> And proudly we will cherish
> His photo on the wall.

(Inserted by his loving father, mother, brothers and sisters.)

244 TROOPER ALEXANDER MADDEN FOOT (1892 – 1915)

5th Light Horse Regiment

Alexander (Alick) Madden Foot was born on the 15th of July 1892 in Townsville. He was the sixth child of eight born to Alfred and Mary Foot (née Madden). Alick Foot's parents lived at Cardington Station via Townsville. He attended Townsville Grammar School (TGS), where he was enrolled as a boarder from 1904 until June 1908 when he left to work as a stockman on the family property, Cardington Station. It was at TGS that he served in the Cadets and achieved the rank of Cadet Corporal. Alick was also a member of the Townsville Rifle Club as were his brothers George and Henry and his sister Couchie.

Six weeks after Australia's involvement in the First World War began on the 4th of August 1914, Alick Foot applied in Townsville to enlist in the Australian Imperial Force (AIF) on the 18th of September 1914. He declared that he was single, employed as a stockman, had previous service as a Cadet Corporal, and was 22 years and two months old. His enlistment medical noted that he was 5'8½" (174 cm) tall, weighed 135 lbs (61 kg), had a medium complexion, dark blue eyes, and brown hair. He identified as Church of England and nominated both parents as his next of kin, giving their address as "Cardington" QNR.

Accepted into the AIF, Alick Foot was allocated service number 244, assigned to the Infantry, and posted to the 5th Light Horse Regiment (LHR). The 5th LHR was founded in Brisbane in September 1914, made up of individuals who had enlisted in Queensland. Light horse troopers were trained as mounted infantry, often participating in dismounted combat. They relied on their horses for battlefield transport and quick withdrawal during retreats or retirements. Three months of training followed enlistment for Trooper Alick Foot. Preparations completed, the 5th LHR sailed from Sydney, bound for Egypt, on the 21st of December 1914. They sailed as part of Convoy No. 2 aboard HMAT A34 *Persic*, which carried 532 troops. The entire convoy of 17 ships carried 11,154 military personnel, 129 civilians (including 12 stowaways), and 4,643 horses on the voyage. Six weeks later, the regiment disembarked in Egypt on the 1st of February 1915.

In September 1915, Trooper Alick Foot was admitted to the hospital in Cairo, suffering from diphtheria. He was released in October 1915 and returned to the 5th LHR, where he was posted to Headquarters as Assistant Groom to the Commanding Officer, Lieutenant Colonel Hubert Harris.

The terrain of Gallipoli was assessed as unsuitable for cavalry operations. Consequently, the light horse regiments, including the 5th LHR, were retained in Egypt while the Australian infantry units were deployed to the Gallipoli Peninsula.

The infantry suffered substantial casualties, and in response, the 2nd Light Horse Brigade, including the 5th LHR, was dispatched as dismounted reinforcements in May 1915. The regiments' horses remained in Egypt. Trooper

Alick Foot's brother, Lance Corporal Henry Daintree Foot, was one of the infantry casualties, Killed in Action at Gallipoli on the 1st of May 1915.

Trooper Alick Foot found himself in Egypt for much of 1915, tending to the horses while the remainder of the Light Horse troopers were sent to fight on foot at Gallipoli. Several months after the event, Alick learned that his older brother Henry had been killed on the 1st of May 1915 at Quinn's Post. Members of the 5th Light Horse continued to reinforce Gallipoli, but the grooms were exempt as the horses required care.

On the 31st of July 1915, Lieutenant Colonel Harris was killed at Gallipoli, and Alick Foot expected that he might finally be released from his groom's duties to join the fight. However, this did not occur, and he was later advised that it was because he had an exemplary record.

On the 14th of November 1915, Trooper Alick Foot finally got his wish and deployed to Gallipoli, where he was fortunate enough to catch up with his brother, Private George Foot.

The 5th LHR was assigned to the 1st Division, which was already engaged in a defensive battle to protect the beachhead established at Anzac Cove during the campaign. The regiment engaged in several small-scale attacks. Tragically, Trooper Alick Foot only survived in the battle for a week and was Killed in Action on the 22nd of November 1915. He was 23 years and four months old and had been a member of the AIF for fourteen months.

Trooper Alick Foot was buried on Gallipoli at the Shell Green Cemetery (Plot I, Row H, Grave 2). His death was notified in the 125th list of Australian Casualties released on the 20th of December 1915.

Alexander Madden Foot's name is inscribed on The Australian War Memorial Roll of Honour (Panel 4), The City of Townsville WW1 Honour Roll, The Townsville Grammar School Roll of Honour, and The Townsville Cenotaph. Foot Street, West End, Townsville, was named in 1919 after Alick and his brother Henry, who was also Killed in Action during WW1. The street has been closed and is now included as part of Keyatta Park.

244 Trooper Alexander Madden Foot served in WW1 at Gallipoli. He was awarded the 1914/15 Star (no: 17444), the British War Medal (no: 5346), and the Victory Medal (no: 5334). A memorial scroll and a memorial plaque were passed to his father in 1921 and 1922, respectively. Alick's medals were later presented to the Townsville Grammar School and displayed in the Principal's Office.

Trooper Alick Foot had three brothers who also served in WW1; they were: 135 Lance Corporal Henry Daintree Foot - 15th Battalion - Killed in Action Gallipoli on the 1st of May 1915 - no known grave, 74 Private George Pierce Foot - 2nd LHR - returned to Australia on the 22nd of February 1917, and 4064 Private Eric Douglas Foot - 4th Pioneer Battalion - returned to Australia on the 18th of July 1919.

PART 2 – 1916

The Fallen of 1916: Townsville's Sacrifice

The year 1916 marked a devastating chapter for Townsville, as the horrors of World War I continued to reach across oceans, claiming the lives of young men from faraway shores. That year alone, Townsville lost 37 men in some of the fiercest battles of the war. These men, all of them volunteers from the North Queensland region, became part of the nation's collective mourning as the death toll rose on the battlefields of Europe. Among them were sons, husbands, and fathers, whose absence left a lasting void in the town's close-knit community. This chapter recounts their sacrifice, honouring their memory and the heavy price Townsville paid in 1916.

The Soldiers: A Range of Ages and Lives

The men who perished in 1916 came from all backgrounds, reflecting a diverse cross-section of Townsville society. Both the young and the old answered the call to serve. Among them, three had not yet turned 20. The youngest, Sydney Carse, was just 19 years and two months old. Barely out of adolescence, these young men were thrust into the unimaginable horrors of war, their lives full of potential but tragically cut short on the battlefields of Europe.

The oldest was Private Herbert Barnes, aged 46, a man who had lived more than twice as long as the youngest. Among the 37 fallen, three were married, leaving their wives to bear the unbearable weight of loss. Private Augustus Norman, one of the married men, left behind a three-year-old son. These personal tragedies sent ripples of grief through families and communities, leaving scars that would endure long after the war ended.

The Battle of Pozières: A Turning Point

Located in the Somme region of northern France, Pozières became a strategic target for both the Allied and German forces. Over the period 23rd of July to the 3rd of September 1916, the Australian Imperial Force (AIF) was heavily engaged in the battle, enduring relentless shellfire and waves of counterattacks. The AIF lost as many men over these few weeks as it did over eight months on Gallipoli.

Nineteen Townsville soldiers died at Pozières, fighting in the trenches and fields that would eventually become a symbol of Australia's military endurance and sacrifice. Of these, five lost their lives on the 29th of July, when the fighting was particularly fierce. For Townsville, it was a tragic and concentrated loss, as news of their deaths reached their families back home. The scale of the devastation at Pozières deeply affected Townsville, as it did much of Australia, as the community learned of the unimaginable horrors faced by their soldiers in France.

The 25th Battalion

Fifteen of the Townsville men who died in 1916 were members of the 25th Battalion, which had strong ties to the region. These men were among the first to land in France, and they played a crucial role in key battles throughout 1916, including the deadly engagement at Pozières. The 25th Battalion was often thrown into the thick of the fighting, with their tenacity and bravery earning them both praise and a high casualty rate. Sixteen of the soldiers who left Townsville to fight in the battalion never returned. Their names are now etched in history, remembered for their contribution to the war effort, but the cost to Townsville and North Queensland was immense.

No Known Graves: The Loss of Identity

One of the most heartbreaking aspects of the losses in 1916 is that 22 of the 37 soldiers from Townsville have no known graves. These men, like so many others from across Australia and the world, disappeared in the chaos of war, their bodies never recovered or identified. For their families back in Townsville, this added another layer of pain to their grief. There was no grave to visit, no headstone to mark their passing—only their names listed on memorials thousands of miles from home.

The lack of closure for these families was profound. Letters from military authorities often simply informed loved ones that their son or husband was "missing, presumed dead." Without a body to bury or a place to mourn, the families of these men were left with a deep sense of uncertainty and loss that could never fully heal.

3547 PRIVATE CHARLES HENRY CLARKE (1884 - 1916)

49th Infantry Battalion

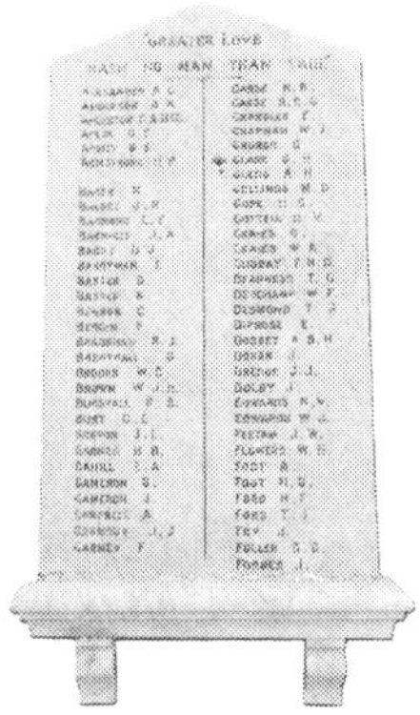

Charles Henry Clarke was born on the 7th of April 1883 in Maryborough, Queensland, to Charles Henry and Emily Clarke (née Gibson, later Mrs. Watson). In 1888, when young Charles was four years old, Emily remarried Robert Joseph Watson, a Master Mariner. The family lived on 6th Street, South Townsville. Charles grew up in Townsville, attended Christian Brothers school, and later followed his stepfather's profession to become a Seaman.

On the 19th of August 1915, Charles Clarke applied to enlist in the Australian Imperial Force (AIF). He declared that he was single, employed as a Seaman, had no previous military experience, and was 31 years and four months old. His medical examination noted that he was 5'5" (166 cm) tall, weighed 10st 3lbs (65kg), had a dark complexion, brown eyes, and black hair. He identified as Roman Catholic and nominated his mother as his next of kin, providing her address as 6th Street, South Townsville.

Accepted into the AIF, Charles Clarke was allocated service number 3547, assigned to the Infantry, and posted to the 9th Infantry Battalion, 11th reinforcements. The 9th Battalion was one of the initial infantry units formed for the AIF in World War I. Along with the 10th, 11th, and 12th Battalions, it formed the 3rd Brigade. On the 25th of April 1915, it was the first battalion to land at Gallipoli and was still fighting there when Charles Clarke enlisted.

Following two months of basic training in Australia, Private Charles Clarke and his fellow reinforcements embarked with Convoy No.13 from Brisbane on the 21st of October 1915, bound for Egypt aboard HMAT A48 *Seang Bee*, which transported 1,013 troops. Convoy 13 transported a combined total of 11,731 troops and 400 horses on the voyage.

Further training took place in Egypt, and in February 1916, Private Clarke was reassigned to the recently formed 49th Battalion as part of the "doubling of the AIF" effort. This initiative aimed to form the 4th and 5th Divisions by handpicking seasoned Officers, NCOs, and soldiers from the 9th Battalion, supplementing them with fresh recruits from Australia, and assigning them to the 49th Battalion. This method allowed for the growth of the AIF while maintaining a core of experienced personnel in the newly established units.

Tragedy struck Private Clarke soon after his transfer to the 49th Battalion. On the 7th of March 1916, he was admitted to the 13th Field Ambulance and then to the 2nd Australian Stationery Hospital at Tel-el-Kebir, suffering from cerebrospinal meningitis. This disease affects the fluid and membranes surrounding the brain and spinal cord, can swiftly propagate throughout the body, and may result in fatality within 24 hours. Tragically, this worst outcome occurred in Private Charles Clarke's case, and he died on the same day he was admitted, the 7th of March 1916. He was 31 years and 10 months old and had served with the AIF for six months and two weeks. He was buried at Tel-el-Kebir War Memorial Cemetery, Egypt, and his death was notified in the 134th list of Australian Casualties released on the 18th of March 1916.

Charles Henry Clarke's name is inscribed on The Australian War Memorial Roll of Honour (Panel 147), the City of Townsville WW1 Honour Roll, and The Townsville Cenotaph. Clarke Street, Garbutt, Townsville, is named in his and two other soldiers' memory.

3547 Private Charles Henry Clarke was awarded the 1914/15 Star (no:1487), the British War Medal (no:21602), and the Victory Medal (no:21526). A memorial scroll and a memorial plaque were passed to his mother in 1921 and 1922, respectively.

The following notice was published in the *Townsville Daily Bulletin* (Qld.) on the 13th of March 1916:

THE WAR IN EUROPE.

Mrs. E. Watson, Main Street, South Townsville, received advice on Saturday of the death of her son, C. H. Clarke (Ninth Battalion), from cerebrospinal meningitis, on the 7th of March at Tel El Kebir.

30108 PRIVATE FRANK HORACE BARNARD (1893 – 1916)

1st Infantry Battalion

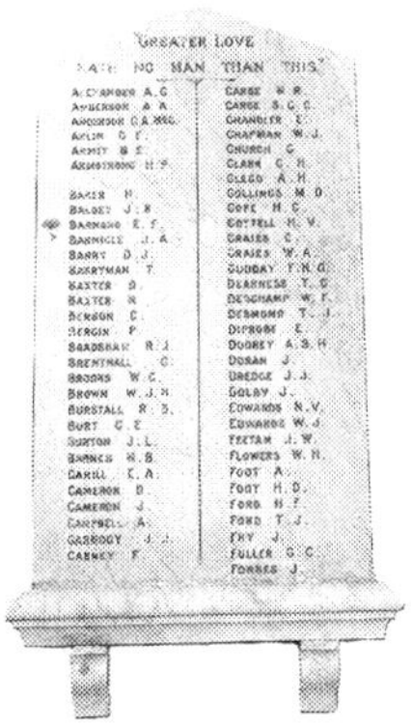

Frank Horace Barnard was born on the 25th of November 1893, in Mount Morgan, Queensland. He was the oldest son of five children born to his father, E. Frank Barnard, and his mother, Louisa (née Neal). He had two brothers and two sisters. There is no record of exactly where Frank Barnard grew up or went to school; however, he qualified as a draper and moved to Sydney in his early twenties.

On the 24th of May 1916, Frank Barnard applied at Victoria Barracks in Sydney to enlist in the Australian Imperial Force (AIF). He declared that he was 22 years and six months old, employed as a draper, and that he had previously served in the militia for two years.

Frank Barnard's enlistment medical examination noted that he was 5'8½" (174 cm) tall, weighed 126 lbs (57kg), had a fair complexion, blue eyes, and fair hair. He identified as Methodist and listed his mother as his next of kin and simply gave her address as Townsville, North Queensland.

Accepted into the AIF, Frank Barnard was allocated service number 30108, assigned to the Infantry, and posted to the 1st Battalion. The 1st Battalion was the first infantry unit recruited for the AIF in New South Wales during WW1 and was raised within a fortnight of the declaration of war in August 1914. The Battalion was fighting on the Western Front when Frank Barnard enlisted.

Three days after enlistment, Frank Barnard was moved to Cootamundra in NSW to complete his initial training prior to joining his unit. He arrived in Cootamundra on the 27th of May and joined "A" Company of the 1st Depot Battalion. Three weeks later, tragedy struck when Private Frank Barnard contracted pneumonia during his training. He died on the 15th of June 1916, at the age of 22, only three weeks after he enlisted in the AIF. He was buried at the Cootamundra General Cemetery.

Frank Barnard's name is inscribed on The Australian War Memorial Roll of Honour (panel 185), The City of Townsville WW1 Honour Roll, and The Townsville Cenotaph. Barnard Street, Aitkenvale, in Townsville, is named in his honour. A memorial scroll was passed to his father in 1922.

874 Captain Claude Elmhurst Benson D.C.M. (1887 – 1916)

9th Infantry Battalion

Claude Elmhurst Benson was born on the 15th of March 1887 at Ryde, New South Wales. He was the eldest son of Edward and Elizabeth Benson (née Jackson). Edward Benson was an orchardist, and Claude spent his childhood in Ryde, attending Ryde Public School. He was a diligent pupil receiving dux of the school in 1897. After completing his studies, Claude relocated to Townsville and secured a position as a clerk in the merchandise department of the Burns Phillip shipping line. He lived in premises on Cleveland Terrace, Melton Hill, in Townsville.

On the 20th of August 1914, Claude Benson applied at Townsville to enlist in the Australian Imperial Force (AIF). He declared that he was single, employed as a clerk, and 27 years and five months old. He also declared that he had previous military service with Senior Cadets.

Claude Benson's enlistment medical record noted that he was 5'9¼" (176 cm) tall, weighed 11st 2lbs (71kg), had a fair complexion, grey eyes, and fair hair. He identified as Church of England and nominated his father as his next of kin, giving his address as "Bensonville," Lane Cove Rd, Ryde, Sydney.

Accepted into the AIF, Claude Benson was allocated service number 874, assigned to the Infantry, and posted as an original member to "H" Company of the 9th Battalion, the first battalion recruited in Queensland. The 9th was raised within weeks of the declaration of war in August 1914 and embarked just two months later.

Within a month, Private Claude Elmhurst was promoted to Sergeant. The 9th Battalion embarked from Brisbane bound for Egypt with the 1st Convoy on the 24th of September 1914. They sailed aboard A5 *Omrah*, which carried 1,142 troops, four civilians, and 15 horses. The 20-ship convoy transported a total of 21,501 military personnel, 28 civilians, and 7,882 horses on the voyage.

Once disembarked in Egypt in December 1914, months of training followed for the 9th Battalion until April 1915, when the 9th Battalion and the bulk of the AIF sailed to Lemnos Island in preparation for the Gallipoli campaign. Sergeant Claude Benson was among the first wave of men ashore on Gallipoli. During the day, *"with great courage and presence of mind, he was known to lead into the firing line men whose officers had all been killed or wounded, and who had suffered very heavy losses. His fine example and devotion to duty were conspicuous."* For these acts, he was awarded the Distinguished Conduct Medal (DCM).

Sergeant Claude Benson was commissioned in the field with the rank of Second Lieutenant on the 28th of April 1915, three days after the landing. In July 1915, 2nd Lieutenant Claude Benson's health deteriorated due to harsh conditions and a poor diet, leading to his evacuation to Lemnos. During this time, he was promoted to the rank of Lieutenant. When his condition did not improve, he was evacuated to Egypt. After making a recovery, he travelled back to Gallipoli on the *Southland* transport. The ship was torpedoed on the voyage, and the troops had to abandon it. Lieutenant Benson and his fellow passengers were stranded in a lifeboat for three hours before being rescued.

Lieutenant Benson returned to his unit on Gallipoli in September 1915 and remained there until the 9th Battalion's evacuation in November of the same year. Returning to Egypt, the Battalion continued training and recovery from the Gallipoli campaign. Four months after the evacuation, the 9th Battalion was deployed to the Western Front in March 1916. Lieutenant Claude Benson was ill when the Battalion sailed, and he did not embark for France until the 8th of May 1916, arriving in Marseilles eight days later. One week after arriving in France, Lieutenant Claude Benson was promoted to Captain.

On arrival at the Western Front, the 9th Battalion was assigned to the Armentières sector and experienced their first front-line duty in May. In June 1916, Captain Claude Benson participated in a raiding party as the second-in-command to his friend Captain Maurice Wilder-Neligan near Rouge De Bout. The team trained for several weeks and conducted multiple night-time excursions into no man's land to prepare for their attack plan.

On the night of the 1st of July 1916, Captain Benson led the right flank into the German trenches, resulting in intense hand-to-hand combat. The Australian artillery barrage prevented German reinforcements, and the Germans were severely affected. When the order to retreat was issued, the Australians, along with 25 German prisoners, returned to their allied trenches. Captain Benson was the last to leave the German trenches and in the process was fatally shot in the back. Some of his men went back for him, and his body was retrieved on the same day that the 9th Battalion withdrew from the front line. He was buried at midday on the 2nd of July 1916 at the Rue-Du-Bois Military Cemetery, Fleurbaix. He was 29 years and three months old and had served with the AIF and the 9th Battalion for one year and 10 months. His death was notified in the 186th list of Australian Casualties released on the 20th of July 1916.

Claude Elmhurst Benson's name is inscribed on The Australian War Memorial Roll of Honour (Panel 55), Ryde Public School Roll of Honour, The City of

Townsville WW1 Honour Roll, and The Townsville Cenotaph. Benson Street, Rosslea, in Townsville, is named in his memory.

Captain Claude Elmhurst Benson served in Egypt, Gallipoli, and on the Western Front in WW1 and was awarded the Distinguished Conduct Medal, the 1914/15 Star (no:2817), the British War Medal (no:325), and the Victory Medal (no:281). These medals, as well as a memorial scroll and a memorial plaque, were passed to his father.

The following tribute was published in *The Sun* (Sydney) on the 16th of July 1916:

CAPTAIN BENSON - KILLED

News has been received in Sydney that Capt Claude E. Benson, of Ryde, was Killed in Action in France. He enlisted in August 1914, at Townsville, Queensland, as a private, and before leaving Australia was promoted to sergeant. On April 25, 1915, at Gaba Tepe, he received the D.C.M. for displaying great courage in twice leading forward into the firing line men whose officers had all been killed or wounded, and who had suffered very heavy losses. He was then promoted to the rank of lieutenant. While proceeding to Egypt, after being on furlough, he travelled on the Southland when she was torpedoed. He had charge of 39 men in a collapsible boat, which capsized as soon as it was launched. He and all his men were eventually rescued. He was later promoted to the rank of captain. Through a slight sickness, he had to remain in Egypt in hospital when the rest of his battalion left for France, but he left Egypt later, arriving in Marseilles on May 15.

4613 PRIVATE PERCY HATCHETT (1881 – 1916)

53rd Infantry Battalion

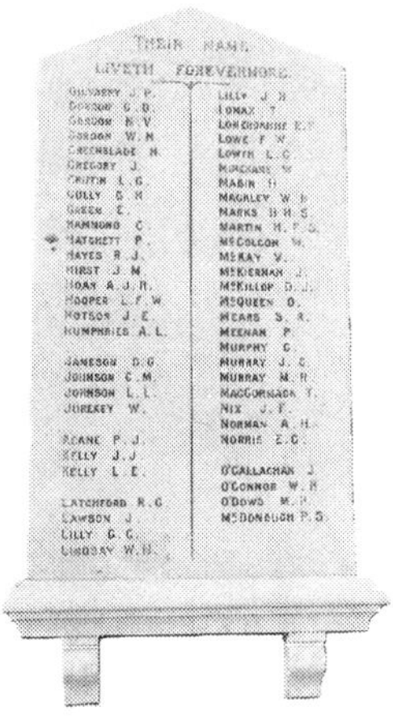

Percy Hanchet was born in Middlesex, England, on the 18th of August 1881, to Thomas and Louisa Hanchet (née Chater). He was the third of eight children and had five brothers and two sisters. He emigrated to Australia and arrived in Townsville on the 24th of October 1913 aboard the *Warilda.* After his arrival, Percy changed his name from Hanchet to Hatchett and lived, and worked as a Steward, at the Tattersalls Club in Townsville.

Two years after arriving in Townsville, Percy Hatchett applied to enlist in the Australian Imperial Force (AIF) on the 11th of October 1915. He declared that he was single, working as a steward, and was 34 years and one month old. He also declared that he had been rejected for service in His Majesty's Navy.

Percy Hatchett's medical examination noted that he was 5'7½" (171 cm) tall, weighed 11st 8lbs (73.4 kg), had a dark complexion, brown eyes, and black hair. He identified as Church of England and nominated his younger brother Bert, who lived in England, as his next of kin.

Accepted into the AIF, Percy Hatchett was allocated service number 4613, assigned to the Infantry, and initially posted to the 9th Battalion, 14th reinforcements. The 9th Battalion was the first battalion recruited in Queensland and was raised within weeks of the declaration of war in August 1914. The 9th Battalion was on the Gallipoli Peninsula when Percy Hatchett enlisted.

Private Percy Hatchett spent the next three months in Enoggera, Brisbane, training for overseas service. He embarked with the other reinforcements from Brisbane on the 31st of January 1916, bound for Egypt as part of Convoy No.17. They sailed aboard HMAT A62 *Wandilla*, which carried 1,052 troops. The entire convoy of eight ships transported 8,410 military personnel, 21 civilians, and 670 horses on the voyage.

After disembarking in Egypt, another three months of training followed at Tel-el-Kebir. In May 1916, Private Hatchett was posted to the 53rd Battalion. This battalion, part of the Australian Imperial Force's "doubling" initiative, was established in Egypt on the 14th of February 1916. Its formation included two distinct cohorts: one comprising recently arrived reinforcements from Australia, and the other consisting of seasoned Gallipoli veterans from the 1st Battalion. Like the 1st Battalion, the 53rd predominantly enlisted soldiers from the suburbs of Sydney. Eventually, the battalion was assigned to the 14th Brigade within the 5th Australian Division.

Private Percy Hatchett, along with the rest of the 53rd Battalion, embarked from Alexandria on the 19th of June 1916 to join the British Expeditionary Force (B.E.F.) in France. They disembarked in Marseilles on the 28th of June 1916. On the 10th of July 1916, the 53rd was deployed to the front line for the first time, and nine days later was committed to their first major engagement on the Western Front, at Fromelles, on the 19th of July 1916. The Battle of Fromelles marked Australia's debut on the Western Front. It aimed to divert German forces from the Somme offensive but ended in disaster. The 53rd Battalion's offensive capability was severely depleted in the battle, with 625 casualties, including their commanding officer. The battalion was reduced to one quarter of its normal strength. The battle highlighted the inevitable waste of lives when troops were committed to frontal assaults on well-prepared enemy positions.

Private Percy Hatchett was one of the casualties of the Battle of Fromelles. He was Killed in Action on the 19th of July 1916. He was 34 years old and had been a member of the AIF for nine months at the time of his death. His body was not recovered, and he has no known grave. His death was notified in the combined 206th and 207th lists of Australian Casualties released on the 16th of September 1916.

Percy Hatchett's name is inscribed on The Australian War Memorial Roll of Honour (Panel No.157), V.C. Corner (Panel No.8), Australian Cemetery Memorial, Fromelles, France, The City of Townsville WW1 Honour Roll, and The Townsville Cenotaph. Hatchett Street, Cranbrook, Townsville, was named in his memory in 1926.

Private Percy Hatchett was awarded the British War Medal (no: 31155) and the Victory Medal (no: 30932). A memorial scroll and a memorial plaque were passed to his brother in 1922.

938 PRIVATE DONALD JOHN DUNCAN McKILLOP (1886 – 1916)

30th Infantry Battalion

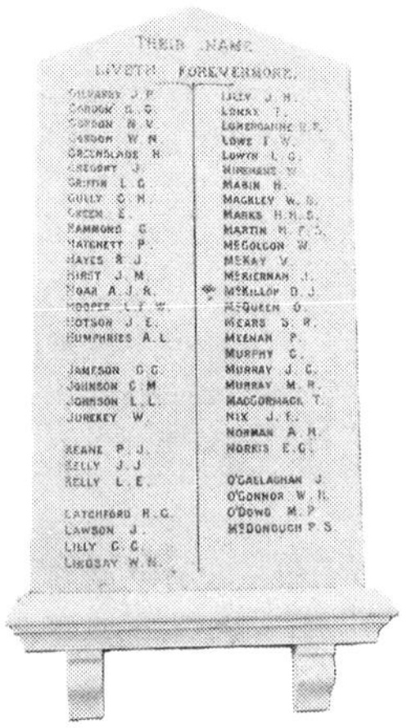

Donald (Don) John Duncan McKillop was born on the 29th of January 1886 in Cunnamulla, Queensland. He was the only son of John and Mary McKillop (née McLeod) and was named after Mary's father. The marriage broke down and John and Mary separated. After the separation, Mary had three children with John George Merrifield, publican of the Tallyhoo Hotel near Kyabra. Merrifield died in 1890. Mary met and married her final husband, Richard Brinsley Pitt, and had seven more children. Don attended a public school and had a connection with Glen Innes in New South Wales. After his school years, he started work as a farrier.

On the 20th of July 1915, Don McKillop applied at Liverpool in NSW to join the AIF, listing his address as Highgrove via Armidale. He declared that he was 30 years old and had no previous military experience. His medical examination noted that he was 5'3½" (161 cm) tall, weighed 126 lbs (57 kg), had a sallow complexion, brown eyes, and black hair. He identified as Presbyterian and nominated his mother as his next of kin, giving her address as Mrs. Mary Pitt, Charters Towers, Qld. This address was changed several times until it reflected Don McKillop's mother living in Townsville.

Accepted into the AIF, Don McKillop was allocated service number 938, assigned to the Infantry, and posted to "D" Company of the freshly minted 30th Battalion. The 30th was raised on the 5th of August 1915 as part of the 8th Brigade at Liverpool in NSW. Most of its troops were recruited from the Newcastle region and other parts of country NSW.

After three months of training in Australia, Private Don McKillop embarked with the 30th Battalion on the 9th of November 1915 from Sydney, bound for Egypt. They sailed as part of Convoy No.14, aboard HMAT A72 *Beltana*, which carried 1,417 troops. The 14-ship convoy transported a total of 12,971 troops and 1,561 horses on the voyage.

Private McKillop disembarked with his battalion in December 1915 and commenced training in Egypt. In June 1916, the 30th Battalion joined the newly formed 5th Australian Division in Egypt and embarked for France, where they were destined for the Western Front.

The 30th Battalion's first major engagement was at Fromelles on 19 July 1916. Initially tasked with carrying supplies, they were quickly drawn into intense combat. This battle was a harsh introduction to the brutal conditions of the Western Front for Australian forces.

The offensive aimed to divert German reserves from the Allied push at the Somme, which had begun on the 1st of July 1916. The 5th Australian Division, along with the British 61st Division, was ordered to assault the fortified German positions along Aubers Ridge in French Flanders.

The operation ended in a catastrophic defeat. Exposed to relentless German artillery in daylight, Australian and British troops suffered heavy casualties. Australian losses were severe, with over 5,500 casualties, including 2,000 killed or dying from their wounds, and 400 captured. This battle marked the worst loss for a single division in a 24-hour period during World War I and remains one of Australia's darkest military moments.

It was during the Battle of Fromelles that Private Don McKillop was Killed in Action. The circumstances of his death are reproduced below, inter alia, from the statement of Captain T.C. Barbour made in January 1917:

"The above Pte fought in the engagement on the attack of Fromelles 19/20 July and did some excellent work under heavy fire on the above date. He reached the German support lines carrying ammunition time after time and came up to me inquiring what other duties he could carry out. He personally helped me to clear a sap which had been blocked up with stores and ammunition boxes; this sap was being dug to connect with the enemy front lines. As that was finished, he volun-

teered to lead a party over to the enemy support lines again; several men went with him only one of whom to my knowledge returned, and he has since been posted as missing and nothing further has been heard of him."

2984 Pte F.L. Murphy stated the following in February 1917 about Private McKillop:

"I knew him quite well; we were making an attack on the German trench from Fleurbaix. He got wounded during the attack by M.G. fire. Pte J. Robinson saw him fall to the ground and said that he was badly wounded. There was a lot of shelling as well."

Initially declared missing, Private Don McKillop was formally declared Killed in Action by a Court of Enquiry held two days after his disappearance. He was 30 years and six months old and had served with the AIF for one year at the time of his death. Private Don McKillop was buried at Ration Farm Military Cemetery (Plot VI, Row F, Grave No.46), La Chapelle d'Armentières, France. His headstone is incorrectly inscribed "*McKullop*." His death was notified in the 334th list of Australian Casualties released on the 1st of September 1917.

Donald John Duncan McKillop's name is inscribed on The Australian War Memorial Roll of Honour (Panel 117), The City of Townsville WW1 Honour Roll, and The Townsville Cenotaph. McKillop Street, Belgian Gardens, Townsville is named in his memory.

938 Private Donald John Duncan McKillop was awarded the 1914/15 Star (no:2447), British War Medal (no:26955), and the Victory Medal (no:26780). A memorial scroll and a memorial plaque were passed to his mother in 1921 and 1922, respectively. Don McKillop's mother was granted a war pension of 14 shillings per fortnight from the 17th of October 1916 for the loss of her son. This was one fifth of the wage paid to AIF soldiers in 1916.

2nd LIEUTENANT RICHARD STEWART BURSTAL (1891 – 1916)

1st Infantry Battalion

Richard (Dick) Stewart Burstal was the first of four children born to Richard and Maria Burstal (née Soares) at Temora, NSW, on the 14th of October 1891. Richard Burstal Sr. was the Manager of the Bank of New South Wales in Young, NSW, and it was in this district that young Dick Burstal spent his early years. Following his school years, Dick passed the Bankers' Institute Exam at the University of Sydney.

Richard Burstal Sr. was appointed to the position of Manager of the Bank of NSW in Townsville, Queensland after spending 12 years as Bank Manager at the township of Young in NSW. Dick Burstal moved to Townsville with his parents, and the family lived at premises on Melton Hill. Dick Burstal pursued a career as a Stock and Station Agent until the 13th of April 1915, when he applied at Townsville to enlist in the Australian Imperial Force (AIF). He declared that he was single, a Stock and Station Agent, had no previous military experience, and was 23 years and seven months old. His medical examination noted that he was 5'8½" (174 cm) tall and weighed 158 lbs (72 kg). Dick Burstal identified as Church of England and nominated his father as his next of kin, giving his address as Melton Hill, Townsville. This address was later changed to Inspector in Charge, Bank of NSW, Brisbane.

Accepted into the AIF, Dick Burstal was assigned to the Infantry and posted to the 1st Battalion, 15th reinforcements. The 1st Battalion was the initial infantry unit raised in the AIF. The battalion was in Egypt, preparing for the forthcoming battles on the Gallipoli Peninsula, when Dick Burstal enlisted. In the first few months following enlistment, he applied for a commission as an Officer in the AIF. He was successful, and after completing the necessary training in Australia, he was appointed as a Second Lieutenant on the 1st of December 1915.

Three months of further training followed for the newly commissioned 2nd Lieutenant Burstal until he embarked from Sydney bound for Egypt on the 8th of March 1916. He sailed as part of Convoy No.18 aboard HMAT A15 *Port Sydney* (ex-*Star of England*), which carried 1,508 troops. The entire convoy of 13 ships carried 10,772 military passengers, 54 civilians, and 3,840 horses on the voyage. 2nd Lt Burstal disembarked in Suez on the 2nd of May 1916 and, one week later, embarked for the Western Front on the 9th of May 1916. Arriving in Marseilles eight days later, 2nd Lieutenant Burstal was taken on strength at the 1st Australian Division Base Depot, where he undertook training prior to joining the 1st Battalion on the 7th of June 1916.

Soon after his arrival, the 1st Battalion was committed to the Battle of Pozières, which was a major engagement that commenced on the 23rd of July 1916 and endured for more than six weeks. Allied forces in the battle included the AIF and the British Army. The primary objective was the capture of the strategic village of Pozières in northern France, a crucial point in the German defensive line. This fiercely contested battle exacted a heavy toll on both sides, marked by intense artillery barrages and steadfast German resistance. The Australians successfully captured Pozières on the 6th of August 1916. However, this triumph came at a significant price, with more than 23,000 Australian casualties.

The Battle of Pozières underscored the determination and skill of the Australian soldiers and their substantial contribution to the overall war effort. It was during this battle that 2nd Lieutenant Richard Burstal was Killed in Action on the 25th of July 1916. A witness to his death, Sergeant TJ Smith, stated that he saw 2nd Lt Burstal killed and that he was a *"recklessly brave man, quite without fear."* Smith went on to state that Richard was buried behind the trench in which he died.

2nd Lt Richard Burstal was 24 years and nine months old, had served with the AIF for 15 months, and was a member of the 1st Battalion for two months at the time of his death. He has no known grave. His death was notified in the 196th list of Australian Casualties released on the 25th of August 1916.

Richard Burstal's name is inscribed on The Australian War Memorial Roll of Honour (Panel 28), The Australian National Memorial, Villers-Bretonneux, France, The City of Townsville WW1 Honour Roll, and The Townsville

Cenotaph (n.b. misspelt as Burstall). Burstall Street, Hyde Park, Townsville is named in his memory.

2nd Lieutenant Richard Stewart Burstal was awarded the British War Medal (no: 33060) and the Victory Medal (no: 32834). A memorial scroll and a memorial plaque were passed to his father in 1922.

The following notice was published in the *Townsville Daily Bulletin* 14th August 1916:

DEATHS. BURSTAL. —Killed in Action at Pozières, France, July 22nd, Lieut. R. S. Burstal, 1st Battalion A.I.F. Aged 25. Elder son of R. W and Maria Burstal, of Townsville.

The following notice was published in the *Sydney Morning Herald* 19 August 1916: BURSTAL. -July 22, 1916, Killed in Action in France, Lieut. R.S. Burstal, aged 23 years, elder son of R.W. Burstal, of Townsville, Queensland, late of Young.

The following article was published in *The Young Witness* (NSW) 22nd August 1916:

"LIEUT. R. S. BURSTAL. Another local boy was also Killed in Action on 22nd July. The sad news was published in the casualty list on Saturday that Lieutenant Richard S. Burstal, met his death during the first dash on the Somme. He also only left Sydney in April last. "Dick" was the elder son of Mr. and Mrs. Burstal, now of Townsville (Queensland) where Mr. Burstal is inspector for the Bank of New South Wales, and prior to receiving that appointment was manager of the Young branch of that bank for about 12 years. Deceased was a grandson of Rev. G. Soares, formerly rector of Lake Bathurst, near Goulburn. He was prior to enlisting on the land and was well known throughout the Young district. His younger brother, Mr. Bert Burstal who has just been through the Sydney University studying medicine, intends, we understand, to go into camp almost immediately. Mr. and Mrs. Burstal's many friends at Young will join in sympathy with them in the loss of their eldest son. Prior to the usual war intercessions in St. John's Church on Sunday night, the rector made a touching reference to the sad news which had come through during the past few days. In doing so, he said, that all of us had been thrilled a few weeks back at the splendid achievements of our boys at the front.

Now, alas, will come the cost. He urged the people to be more earnest with their prayers during the next three or four weeks, for many homes will be saddened, and the bereaved will need consolation. In referring to the death of Lieut. Burstal the rector said that whilst at Young the family had been great workers for the church, and he knew the people would regret to learn of Lieut. Burstal's death."

3043 PRIVATE NORMAN VICTOR EDWARDS (1896 – 1916)

12th Infantry Battalion

Norman Victor Edwards (Vic) was born at Bluff Downs on the 3rd of September 1896 in Charters Towers, Queensland, to Henry and Harriet Edwards (née West). He was the second child and first-born son and had five brothers and two sisters. The Edwards family moved to Wills St in Townsville prior to 1903, and Henry Edwards worked as a cordial maker. By 1913, the family had moved to Stagpole Street, and Henry Edwards was employed as a wheelwright. Vic Edwards grew up in Townsville and attended the Townsville West State School prior to qualifying as a motor mechanic. During this time, he served for three years in the Senior Cadets and one year in the 2nd Infantry. Vic Edwards also served on the expedition to New Guinea in the *Kanowna* with the 2nd Battalion, Australian Naval and Military Expeditionary Force (AN&MEF).

The AN&MEF was tasked with the Australian takeover and occupation of the Pacific colony of German New Guinea in September – November 1914. The contingent boarded the *Kanowna* and started the journey towards the eventual capture of German New Guinea. As events turned out, they did not participate in the capture of German New Guinea on account of *"trouble caused by the firemen on the troopship."* Private Vic Edwards returned to Townsville on the 18th of September 1914, and upon arrival, was discharged from the AN&MEF.

Five weeks later, Vic Edwards applied to enlist in the Australian Imperial Force (AIF) at Townsville on the 23rd of August 1915. He declared that he was 18 years, 11 months old, single, qualified as a mechanic, and working as a Blacksmith with his father in Flinders Street. He also declared his previous military service. Vic Edwards' enlistment medical record noted that he was 5'7½" (171 cm) tall, weighed 9st 11lbs (62kg), had a fair complexion, grey eyes, and dark brown hair. He identified as Methodist and nominated his father as his next of kin and gave his address as Flinders St, Townsville.

Accepted into the AIF, Vic Edwards was allocated service number 3043, assigned to the Infantry, and was posted to the 26th Battalion, 7th reinforcements. The 26th Battalion was raised at Enoggera, Queensland, in April 1915 from recruits enlisted in Queensland and Tasmania, and formed part of the 7th Brigade. The 26th was fighting on the Gallipoli Peninsula when Vic Edwards enlisted.

The 7th reinforcements spent the period from August to December 1915 completing Recruit and Basic Training in Australia. On the 30th of December 1915 Private Vic Edwards and his fellow reinforcements embarked from Brisbane bound for Egypt. They sailed with Convoy No.15 aboard HMAT A50 *Itonus*, which carried 783 troops. The entire convoy of 11 ships transported 8,994 troops and 128 civilians on the voyage. Private Vic Edwards was promoted to Acting Sergeant prior to the voyage. He relinquished his Acting Sergeant Rank and reverted to Private when he joined the 12th Battalion from the 26th Battalion at Zeitoun near Cairo Egypt in March 1916.

The 12th Battalion sailed for France and the Western Front in March, arriving in the first week of April 1916. Soon after his arrival, the Battalion was committed to the Battle of Pozières which was a major engagement that commenced on the 23rd of July 1916 and endured for more than six weeks. Allied forces fighting in the battle included the AIF and the British Army. The primary objective was the capture of the strategic village of Pozières in northern France, a crucial point in the German defensive line.

This fiercely contested battle exacted a heavy toll on both sides, marked by intense artillery barrages and steadfast German resistance. The Australians successfully captured Pozières on the 6th of August 1916. However, this triumph came at a significant price, with more than 23,000 Australian casualties. The Battle of Pozières underscored the determination and skill of the Australian soldiers and their substantial contribution to the overall war effort.

Private Norman Victor Edwards was one of those casualties when he was Killed in Action on the 25th of July 1916. There are no recorded details of his death, and he has no known grave. On the date of his death, Private Edwards was 19 years, 10 months old, had been a member of the AIF for 11 months, and a member of the 12th Battalion for four months and two weeks. His death was notified in the combined 200th and 201st list of Australian Casualties released on the 29th of August 1916.

Norman Victor Edwards served in Egypt and on the Western Front in WW1. His name is inscribed on The Australian War Memorial Roll of Honour (Panel 65), The Australian National Memorial, Villers-Bretonneux, France, The Townsville West State School Honour Roll, The City of Townsville WW1 Honour Roll, and The Townsville Cenotaph. Edwards Street, West End, Townsville, is named in his memory.

3043 Private Norman Victor Edwards was awarded the British War Medal (no:28942) and the Victory Medal (no:28761). These medals were passed to his father, along with a memorial plaque and a memorial scroll, which were also passed to him in 1922.

The following tributes were published in the *Townsville Daily Bulletin* on the 25th of July 1918:

EDWARDS. —In loving memory of our dearly beloved eldest son and brother, Sergeant Norman Victor, who was Killed in Action at Pozières, France, on July 25th, 1916. (Inserted by his loving parents, sisters and brothers.)

EDWARDS. —In loving memory of Sergeant Norman Victor Edwards, 12th Batt., who was killed at Pozières, on the 25th of July 1916. To memory ever dear. (Inserted by his loving grandmother, aunts, uncles and cousins.)

EDWARDS. —In loving remembrance of Sergeant Norman Victor Edwards, 12th Batt., who was killed in action, at Pozières, France, on July 25th.

(Inserted by his loving grandmother, aunts, uncles and cousins.)

EDWARDS. — A tribute to the memory of my dear mate, Vic., who was Killed in Action at Pozières on the 25th July, 1916.

He was fighting for his country,
He answered duty's call,
His friends, his home, his comforts,
He sacrificed them all.

(Inserted by his mate Corporal R. E. Burns, 48th Batt.)

365 CORPORAL HENRY FRANCIS FORD (1895 - 1916)

25th Infantry Battalion

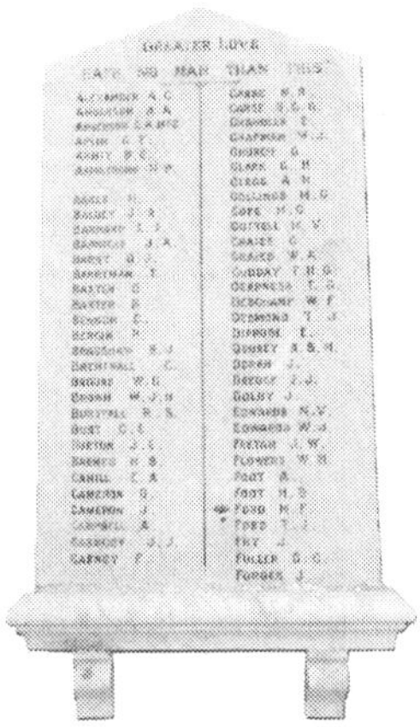

Henry Francis Ford was born in Bowen, Queensland, on the 21st of August 1895, to Henry and Christina Ford (née Roche). He was one of five surviving children and had one sister and three brothers. Henry Ford attended St. Mary's Roman Catholic School, West End, Townsville, and Christian Brothers College, Charters Towers. He undertook a three-year apprenticeship in Townsville and qualified as a cycle/motor mechanic. He lived in Mitchell St North Ward, Townsville, and served for three years in the Senior Cadets. After Cadets, he joined the Kennedy Infantry Regiment of the Citizen Forces.

Henry Ford served on the expedition to New Guinea as a member of the Australian Naval and Military Expeditionary Force (AN&MEF). The AN&MEF was tasked with the Australian takeover and occupation of the Pacific colony of German New Guinea in September – November 1914. The contingent boarded the SS *Kanowna* on the 8th of August 1914 and started the journey towards the eventual capture of German New Guinea. As events turned out, they did not participate in the capture of German New Guinea on account of *"trouble caused by the firemen on the troopship."* Henry Ford returned to Townsville on the 18th of September 1914, and upon arrival, was discharged from the AN&MEF.

On the 15th of February 1915, Henry Ford applied to join the Australian Imperial Force (AIF) at Townsville. He declared that he was 19 years, six months old, single and was employed as a cycle mechanic. He also declared his previous military experience. His enlistment medical notes that he was 5' 7½" (171 cm) tall, weighed 8st 4lbs (53kg), had a fresh complexion, brown eyes, and dark hair. He identified as Roman Catholic and, because his father was deceased, listed his mother as his next of kin. His brother, Thomas, had enlisted two weeks earlier, and both brothers were allocated to the Infantry and posted to the newly raised 25th Battalion.

After four months training in Australia, the 25th Battalion embarked from Brisbane bound for Egypt on the 29th of June 1915. They sailed aboard HMAT A60 *Aeneas*, which carried 1,720 troops on the voyage. Their ship was one of four that made up Convoy No.7. The convoy transported 5,052 men and 20 horses.

In August 1915, the 25th Battalion underwent training in Egypt. In September they deployed to Gallipoli as part of the 7th Brigade and provided reinforcement to the diminished New Zealand and Australian Division. The last major Allied offensive had been launched and turned back in August 1915, and consequently, the 25th Battalion had a quiet time. Private Henry Ford was promoted to Acting Corporal at the end of October 1915. He left the Gallipoli peninsula with the 25th Battalion on the 18th of December 1915. The 25th underwent further training in Egypt after their evacuation from Gallipoli. Acting Corporal Ford's promotion to substantive Corporal was confirmed on the 27th of January 1916.

In mid-March 1916, the 25th Battalion departed Alexandria to join the British Expeditionary Force on the Western Front in France. Arriving on the 19th of March, it became the first AIF battalion to land there. The battalion's first major engagement was at Pozières, a key battle in the larger Battle of the Somme, which took place between July and September 1916. Their objective was to capture the strategically important village of Pozières, a vital position in the German defensive line.

As the Australian troops fought to capture Pozières, the German Army launched a bombardment of the village and its surrounding areas. The two sides engaged in fierce shelling, with the intensity reaching its peak on the 26th of July. This period of bombardment was one of the most severe ever faced by an Australian Division. On the night of the 28th of July 1916, the 25th Battalion attacked the German lines north and east of the village. The next day, 29th of July 1916, Henry Ford was reported as missing. This was amended to Killed in Action following a board of enquiry held one year later in July 1917. In a statement made to the Australian Red Cross in 1917, 2149 Private Arthur Gabbert said: *"I saw Cpl. Ford No. 365, killed at Pozières on the 29th July 1916 by a machine gun. It was dark, and I was about 2 yards away from him. We were making an attack on Pozières Ridge and had to withdraw owing to the wire being uncut. I knew Ford well having been with him in Gallipoli. I do not know what became of his body; it was left behind when we returned to our trenches. Ford was in my Company and Platoon."*

Corporal Henry Ford was one month short of his 28th birthday at the time of his death and has no known grave. He had served in the AIF for 17 months and his death was notified in the 331st list of Australian Casualties released on the 18th of August 1917.

Henry Francis Ford's name is inscribed on The Australian War Memorial Roll of Honour (Panel 104), The City of Townsville WW1 Honour Roll, The Townsville Cenotaph, and The Villers-Bretonneux Memorial (Australian National Memorial-France). Ford Street, Hermit Park, Townsville is named in his and his brother Thomas' honour.

365 Corporal Henry Francis Ford served in WW1 in Gallipoli and on the Western Front. He was awarded the 1914/15 Star (no:20797), the British War Medal (no:16622), and the Victory Medal (no:16555). These medals, as well as a memorial scroll and a memorial plaque, were passed to his mother.

Two of Henry Ford's brothers, John and Thomas, also served in WW1. 862 Private Thomas James Ford was Killed in Action on the 14th of November 1916. 1652 Private John Henry Ford survived the war and returned to Australia on the 30th of June 1918.

2108 PRIVATE GERALD EDWARD APLIN (1883 – 1916)

25th Infantry Battalion

Gerald Edward Aplin was born in Townsville, Queensland, on the 22nd of July 1883, to William and Mary Jane Aplin (née Bristol). He was one of nine children. William Aplin arrived in Australia in December 1862 and moved to Bowen in 1863. He worked as a pastoralist & politician from 1865 to 1901 and was the Mayor of Townsville in 1869. He became the Chairman of Thuringowa Shire Council and was a Member of the Queensland Legislative Council from 1880 until his death in 1901.

Gerald Aplin grew up in Townsville and attended the Townsville Grammar School. His father was a member of the first board of trustees of Townsville Grammar School in 1888. After his school years were completed, Gerald Aplin became a grazier on "Woolgarang" Station, Burketown, Northern Territory. He married Margaret Constance Knight-Eaton on the 7th of July 1915 in Brisbane, and 20 days after his wedding, Gerald Aplin applied at Townsville to enlist in the Australian Imperial Force (AIF). He declared that he was 32 years old, married, and a grazier. His medical examination noted that he was 5'7" (170 cm) tall and weighed 132 lbs (60kg). He had a fresh complexion, blue eyes, fair hair, and identified as Church of England. Accepted into the AIF, Gerald Aplin was allocated service number 2108, assigned to the Infantry, and posted to the 25th Battalion, 4th reinforcements.

After six weeks of basic training in Australia, Private Gerald Aplin and his fellow reinforcements embarked with Convoy No.11 from Brisbane bound for Egypt. They sailed aboard HMAT A26 *Armadale*, which carried 244 troops. The entire convoy of nine ships transported 7,336 troops on the voyage. Private Aplin was taken on strength in the 25th Battalion on the 4th of February 1916 in the town of Ismailia on the Egyptian Suez Canal. Training in Egypt followed until the 25th Battalion embarked from Alexandria on the 14th of March 1916 to join the British Expeditionary Force fighting in France on the

Western Front. Landing in France on the 19th of March 1916, the 25th was the first AIF battalion to arrive there.

The battalion took part in its first major battle at Pozières, fought between July and September 1916, as part of the larger Battle of the Somme. The primary objective was the capture of the strategic village of Pozières in northern France, a crucial point in the German defensive line. This fiercely contested battle exacted a heavy toll on both sides, marked by intense artillery barrages and steadfast German resistance.

On July 23rd, the Australian forces seized control of Pozières, prompting the German Army to retaliate with a bombardment of the village and its surroundings. The ensuing shelling was one of the most intense ever encountered by an Australian Division, culminating in its peak on July 26th. The Australians successfully captured Pozières; however, this triumph came at a significant price, with more than 23,000 Australian casualties. It is remembered as one of the most ferocious battles of the war.

Private Gerald Aplin was one of the Australian casualties at Pozières. He was reported missing in France on the 29th of July 1916, and it was not until July 1917 that a Court of Enquiry found that he was Killed in Action by gunshot on the day he was reported missing. Private Gerald Aplin was 33 years old at the time of his death and had served in the AIF for one year. He was a member of the 25th Battalion for six months. He has no known grave. His death was notified in the 331st list of Australian Casualties released on the 18th of August 1917.

In an Australian Red Cross investigation of Private Gerald Aplin's death, 439 Sergeant Frederick Mortyn of the 25th Battalion stated, in October 1916:

"I know about Aplin from a man called Fennelly - who as he came in on July 29th found Aplin wounded in a shell hole - Fennelly was just giving him a cup of water, when Aplin was struck by another bullet, and he was killed instantly. He practically died in the arms of Fennelly, who told me about it after that when we met at Étaples, where I was on my way to the Lewis Gun School."

Gerald Aplin's name is inscribed on The Australian War Memorial Roll of Honour (Panel 103), The Villers-Bretonneux Memorial (Australian National

Memorial-France), The Darwin War Memorial Cenotaph, The Burketown Burke Shire Council WW1 Roll of Honor, The Townsville Grammar School Roll of Honour, The City of Townsville Honour Roll, and The Townsville Cenotaph. The Aplin Weir is named after the family, and Aplin Road in Girraween, Litchfield Council area, in the Northern Territory is also named in Gerald Aplin's honour.

2108 Private Gerald Edward Aplin served in WW1 in Egypt and on the Western Front. He was awarded the 1914/15 Star (no:21048), the British War Medal (no:19779), and the Victory Medal (no:19707). These medals, as well as a memorial scroll and a memorial plaque, were passed to his widow.

The following was published in *The Telegraph* (Brisbane) on the 22nd of September 1916:

PERSONAL NOTES Mrs. Gerald Aplin, of Wallogorang Station, Burketown, has received word that her husband, Pte G. E. Aplin, has been missing since 29th July. Mrs. Aplin is at present in Brisbane, prior to her departure for England shortly.

8769 PRIVATE HAROLD HARRY BAKER (1888 - 1916)

25th Infantry Battalion

Harold Harry Baker was born in Brisbane in about 1887 to Alfred and Emily Baker (née Louden). Harold, Emily, and Annie arrived in Rockhampton from Birmingham, England, in December 1886. His infant brother Henry died on the voyage. Harold was the third child born; however, his brother's death meant that he only had one surviving older sibling, his sister Annie.

In 1903, Annie Baker is recorded as living at Ma Ma Creek, located about 84 km from Brisbane. She later moved to Grantham with her husband, and Harold lived with them. In 1913, Harold Baker was working as a carpenter and living at the Tower

Hotel in Townsville. The details of his parents are not recorded; however, letters from his sister confirm that they were deceased prior to 1915.

After working as a carpenter and labourer, Harold Baker became a fireman. On the 29th of July 1915, he was living in Grantham when he applied, in Brisbane, to join the AIF. Harold Baker declared that he was 28 years old, single, employed as a fireman, and that he had no previous military experience. His medical examination noted that he was 5'4" (162 cm) tall, weighed 151 lbs (68 kg), had a dark complexion, blue eyes, and black hair. He identified as Methodist and nominated his sister as his next of kin, giving her address as Mrs. G. Metcalf, Grantham, Queensland.

Accepted into the AIF, Harold Baker was allocated service number 8769, assigned to the Infantry, and posted to the 25th Infantry Battalion. The 25th, also known as "The Darling Downs Battalion," was formed in Queensland in March 1915. In November 1915, Private Harold Baker was reposted to the 2nd Divisional Ammunition Column (2nd DAC), and his rank was changed to Gunner. The DAC provided essential logistical support for the 2nd Division, responsible for the transportation and supply of ammunition to the front lines. The unit consisted of over 1,000 personnel, including drivers, mechanics, and other support staff, as well as horses and wagons.

After completing basic training in Australia, Gunner Harold Baker embarked with his unit from Melbourne bound for Egypt on the 16th of November 1915. They joined Convoy No.14 and sailed aboard HMAT A39 Port Macquarie, which boarded 396 troops and 300 horses. The 14 ships that made up the convoy carried a total of 12,190 troops and 1,561 horses on the voyage to Egypt.

After arriving in Egypt in December 1915, Gunner Baker spent the next three months in training. During that period, he was transferred to the 5th Field Artillery Battery and then the 22nd Howitzer Brigade. In March 1916, he rejoined the 25th Battalion, was redesignated as a Private, and sent with the Battalion to France to fight on the Western Front with the British Expeditionary Force. The 25th Battalion was the first Australian Battalion deployed to France, arriving on the 19th of March 1916.

As a unit of the 2nd Division, the 25th Battalion engaged in their first significant encounter at Pozières from the 25th of July to the 7th of August. On the 28th of July 1916, during a night assault on the heights, the battalion suffered substantial casualties, with 12 officers and 350 other ranks killed or wounded. Many soldiers found themselves entangled in dense wire entanglements that had survived the preceding artillery barrage. A subsequent offensive on the 4th of August led to additional losses, resulting in a total of 785 casualties for the 25th Battalion's inaugural battle on the Western Front.

Private Harold Baker was one of these casualties, having been Killed in Action on the 29th of July 1916. In a statement made to the Australian Red Cross in 1918, 1923 Private Rupert Hill said regarding Private Prichard, Private Baker, and Corporal Macdonald:

"These three men I saw dead about halfway between the lines, just in front of Contalmaison. They all fell in front of me. I am quite certain that Cpl. Macdonald was dead, as he was shot in the head and there is very little doubt about the other two men being shot dead also. The attack was a failure, and I do not think it in the least likely that the Germans would have come out in 'No Man's Land' to look for such men as these, and I am sure we never got them in. They were too far out from both sides. This happened on the night of July 28th or early the following morning. The Germans were shelling the front heavily, so anything might have happened to their bodies."

Private Harold Baker was 29 years old at the time of his death and had served in the AIF for exactly one year. He had been a member of the 25th Battalion for four months. He has no known grave. Private Harold Baker's death was notified in the 218th list of Australian Casualties released on the 30th of September 1916.

Harold Harry Baker's name is inscribed on The Australian War Memorial Roll of Honour (Panel 103), The Australian National Memorial, Villers-Bretonneux, France, The City of Townsville WW1 Honour Roll, and The Townsville Cenotaph. Baker Street, Hermit Park, in Townsville was named in his memory.

8769 Private Harold Harry Baker served in WW1 in Egypt and on the Western Front. He was awarded the 1914/15 Star (no:24646), the British War

Medal (no:24032), and the Victory Medal (no:23864). These medals, as well as a memorial scroll and a memorial plaque, were passed to his sister.

34 SERGEANT ALFRED HENRY CLEGG (1894 – 1916)

26th Infantry Battalion.

Alfred Henry Clegg was born in Charters Towers on the 15th of July 1894. He was the first son and eldest of five children born to Alfred and Prudence Clegg (née Norris). Alfred Clegg Sr. was a watchman in Charters Towers.

Alfred Clegg Jr. grew up in Charters Towers in his early childhood, where he attended Charters Towers State School. By 1913, the family had moved to Townsville and lived on Allen Street. Alfred Clegg then attended Townsville State School and went on to Technical College to become a Clerk. He was also an active member of the church, serving as a choir member and Sunday School teacher.

After finishing school, Alfred Clegg gained joined the Railways, working as a clerk in Townsville and Cairns. He also served in the Senior Cadets, and in 1914, he volunteered for the 2nd Australian Naval and Military Expeditionary Force (AN&MEF), raised to take part in the capture of German New Guinea.

On the 8th of August 1914, Alfred Clegg was sent to Thursday Island for garrison duty. After enlisting for service outside Australia, he left Thursday Island on the 16th of August 1914 on the troopship S.S. *Kanowna* and was formally taken on strength of the AN&MEF on the 6th of September 1914. The ship returned to Townsville without reaching New Guinea on the 18th of September due to a dispute with the vessel's firemen. The soldiers aboard S.S. *Kanowna* were discharged upon their return, and Alfred Clegg returned to Cairns.

Six months later, on the 13th of March 1915, Alfred Clegg applied in Cairns to join the Australian Imperial Force (AIF). He declared that he was 20 years, seven months old, single, qualified as a clerk, and still serving as a Corporal in "E" Company of the Citizen Forces. His enlistment medical examination noted that he was 5'9" (175 cm) tall, weighed 9st 4lbs (59kg), had a fair complexion, blue eyes, and dark brown hair. He identified as Church of England and listed his mother as his next of kin, giving her address as Allen St, South Townsville.

Accepted into the AIF, Alfred Clegg was allocated service number 34, assigned to the Infantry, and posted as a Corporal and an original member of "A" Company of the 26th Battalion. The 26th Battalion was raised at Enoggera, Queensland, in April 1915 from recruits enlisted in Queensland and Tasmania and formed part of the 7th Brigade.

After three months of training, Corporal Alfred Clegg, as a member of the 26th Battalion, joined Convoy No.7 at Brisbane on the 24th of May 1915, bound for Egypt. The battalion sailed aboard HMAT A11 *Ascanius*, which carried 1,748 troops and 10 horses on the voyage. The four ships of Convoy 7 transported a total of 5,052 troops and 20 horses to Egypt.

Following further training in Egypt, the 26th Battalion was deployed to the Gallipoli Peninsula as part of the 7th Brigade on the 4th of September 1915. As the Battalion arrived at Gallipoli after the conclusion of the August offensive, their role was purely defensive, protecting the ANZAC perimeter at Courtney's and Steele's Posts, and Russell's Top. They served in that role until the general evacuation in December 1915.

Corporal Alfred Clegg was promoted to Sergeant and left the Gallipoli Peninsular on the 11th of December 1915, headed for Alexandria, Egypt. Three months later, in March 1916, the 26th Battalion was redeployed to the Western Front, France. The battalion's first major action in France was at Pozières in the Somme valley.

Pozières was a brutal battle. Charles Bean's quote, now inscribed on a plinth near the ruins of the Windmill, states: *"The ruin of Pozières Windmill which lies here was the centre of the struggle in this part of the Somme battlefield in July and August 1916. It was captured on August 4 by Australian troops who fell more thickly on this ridge than on any other battlefield of the war."*

It was during the battle of Pozières that Sergeant Alfred Clegg was Killed in Action on the 29th of July 1916. He was 21 years, 11 months old and had served with the AIF for one year and four months. He has no known grave, and his death was notified in the 223rd list of Australian Casualties released on the 28th of September 1916.

Alfred Henry Clegg's name is inscribed on The Australian War Memorial Roll of Honour (Panel 107), The Villers-Bretonneux Memorial (Australian National Memorial-France), The Millchester State School Honour Board, The Ross Island State School Honour Board, Townsville, The Townsville Railway Honour Board, The Cairns War Memorial, The City of Townsville WW1 Honour Roll, and The Townsville Cenotaph.

34 Sergeant Alfred H. Clegg served in WW1 at Gallipoli and on the Western Front. He was awarded the 1914/15 Star (no:22120), the British War Medal (no:12685), and the Victory Medal (no:12627). A memorial scroll and a memorial plaque were passed to his father in 1921 and 1922, respectively.

The following notice was published in the *Townsville Daily Bulletin* (Qld.) on the 4th of October 1916:

"The Great War.

At St. John's Church, South Townsville, a memorial service was held on Sunday night, September 24, in memory of the late Sergeant A. H. Clegg, who was killed in France. The service was conducted by the Rev. J. E. Dale. The large congregation consisted chiefly of the late sergeant's friends. The 'Dead March' was played by Miss E. Clegg. Sergeant A. H. Clegg was an ardent church worker, being a member of the church choir and a teacher of one of the Sunday school classes. Enlisting at the beginning of the war, Sergeant Clegg was one of the S.S. Kanowna boys, and on returning from Thursday Island joined the Expeditionary Forces."

The following notice was published in *The Northern Miner* (Charters Towers, Qld.) on the 20th of September 1916:

"Sergeant A. H. Clegg, whose death was reported yesterday, is a son of Mr. A.V. Clegg, who, for so many years was such a prominent member of the Millchester Cricket Club, and whose left-handed bowling was such a tower of strength to his

side. Young Clegg, who was 22 years of age, was born in Charters Towers, but enlisted in Cairns, where he was employed in the Railway Department. He went with the 26th Battalion, and fought at Gallipoli, where he spent his 21st birthday, and his 22nd birthday found him in the trenches in France. He was a sunny open-hearted boy, a good sportsman, and a brave soldier. Great sympathy is felt by his parents, who now reside in South Townsville."

200 SERGEANT WILLIAM GEORGE TRIX (1885 – 1916)

26th Infantry Battalion

William George Trix was born on the 20th of January 1885 in Southport, Queensland. He was the eldest of four children born to George and Susan Trix (née Jones). George Trix was a Sugar Planter, and the family lived in Skyring Creek, south of Gympie, near Pomona in Queensland.

William Trix attended Pomona State School, and after finishing his education, he became a dairyman. In 1913, he lived at Rosemount, Ithaca Creek, in Brisbane.

William Trix was also a member of the Citizen Military Forces, and on the 8th of August 1914, he was mobilized for war service and embarked at Cairns for Thursday Island (War Station) Garrison Duty. One week later, he enlisted for service outside Australia and volunteered for the Australian Naval and Military Expeditionary Force (AN&MEF), raised to participate in the capture of German New Guinea.

He left Thursday Island on the 16th of August 1914 on the troopship S.S. *Kanowna* and was formally taken on strength of the AN&MEF on the 6th of September 1914. However, the ship returned to Townsville without reaching New Guinea on the 18th of September due to a dispute with the vessel's firemen.

William Trix, along with the rest of the AN&MEF, was discharged from the Army on the day they disembarked in Townsville. Following his discharge,

William assumed employment in Townsville as a storeman, working for Thos Browns & Sons, and in 1915 was living at the YMCA. On the 5th of April 1915, William Trix applied to join the Australian Imperial Force (AIF). He was 30 years, two months old, and he declared that he was single, qualified as a despatch clerk, and had previous military service with the Citizen Forces. His medical examination noted that he was 5'7½" (171 cm) tall, weighed 10st (64kg), had a fair complexion, hazel eyes, and light brown hair. He identified as Church of England and nominated his father as his next of kin, giving his address as Annerley St., Toowong, Brisbane.

Accepted into the AIF, William Trix was allocated service number 200, assigned to the Infantry, and posted as an original member to "A" Company, 26th Infantry Battalion, at the rank of Corporal. The 26th Battalion was raised at Enoggera, Queensland, in April 1915 from recruits enlisted in Queensland and Tasmania and formed part of the 7th Brigade.

After a brief period of basic training, the Battalion's first draft, consisting of two companies of Queenslanders, embarked from Brisbane as part of Convoy No. 7. They sailed aboard HMAT A11 *Ascanius* on the 24th of May 1915. The ship was bound for Egypt and made a stop in Sydney during the voyage. While the *Ascanius* was at port in Sydney Harbour, the soldiers were camped at the Sydney Show Grounds. During this stay, Corporal Trix married Agnes Gordon on the 28th of May 1915 at Marrickville, Sydney. A short honeymoon followed before the *Ascanius* resumed its voyage.

Once disembarked in Egypt, the first draft of the 26th Battalion undertook further training and was joined by its second draft of two Tasmanian companies in late June. Once formed up as a complete battalion, the 26th was deployed to the Gallipoli Peninsula, landing there on the 12th of September 1915.

Having arrived after the conclusion of the August Offensive, the 26th's role was purely defensive, protecting the ANZAC perimeter at Courtney's and Steele's Posts, and Russell's Top. They served in that capacity until the general evacuation in December 1915.

Corporal Trix developed bronchitis while at Gallipoli and was evacuated back to medical facilities at Mudros on the 4th of December 1915. He subsequently developed jaundice and was evacuated to a hospital in Alexandria. After treatment and recovery, he rejoined the 26th Battalion at Tel-el-Kebir in early February 1916.

Nine days later, he suffered a bout of rheumatism and spent five days in the 7th Field Ambulance. On the 11th of March 1916, William Trix was promoted to Sergeant. Four days after his promotion, Sergeant Trix and the 26th Battalion boarded a ship in Alexandria to join the British Expeditionary Force on the Western Front in France.

The battalion's first major action in France was at Pozières in the Somme valley. Pozières was a brutal battle. Charles Bean's quote, now inscribed on a plinth near the ruins of the Windmill, states: *"The ruin of Pozières Windmill which lies here was the centre of the struggle in this part of the Somme battlefield in July and August 1916. It was captured on August 4 by Australian troops who fell more thickly on this ridge than on any other battlefield of the war."*

It was during this battle that Sergeant Trix was Killed in Action on the 29th of July 1916. In a statement recorded by the Australian Red Cross in 1917, Lance Corporal A.G. Salmond said, *"I was alongside him when he was killed near the German front-line trench at Pozières on the night of 28th July. He was shot – in the body, I think- and fell, killed outright. He had just called to us to come on. The last words he spoke were: 'Come on No.1 we've never been beaten yet!' He fell almost immediately. He was Sgt. of No.1 Pl. A. Coy."*

Sergeant William Trix was 31 years, six months old and had served with the AIF for one year and four months when he was Killed in Action. His body was not recovered, and he has no known grave.

William George Trix's name is inscribed on The Australian War Memorial Roll of Honour (panel 109), The Villers-Bretonneux Memorial (Australian National Memorial-France), The City of Townsville WW1 Honour Roll, The Pomona State School Roll of Honor, and The Townsville Cenotaph. Trix Street, Rosslea, Townsville, is named in his memory.

200 Sergeant William George Trix served at Gallipoli and on the Western Front in WW1. He was awarded the 1914/15 Star (no:21717), the British War Medal (no:12882), and the Victory Medal (no:12824). For her husband's

sacrifice, his widow, Agnes, received a war pension of two pounds, 12 shillings and three pence (£2/12/3d) per fortnight (which was about half of the average national wage in 1916). A memorial scroll and a memorial plaque were passed to her in 1920 and 1921, respectively.

The following notices were published in The Brisbane Telegraph on the 28th of July 1917:

TRIX. — Killed in action, in France, 29th July, 1916, Sergeant William George Trix, aged 31 years, dearly beloved husband of Agnes Trix.

> There is a lonely grave, so far away,
> Where my beloved sleeps in peace today,
> Yet not more lonely than the sorrowing heart
> Which still mourns for the hero who played his part.

TRIX.— In loving memory of Sergeant W.G. Trix, killed in action, Pozières, France, 29th July, 1916.

Inserted by his loving parents.

777 PRIVATE GEORGE MERVIN WATSON (1897 – 1916)

25th Infantry Battalion

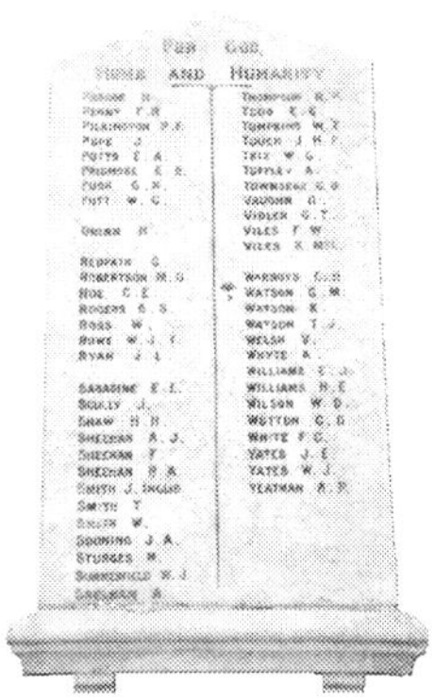

Mervin Watson was born John Henry William Ogier in about 1897, and he was the child of William Ogier. There is no record of his mother's name. William Ogier is recorded as living in New Farm for several years. Other residents of Kent Street stated that young John Ogier lived with him.

In 1922, the Army requested the Queensland Police to investigate the relationship between George Watson and William Ogier. The request was made because George had nominated William Ogier as the sole beneficiary in his Will but nominated a female friend as his next of kin.

It was important to clarify the exact relationship as legislation at the time required that medals of deceased soldiers be sent to the recipient's father if possible. The Police determined that William Ogier was George's father. They also determined that George was underage and that his father had refused him permission to enlist. As a result, the young John Ogier left home and enlisted in Townsville, using the name 'George Mervin Watson'.

In early 1915, it was necessary to have attained the age of 19 years to enlist in the Australian Imperial Force (AIF). This changed in June 1915 to 18 years. John Henry William Ogier was, therefore, less than 19 years old when he applied to join the AIF on the 20th of February 1915. On his Attestation Paper, he declared that his name was George Mervin Watson and that he was 23 years old. He also declared that he was born in Brisbane, that he was single, a labourer, and that he had no previous military experience.

George Watson's medical examination noted that he was 5' 4" (163 cm) tall, which should have excluded him from enlistment as the minimum height requirement was 5'6" (168 cm). However, the enlistment proceeded. He weighed 9st 3lbs (59kg), had a brown complexion, brown eyes, and black hair. He identified as Presbyterian and nominated Miss Muriel Ward as his next of kin, giving her address simply as Einasleigh.

George Watson was accepted into the AIF and was allocated service number 777, assigned to the Infantry, and posted as an original member to "C" Company of the newly formed 25th Battalion. The 25th was raised at Enoggera in Queensland in March 1915 as part of the 7th Brigade. The Battalion was composed of men recruited in Queensland but also included a small contingent of men from Darwin. The battalion underwent four months of training in Brisbane in preparation for their deployment overseas.

On the 29th of June 1915, Private George Watson, and the rest of the 25th Battalion embarked from Brisbane bound for Egypt. They sailed aboard HMAT A60 *Aeneas,* which carried 1,720 troops on the voyage. Their ship was one of four that made up Convoy No.7. The entire convoy transported 5,052 men and 20 horses. Once landed in Egypt, the 25th underwent further training until early September 1915, when the Battalion was deployed to the

Gallipoli Peninsula. They remained on Gallipoli in a defensive role until the evacuation back to Egypt in mid-December.

Private George Watson's conduct record was blemished, as he was in the habit of going Absent Without Leave and Disobeying Superior Officers. In December 1915, at Mudros, he was charged for absenting himself without leave and disobeying a lawful command from his superior officer. Private Watson was found guilty and awarded 28 days Field Punishment Number 2 (F.P.No.2). There were two categories of field punishment. Field punishment No.1 consisted of heavy labouring duties, possibly being restrained in handcuffs or fetters, and being tied to a post or wheel. Field punishment No.2 differed, in that the offender was not liable to be attached to a fixed object. In both forms of field punishment, the soldier was also liable for loss of pay. In January 1916, at Tel el Kebir, Private George Watson was again charged for absenting himself without leave and disobeying a lawful command from his superior officer. This time he received 28 days F.P. No.2.

In March 1916, the 25th was redeployed to the Western Front, France. Landing on the 19th of March 1916, it was the first AIF battalion to arrive in France. In May 1916, in France, George faced another charge, this time for disobedience of an order. Another 14 days F.P. No.2 followed.

The 25th Battalion, fighting as part of the 2nd Division, was engaged in its first major action in the Battle of Pozières on the 28th of July 1916. During the battle, the battalion suffered heavy losses during a night attack on the high ground, with 12 officers and 350 other ranks killed or wounded. Many soldiers were caught in thick wire obstacles that had not been cleared by the pre-attack artillery barrage. A follow-up attack on the 4th of August resulted in additional casualties, bringing the total loss for the 25th Battalion to 785 in its first encounter on the Western Front.

It was during the Battle of Pozières that Private George Watson was initially reported as missing on the 29th of July 1916. This was later confirmed as Killed in Action. His exact age at the time of his death is unknown but is recorded as 25 years. It is more likely that he was closer to 20 years old and had served with the AIF for one year, five months. He has no known grave, and his death was notified in the 226/227th list of Australian Casualties released on the 2nd of October 1916.

George Watson's name is inscribed on The Australian War Memorial Roll of Honour (Panel 106), The Villers-Bretonneux Memorial (Australian National Memorial-France), Georgetown Etheridge War Memorial, The City of Townsville WW1 Honour Roll, and The Townsville Cenotaph.

777 Private George Mervin Watson served in WW1 at Gallipoli and on the Western Front. He was awarded the 1914/15 Star (no:21445), the British War Medal (no:16806), and the Victory Medal (no:16740). A memorial scroll and a memorial plaque were passed to his father in 1921 and 1922, respectively.

The following tributes were published in the *Townsville Daily Bulletin* (Qld.) on the 17th of October 1916:

CHURCH — In loving memory of my dear son, Pte George Church, aged 30 years, and also his true comrade, Pte George M. Watson, aged 28, who both fell and died in France after one year and nine months' fighting for home and freedom.

> George, the parting day was a sad one,
> And it nearly broke my heart:
> And only a mother can feel the pain
> That this message to-day imparts.
>
> (Inserted by his loving mother, Elizabeth Church.)

194 CORPORAL WILLIAM HENRY O'Connor (1893 – 1916)

25th Infantry Battalion

William Henry O'Connor was born in Charters Towers, Queensland, on the 3rd of May 1893 to Thomas and Hannah O'Connor (née Thomas). He was the sixth of 13 children and had five brothers and seven sisters. Thomas O'Connor was a miner, and the family lived at Millchester Rd, Charters Towers. William O'Connor grew up in Charters Towers and attended Queenton State School. When his school years were finished, he worked as a labourer and ended up as a crane driver. In his spare time, he served in the senior cadets and joined the Kennedy Regiment when he turned 18. William O'Connor was also a member of the Townsville Rifle Club

Australia entered the war on the 4th of August 1914, and four days later, Private William O'Connor was mobilised for war service. He embarked along with other members of the Kennedy Regiment at Cairns, bound for Thursday Island (War Station) Garrison Duty. One week later, he enlisted for service outside Australia and volunteered for the Australian Naval and Military Expeditionary Force (AN&MEF), raised to take part in the capture of German New Guinea. He left Thursday Island on the 16th of August 1914 on the troopship S.S. *Kanowna* and was taken on strength of the AN&MEF on the 6th of September 1914. The ship returned to Townsville without reaching New Guinea on the 18th of September, on account of a dispute with the vessel's firemen. William O'Connor, along with the rest of the AN&MEF, was discharged from the Army on the day they disembarked in Townsville.

Three months later, on the 21st of December 1914, William O'Connor applied at Townsville to enlist in the Australian Imperial Force (AIF). He declared he was 21 years, 7 months old, single, and a labourer. His medical examination noted that he was 5'5" (165 cm) tall and weighed 9st 5lbs (59kg), had a dark complexion, grey eyes, and black hair. He identified as Church of England and nominated his father as his next of kin.

Accepted into the AIF, William O'Connor was allocated service number 194, assigned to the Infantry, and posted as an original member of "A" Company in the newly formed 25th Battalion. This battalion was established in March 1915 at Enoggera in Queensland as a part of the 7th Brigade. While most of the battalion's recruits were from Queensland, a few men from Darwin were also included in its formation. The 25th spent six months forming up the battalion and completing basic training in Australia. On the 29th of June 1915, Private William O'Connor, with the rest of the 25th Battalion, embarked from Brisbane bound for Egypt. They sailed aboard HMAT A60 *Aeneas*, which carried 1,720 troops on the voyage. Their ship was one of four that made up Convoy No.7. The entire convoy transported 5,052 men and 20 horses.

Once landed in Egypt, the 25th underwent further training in Egypt during August and by early September was manning trenches at Gallipoli. At Gallipoli, the 7th Brigade reinforced the depleted New Zealand and Australian Division. The 25th Battalion, however, had a relatively quiet time because the last major Allied offensive had been launched, and turned back, in the previous month.

During his time on Gallipoli, Private William O'Connor was appointed as an Acting Sergeant in November 1916. He relinquished the acting rank when he was transferred to a hospital ship at the end of November 1916. He had contracted hepatitis and was evacuated back to the 2nd Australian General Hospital in Cairo. As his condition improved, he was transferred to other medical units until he was finally discharged from Ras-El-Tin on the 21st of December 1915. By this time, the 25th Battalion had left the Gallipoli Peninsula, and Private William O'Connor re-joined the unit at Tel-el-Kebir on the 20th of January 1916. He was promoted to Corporal on the 3rd of March 1916.

The 25th Battalion proceeded to France after completing further training in Egypt. Landing on the 19th of March 1916, it was the first AIF battalion to arrive on the Western Front. The Battalion soon learned the difference between fighting on Gallipoli and the trench warfare of France.

From late July to early September 1916, the 25th was committed to the Battle of Pozières. On the 23rd of July, as Pozières was gained by Australian troops, the German Army bombarded the village and its approaches. The two opposing forces increased their shelling, reaching a peak on the 26th of July. It was recorded as some of the worst shelling endured by an Australian division.

On the night of the 28th of July, the 25th Battalion attacked the German lines north and east of the village. It was during this action that Corporal William O'Connor was Wounded in Action on the 29th of July and admitted to the 44th British Casualty Clearing Station. The exact nature of his wounds is not recorded. Tragically, he Died of Wounds the next day, the 30th of July 1916. He was laid to rest the same day at the Puchevillers British Cemetery (Plot II, Row C, Grave No.36), France. Corporal William O'Connor was 23 years and three months old and had served in the AIF for 19 months when he Died of Wounds.

William Henry O'Connor's name is inscribed on The Australian War Memorial Roll of Honour (Panel 105), the Queenton State School Roll of Honour, the Richmond Hill Teachers and Ex-Pupils Great War Honour Roll, The City of Townsville WW1 Honour Roll, and The Townsville Cenotaph. O'Connor St, Rosslea, Townsville, is named in his memory.

194 Corporal William Henry O'Connor served in WW1 in Gallipoli and on the Western Front. He was awarded the 1914/15 Star (no:20904), the British War Medal (no:20677), and the Victory Medal (no:20599). These medals, as well as a memorial scroll and a memorial plaque, were passed to his father in 1922.

William O'Connor's younger brother, 2704/5948 Company Sergeant Major, Victor David O'Connor, also enlisted in the AIF and served in the 25th Battalion. He Died of Wounds received in battle in France on the 2nd of September 1918, aged 22 years.

447 2nd LIEUTENANT JAMES CASSIE MURRAY (1889 – 1916)

25th Infantry Battalion/2nd Machine Gun Battalion

James Cassie Murray was born in Townsville on the 12th of December 1889. He was the oldest of two sons, born to George and Margaret Murray (née Macgregor). George Murray was a contractor and a previous Mayor of Townsville. The family lived in premises on Alexander St, Stanton Hill for many years. James Murray attended the Central State School, Townsville, and then qualified as a Joiner.

On the 15th of March 1915, James Murray applied at Townsville to enlist in the Australian Imperial Force (AIF). He declared that he was 25 years, five months old, employed as a joiner, and single. He also declared that he had previous military experience: three years and seven months in the Garrison Artillery and two years in the Naval Reserve. James Murray's medical examination noted that he was 5'10" (178 cm) tall, weighed 164 lbs (74 kg), had a dark complexion, brown eyes, and black hair. He identified as Presbyterian and listed his father as his next of kin giving his address as Stanton Hill, Townsville.

Accepted into the AIF, James Murray was allocated service number 447, assigned to the Infantry, and posted as a Corporal and an original member to the 25th Battalion, "D" Company. The 25th Battalion was raised at Enoggera in Queensland in March 1915 as part of the 7th Brigade and was predominantly composed of men recruited in Queensland plus a small contingent of men from Darwin. After enlistment, Corporal James Murray spent three and a half months in Enoggera, Brisbane, training for overseas service.

On the 29th of June 1915, the 25th Battalion embarked from Brisbane for Egypt aboard HMAT A60 *Aeneas*, which carried 1,720 troops. The ship was one of four in Convoy No.7, which transported a total of 5,052 men and 20 horses.

Once landed in Egypt, the 25th underwent further training in August and by early September was manning trenches at Gallipoli. The 7th Brigade reinforced the depleted New Zealand and Australian Division; however, the 25th Battalion experienced a relatively quiet time as the last major Allied offensive had been launched and turned back the previous month. This quiet period continued until mid-December when the general evacuation occurred.

On the 5th of October 1915, Corporal James Murray was promoted to Sergeant. While on the Peninsula, he also contracted jaundice (hepatitis), which resulted in his evacuation to a hospital in Alexandria in December 1915, just before the general evacuation. After being discharged from the hospital at the end of January 1916, Sergeant Murray re-joined the 25th Battalion in Egypt.

After further training in Egypt, the 25th Battalion was deployed to France to fight on the Western Front. Landing on the 19th of March 1916, it was the first AIF battalion to arrive on the Western Front. On the 14th of March 1916, Sergeant Murray was promoted to 2nd Lieutenant and seconded for duty with the 7th Machine Gun (MG) Company, which was allocated to the 7th Brigade. The elements of the MG Company were allocated to sections of the line in detachments of varying sizes depending on the task, but normally in pairs as a 'section'. A section would be two guns and their crews (three men per gun). They were equipped with the legendary Vickers Medium Machine Gun.

The 25th Battalion, now fighting as part of the 2nd Division, took part in its first major battle at Pozières between the 25th of July and the 7th of August 1916. During this battle, the AIF lost as many men over a few weeks as it did over 8 months on Gallipoli. One of these casualties was 2nd Lieutenant James Murray, who was Killed in Action on the 5th of August 1916. 1702 Pte Percy Kemp said in a statement made in January 1917, *"On Aug. 5th. 1916 I saw Lieut. Murray shot by a bullet near the heart about 20 yards in front of the first trench taken that morning at Pozières. I spoke to him, and he said he was feeling pretty bad and asked about the guns and told me to carry on with the work. When I returned, I was told he had died of his wounds about an hour afterwards. I think he would have been buried behind the lines at Pozières."*

2nd Lieutenant James Murray was 26 years, eight months old and had been a member of the AIF for 17 months at the time of his death. His body was not recovered, and he has no known grave. His death was notified in the 216th list of Australian Casualties released on the 20th of September 1916.

James Cassie Murray's name is inscribed on The Australian War Memorial Roll of Honour (Panel 179), The Australian National Memorial, Villers-Bretonneux, France, The City of Townsville WW1 Honour Roll, and The Townsville Cenotaph. Murray Street, Nelly Bay is named in the memory of James and his brother Murdoch.

2nd Lieutenant James Cassie Murray was awarded the 1914/15 Star (no:23424), the British War Medal (no:16708), and the Victory Medal (no:16642). A memorial scroll and a memorial plaque were passed to his father in 1921 and 1922, respectively.

James Murray's brother, Lance Corporal Murdoch Ross Murray, also served in the AIF. He was Killed in Action with the 15th Battalion on the 18th of September 1918.

600 PRIVATE GEORGE CHURCH (1886 – 1916)

25th Infantry Battalion

George Church was born on the 17th of October 1886 in Drouin, Victoria. He was the eighth of 10 children born to George and Elizabeth Church (née Meyer). He had four surviving brothers and three sisters. Elizabeth Church remarried Alan Barr at Rockhampton in January 1900. By 1903, the Barr family was living in Flinders St, Townsville, and Alan Barr was an engineer.

George Church worked as a labourer and lived at Towers Road, Townsville, when he applied to join the Australian Imperial Force (AIF) on the 22nd of February 1915. He declared that he was single, 28 years old, and had no previous military experience. He nominated his mother as his next of kin and listed her address as Hermit Park, Townsville. He later amended his mother's address and name in his service records to Mrs. E. Barr of 6 Underwood St, Paddington, Sydney, NSW.

George Church's medical examination noted that he was 5'5½" (166 cm) tall, weighed 11st 2lbs (71 kg), had a brown complexion, grey eyes, and black hair. He identified as Church of England. Accepted into the AIF, he was allocated service number 600, assigned to the Infantry, and was posted as an original member to "C" Company of the 25th Battalion, which was raised at Enoggera in Queensland in March 1915 as part of the 7th Brigade.

The 25th Battalion was predominantly composed of men recruited in Queensland but also included a small contingent of men from Darwin. After the battalion was formed up and basic training was completed, Private George Church and the rest of the 25th Battalion embarked from Brisbane bound for Egypt. They sailed aboard HMAT A60 *Aeneas*, which carried 1,720 troops on the voyage. Their ship was one of four that made up Convoy No.7. The entire convoy transported 5,052 men and 20 horses.

Once landed in Egypt, the 25th underwent further training during August, and by early September was manning trenches at Gallipoli. At Gallipoli, the 7th Brigade reinforced the depleted New Zealand and Australian Division. The 25th Battalion, however, had a relatively quiet time because the last major Allied offensive had been launched, and turned back, in the previous month. This continued until mid-December when the general evacuation took place.

Three months later, in March 1916, the 25th was redeployed to the Western Front, France. In early May 1916, Private George Church was charged for disobeying an order, found guilty, and awarded 14 days Field Punishment No.2 (FP2), which consisted of heavy labouring duties without pay. After a brief period in 'The Nursery' near Armentieres/Fleurbaix/Fromelles, the Battalion headed south for the great Somme offensive.

The focus of the AIF's endeavours was to capture Pozières on the right flank of the British front. The battle, which took place between July and August 1916, was part of the larger Somme Offensive, a joint British and French operation to break through German lines and push towards the city of Bapaume. The 25th Battalion was one of several Australian units tasked with taking the village of Pozières, a strategically critical position held by the Germans.

On the 23rd of July, Private George Church advanced across no man's land with the rest of the battalion which had been ordered to capture the German front line. Despite heavy resistance, the 25th successfully took the position and continued to push forward. Over the next several days, the battalion engaged in fierce fighting against German counterattacks and endured constant artillery fire. Despite heavy losses, the 25th Battalion managed to hold their ground and successfully repel enemy attacks.

Pozières was a brutal battle. Charles Bean's quote, now inscribed on a plinth near the ruins of the Windmill, states: *"The ruin of Pozières Windmill which lies here was the centre of the struggle in this part of the Somme battlefield in July and August 1916. It was captured on August 4 by Australian troops who fell more thickly on this ridge than on any other battlefield of the war."*

It was during this battle that Private George Church was reported as Wounded in Action on the 5th of August 1916. This was changed on the 21st of September 1916 to Died of Wounds on the 5th of August 1916. He was 30 years old and had served with the AIF for 17 months. He has no known grave; however, a headstone bearing his name was erected by his family in the Belgian Gardens Cemetery in Townsville. His death was notified in the 238th list of Australian Casualties released on the 26th of October 1916. He had previously been notified as wounded in the combined 220th/221st lists released on the 26th of September 1916.

George Church's name is inscribed on The Australian War Memorial Roll of Honour (Panel 104), The Villers-Bretonneux Memorial (Australian National Memorial-France), The City of Townsville WW1 Honour Roll, and The Townsville Cenotaph.

600 Private George Church served in WW1 at Gallipoli and on the Western Front. He was awarded the 1914/15 Star (no:20925), the British War Medal (no:16588), and the Victory Medal (no:16521). A memorial plaque was passed to his mother in 1923.

The following tributes were published in the *Townsville Daily Bulletin*:

IN MEMORIAM ROLL OF HONOR.

CHURCH — In loving memory of my dear son, Pte George Church, aged 30 years, and also his true comrade, Pte George M. Watson, aged 28, who both fell and died in France after one year and nine months' fighting for home and freedom.

George, the parting day was a sad one,
And it nearly broke my heart:
And only a mother can feel the pain
That this message to-day imparts.

(Inserted by his loving mother, Elizabeth Church.)

CHURCH — In loving memory of our dear brother and uncle, Pte George Church, who died of wounds, and also his true and trusted friend, George Mervin Watson, who was Killed in Action on the same day, August 5th, 1916.

"Greater love can no man give than to lay down his life for others."

(Inserted by his loving sister and brother and nieces, Mr and Mrs Stagg and E. and L. Lowe.)

CHURCH — In loving memory of our dear brother, Pte George Church, who Died of Wounds on the 5th of August 1916.

(Inserted by his loving brother and sister, Mr and Mrs W. Church.)

CHURCH — In loving memory of our dear brother, Pte George Church, who Died of Wounds on the 5th of August 1916. May his soul rest in peace.

(Inserted by his bereaved sister and brother, Mr and Mrs Samuel Church.)

4519 PRIVATE AUGUSTUS HARRY NORMAN (1889 – 1916)

25th Infantry Battalion

Augustus Harry Norman (Harry) was born in October 1888 in Woodburn, New South Wales, to Augustus and Caroline Norman (née Lane). He was the fifth of eight children and had four brothers and three sisters. The Norman family moved to Townsville before WW1, where Harry Norman worked as a labourer.

He married Florence Louisa Nimmo on the 19th of August 1911 in Queensland, and they had one son, Lewis, born on the 5th of May 1913. On the 4th of October 1915, Harry Norman applied to join the Australian Imperial Force (AIF). He lied on his application, declaring that he was single. It was not until February 1916, when he was in the Army, that he "confessed."

Harry Norman outlined in a letter that he had declared he was single because the recruiters in Townsville would not accept his application without a letter from his wife stating that she had no objection to him joining. Harry Norman could not get that letter because Florence was not in Townsville at that time, so he simply declared that he was single. He included the letter from Florence in his "confession" and applied for the separation allowance paid to married men.

The rest of Harry Norman's application to join the AIF was correctly filled out. He declared that he was 27 years, two months old, a labourer, and had no previous military experience. Because he had lied about his marital status, he listed his mother as his next of kin. This was later changed to his wife, whose address by that time was Orchard Street, Enoggera.

Harry Norman's medical examination noted that he was 5'7" (170 cm) tall and weighed 9st 4lbs (59 kg). He had a fair complexion, brown eyes, and light brown hair. He identified as Church of England.

Accepted into the AIF, Harry Norman was allocated service number 4519, assigned to the Infantry, and posted to the 25th Battalion, 11th reinforcements. The 25th Battalion was manning the trenches at Gallipoli when Harry enlisted. For the next six months, Private Harry Norman completed his basic training in Australia.

On the 31st of March 1916, Private Harry Norman and the other 11th reinforcements embarked from Sydney bound for Egypt as part of Convoy No.19. They sailed aboard HMAT A16 *Star of Victoria* (later renamed HMAT *Port Melbourne*), which carried 1,552 troops on the voyage. The 10-ship convoy transported a total of 8,075 personnel and 1,168 horses to Egypt. Disembarking at Alexandria on the 10th of May 1916, the contingent re-embarked eight days later for movement to France.

After further training in France, the reinforcements joined the 25th Battalion on the 2nd of August 1916, in time to take part in the 25th's first major battle at Pozières (25th of July - 7th of August 1916). The Somme offensive, directed at diverting the attention of the German army from Verdun, included the battle for Pozières. This strategically significant location commanded the high ground in the British advance towards Thiepval.

Although the 1st, 2nd, and 4th Australian divisions managed to capture a small, ruined area from the enemy, the cost was immense. Over 42 days, the Australians launched up to 19 assaults against the Germans, resulting in a devastating 23,000 casualties, including the dead, wounded, and captured. These heavy losses were nearly unbearable for a volunteer army. The 25th Battalion alone endured 785 casualties during this gruelling period.

It was during these attacks that Private Harry Norman was initially reported as missing. However, this report was later changed to Killed in Action on the 5th of August 1916, three days after joining the 25th in the field. While he was buried on the Pozières battlefield soon after being killed, he has no known grave today. He was 28 years old and had served with the AIF for 10 months at the time of his death. His death was notified in the 238th list of Australian Casualties released on the 26th of October 1916.

Augustus Harry Norman's name is inscribed on The Australian War Memorial Roll of Honour Panel 105, The Australian National Memorial, Villers-Bretonneux, France, The City of Townsville WW1 Honour Roll, and The Townsville Cenotaph.

4519 Private Augustus Harry Norman was awarded the British War Medal (no:35833) and the Victory Medal (no:35565). A memorial scroll and a memorial plaque were passed to his widow in 1921 and 1922, respectively. Florence Norman was granted a war widow's pension of two pounds (£2/-/-) per fortnight, about half of the wage paid to AIF soldiers in 1916.

The following notice was published in the *Townsville Daily Bulletin* on the 5th of August 1926:

NORMAN — In loving memory of my dear son, Private A. H. Norman, 25th Battalion, A.I.F., killed in action, August 6th 1916.

> We often sit and think of you,
> When we are all alone;
> For memory is the only thing,
> That grief can call its own.
> We think of you in silence,
> No eyes may see us weep.
> But deep within our hearts,
> Your loving memory we will keep.

Inserted by his loving mother, brothers, and sisters.

2225 PRIVATE HAROLD EWART WILLIAMS (1891 – 1916)

25th Infantry Battalion

Harold Ewart Williams was born in September 1891 in Keynsham, Bristol, England. He was the second child of Walter Williams and his wife Clara (née Hancock). Walter Williams was an insurance agent in 1891 in England when the family lived in Mangotsfield, Gloucestershire.

In the early 1900's, the Williams family left England for Australia. On arrival, the family lived in several locations across North Queensland including Charters Towers, Townsville, Cairns, and Atherton.

On the 16th of August 1915, Harold Williams applied at Townsville to enlist in the Australian Imperial Force (AIF). He declared that he was twenty-five years, one month old, had no military experience, was qualified as a motor driver and single. His enlistment medical noted that he was 5'8" (173 cm) tall, weighed 9st 3lbs (59kg), had a dark complexion, brown eyes, and black hair. He identified as Church of England. He listed his father as his next of kin and gave his address as Main Street, Atherton, North Queensland.

Accepted into the AIF, Harold Williams was allocated service number 2225, assigned to the Infantry, and posted to 25th Infantry Battalion, 4th Reinforcements. The 25th Battalion was raised at Enoggera in Queensland in March 1915 as part of the 7th Brigade. When Harold Williams enlisted, the 25th was training in Egypt in preparation for deployment to Gallipoli in early September 1915.

Private Williams spent the next five weeks in basic training in Australia before his unit embarked from Brisbane, bound for Egypt, on board HMAT A26 *Armadale* on the 18th of September 1915. They sailed as part of Convoy No.11, and their ship carried 244 military personnel. The entire convoy of 10 ships carried 7,336 personnel.

Arriving in Egypt, Private Williams was taken on strength of the 25th Battalion at Tel-el-Kebir on the 10th of January 1916. The 25th had left the Gallipoli peninsula on the 18th of December 1915, and when Private Williams arrived, the Battalion was undergoing further training prior to deployment to the Western Front.

The 25th Battalion deployed to France on the 14th of March 1916 and landed five days later. It was the first AIF battalion to take up positions on the Western Front.

Private Williams became ill with fever on the 8th of July 1916 and was serious enough to be hospitalized. He re-joined the 25th Battalion on the 29th of July 1916. This was the day after the 25th Battalion's first significant battle at Pozières on 28th July 1916. The battalion suffered heavily during a night-time attack on the heights, losing 12 officers and 350 other ranks killed or wounded, many of whom were caught in thick wire obstacles that had been left intact by the pre-attack artillery bombardment.

Private Harold Williams was with the 25th when it launched a follow-up attack on the 4th of August 1916 which added more casualties, bringing the total loss for the 25th Battalion in its first battle on the Western Front to 785.

Private Williams was one of these casualties, initially reported as wounded and then as wounded and missing. Harold Williams' father, Walter, was the Secretary of the Cairns War Council and, later, of the Charters Towers State Recruiting Committee. He wrote several letters to the Army seeking further information and informed them that two returned soldiers had reported to him that his son was killed.

A Court of Enquiry was conducted on the 25th of July 1917, nearly a year after Private Williams was reported missing. The Enquiry determined that Private Harold Ewart Williams was Killed in Action on the 5th of August 1916. He was 26 years old and had served in the AIF for 11 months. He has no known grave. His death was notified in the 331st list of Australian Casualties released on the 18th of August 1917.

Harold Ewart Williams' name is inscribed on The Australian War Memorial Roll of Honour (Panel 106), The Australian National Memorial, Villers-Bretonneux, France, The Atherton War Memorial, The City of Townsville WW1 Honour Roll, and The Townsville Cenotaph.

2225 Private Harold Ewart Williams served in WW1 in Egypt and on the Western Front. He was awarded the 1914/15 Star (no:22469), the British War Medal (no:19809), and the Victory Medal (no:19735). A memorial scroll and a memorial plaque were passed to his father in 1921 and 1922, respectively.

The following notice was published in the *Townsville Daily Bulletin* (Qld.) on the 4th of August 1921:

WILLIAMS. — In affectionate remembrance of Harold Ewart Williams, late 25th Battalion, Killed in Action Pozières, August 4th. 1916. Ever remembered by his loving parents, sisters, and brothers, C. A. and W. Williams, Mrs. and Mr. R. E. Horn, and Jack Williams.

1715 PRIVATE HERBERT BARNES (1871 – 1916)

15th Infantry Battalion

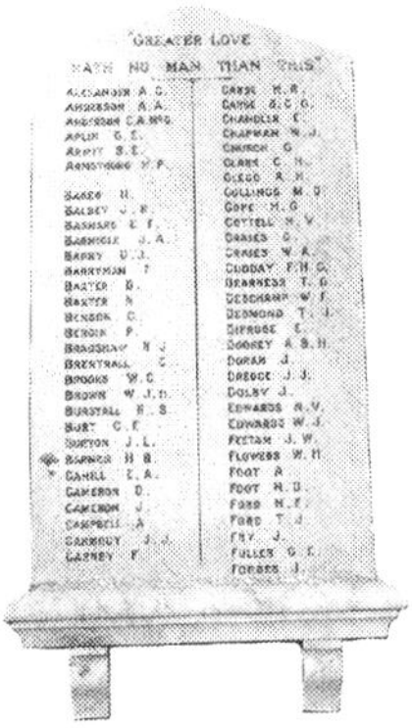

Based on his own declaration in 1915, Herbert Barnes was born in about August 1870 at Mile End, London, England. There is no recorded history of his parents, and it is unknown if he had any other relatives in Australia when he arrived in about 1895.

By 1903, Herbert Barnes lived at Mirani Silver Creek and was employed as a labourer. At some time after 1905, Herbert Barnes moved to Brisbane and lived in Stanley St, South Brisbane, with a friend, Mrs. James Rowe. On the 25th of January 1915, Herbert applied in Townsville to enlist in the Australian Imperial Force (AIF). He declared that he was 44 years and six months old, employed as a cane cutter, and that he had no previous military experience. His enlistment medical examination noted that he was 5'7½" (171 cm) tall, weighed 9st 1lb (62 kg), had a dark complexion, blue eyes, and brown hair. He identified as Church of England, listed Mrs. Rowe as his next of kin, and gave her address as Stanley St, South Brisbane.

Accepted into the AIF, Herbert Barnes was allocated service number 1715, assigned to the Infantry, and posted to the 15th Battalion, 4th Reinforcements. The 15th Battalion was originally raised as part of the all-volunteer AIF in September 1914 and was on its way to Egypt when Herbert Barnes enlisted. Private Herbert Barnes and his fellow reinforcements underwent initial training in Australia until they joined Convoy No. 4 on the 8th of April 1915. They embarked from Brisbane for Egypt and sailed aboard HMAT A15 *Star of England*, which carried 507 troops and 440 horses on the voyage. The entire convoy of five ships transported 3,299 troops and 2,124 horses to Egypt.

After a brief period of training in Egypt, the reinforcements were deployed to the Gallipoli Peninsula, where they joined the main body of the 15th Battalion on the 2nd of June 1915. As part of the August Offensive, the 15th Battalion was tasked to seize Hill 971 on the heights of Sari Bar. Private Herbert Barnes was wounded in this action on the 7th of August 1915. He was evacuated to Mudros and then transferred to the Hospital Ship *Aquitania* for further evacuation to England. He was suffering from dysentery as well as his wound.

After his recovery and time spent at the Convalescent Depot at Abbey Wood in England, Private Barnes re-joined the 15th Battalion in Egypt on the 11th of March 1916. On the 8th of June 1916, the 15th Battalion departed Alexandria for France and the Western Front. After landing at Marseilles, they travelled by rail to Beilleul in northern France and briefly moved into the line around Bois Grenier on the 15th of June 1916. The battalion's first major battle in France was the Battle of Pozières on the 5th of August 1916. During the battle, the battalion encountered intense shelling, followed by a brief German assault once they had deployed in trenches along the sunken road to Courcellette. The assault was thwarted, resulting in the capture of twenty German soldiers. Subsequently, the battalion focused on strengthening their defensive line by excavating saps towards the German trenches and dispatching patrols into no man's land in the following days.

It was during this battle that Private Herbert Barnes was reported as Missing in Action. It was not until the 1st of May 1917 that a Court of Enquiry concluded that he had been Killed in Action on the 7th of August 1916. He was about 46 years old and had been a member of the AIF for 18 months. He has no known grave, and his death was notified in the 303rd list of Australian Casualties released on the 26th of May 1917.

Herbert Barnes' name is inscribed on The Australian War Memorial Roll of Honour (Panel 74), The Australian National Memorial, Villers-Bretonneux, France, The City of Townsville WW1 Honour Roll, and The Townsville Cenotaph.

1715 Private Herbert Barnes served in WW1 on the Gallipoli Peninsula and on the Western Front. He was awarded the 1914/15 Star (no:12743), the British War Medal (no:9978), and the Victory Medal (no:9935). A memorial scroll and a memorial plaque were produced honouring his service. Despite representations made to the Army by his designated next of kin, Private Herbert Barnes' medals, scroll, and plaque were not issued as these could only be passed to *"a blood relative."*

3687 PRIVATE ROBERT PHILIP BERGIN (1888 – 1916)

15th Infantry Battalion

Robert Philip Bergin (Philip) was born, the middle child of a large family, on the 18th of July 1888 in Bundaberg, Queensland. His parents were Stephen and Mary Bergin (née McEnroe). Stephen Bergin was employed as a butcher at the Ross River Meatworks in 1903.

In 1913, the Bergin family lived in Gregory Street. After completing his schooling, Philip Bergin followed in his father's footsteps and worked as a labourer at the Ross River Meat Works.

On the 18th of September 1915, Philip Bergin applied in Townsville to enlist in the Australian Imperial Force (AIF). He declared that he was single, 27 years old, employed as a labourer, and had no previous military experience. His enlistment medical examination noted that he was 5'11¼" (181 cm) tall, weighed 176 lbs (80 kg), had a fair complexion, grey eyes, and auburn hair. He identified as Roman Catholic and nominated his father as his next of kin.

Accepted into the AIF, Philip Bergin was allocated service number 3687, assigned to the Infantry, and posted to the 15th Battalion, 12th reinforcements. The 15th Battalion was raised in September 1914 as a part of the all-volunteer Australian Imperial Force (AIF). Consisting of volunteers from Queensland and Tasmania, it was allocated to the 4th Brigade, alongside the 13th, 14th, and 16th Battalions. The 15th was engaged in combat on the Gallipoli Peninsula at the time of Philip Bergin's enlistment.

After enlistment, the reinforcements underwent three months of initial training in Enoggera, Brisbane. Basic training completed, Private Philip Bergin and the rest of the reinforcements embarked from Sydney, bound for Egypt, on the 30th of November 1915. They sailed as part of Convoy No. 14 aboard HMAT A23 *Suffolk* which carried 973 troops. The entire convoy of 14 ships transported a total of 12,190 troops and 1,561 horses on the voyage.

Disembarking in Egypt, Private Bergin and the other reinforcements joined the 15th Battalion at Tel-el-Kebir on the 6th of March 1916. Further training followed in the desert before the 15th Battalion sailed for France aboard the transport *Transylvania* in June 1916. Disembarking at Marseilles, the battalion moved to northern France via rail to Beilleul, moving into the line around Bois Grenier for a brief period on the 15th of June 1916.

The following month, the 15th Battalion was committed to fighting on the Western Front for the first time, entering the line around Pozières on the 5th of August 1916, as the 4th Division relieved the shattered 2nd Division.

On the 8th of August 1916, the 15th Battalion received orders from the 4th Brigade Headquarters to initiate an offensive to secure designated enemy positions. The assault commenced at 9:20 p.m., supported by heavy friendly artillery fire. The battle continued over the next four days, culminating in the 15th Battalion successfully consolidating the captured enemy positions.

This action exacted a significant toll, with the Battalion sustaining 48 fatalities, 337 injuries, and 45 reported as missing. Among those listed as missing was Private Philip Bergin. There were reports that he had been taken prisoner; however, a statement by an eyewitness (L/Cpl C Norris) in November 1916 confirmed Private Bergin's death from a shell explosion. A Court of Enquiry held at Ribemont on the 1st of May 1917 confirmed that Private Philip Bergin was Killed in Action on the 8th of August 1916. He was 28 years and one month old and has no known grave. He had served with the AIF for 10 months and three weeks and been a member of the 15th Battalion for five months.

Robert Philip Bergin's name is inscribed on The Australian War Memorial Roll of Honour (Panel 74), The Australian National Memorial, Villers-Bretonneux, France, The City of Townsville WW1 Honour Roll, and The Townsville Cenotaph. Bergin Road, Cranbrook in Townsville, is named in his memory.

3687 Private Robert Philip Bergin was awarded the British War Medal (no: 25640) and the Victory Medal (no: 25462). These medals, as well as a memorial scroll and a memorial plaque, were passed to his father. His father was granted a pension of one pound and ten shillings (£1/10/-) per fortnight for the loss of his son. This was about forty percent of the wage paid to AIF soldiers in 1916.

3823 PRIVATE REGINALD GEORGE LATCHFORD (1895 – 1916)

49th Infantry Battalion

Reginald George Latchford (Reg) was born on the 21st of January 1895, in Townsville, Queensland. He was the first-born child and oldest son of Edward and Jane Latchford (née Kennedy). He had two brothers and one sister.

Edward Latchford was a storekeeper and merchant, and the family lived in Charters Towers until about 1913 when they relocated to Toowoomba, Queensland, where Edward established an agricultural merchant company. Reg Latchford grew up in Charters Towers and attended the Christian Brothers School. After completing his school years, he worked as a clerk in his father's produce business in Toowoomba. He also served four years in the Army Cadets and the Citizen Military Forces.

In Toowoomba, Reg Latchford applied to enlist in the Australian Imperial Force (AIF) on the 22nd of September 1915. He stated that he was single, employed as a clerk, and 20 years and eight months old. He also detailed his previous military service. In his medical assessment, it was noted that he was 5'9½" (176.5 cm) tall, weighed 147 lbs (67 kg), had a fair complexion, blue eyes, and light brown hair. He identified as Roman Catholic and he nominated his father as his next of kin, giving his address as Roma St, Toowoomba.

Accepted into the AIF, Reg Latchford was allocated service number 3823, assigned to the Infantry, and posted to the 9th Battalion/12th Reinforcement. The 9th Battalion was one of the initial infantry units formed for the AIF in World War I. Along with the 10th, 11th, and 12th Battalions, it formed the 3rd Brigade. On the 25th of April 1915, it was the first battalion to land at Gallipoli.

After enlisting, Private Reg Latchford underwent three months of basic training in Australia. On the 30th of December 1915, he and the other reinforcements embarked with Convoy No.15 for Egypt from Brisbane. They sailed aboard HMAT A50 *Itonus*, which carried 1,052 troops. The entire convoy, consisting of 11 ships, transported a total of 9,123 troops to Egypt.

Upon reaching the AIF Training Camp in Zeitoun, near Cairo, Private Latchford was promptly transferred to the newly established 49th Infantry Battalion on the 29th of February 1916. This battalion had been formed just two days earlier as part of the AIF's expansion. Roughly half of its members were veterans from the 9th Battalion who had served at Gallipoli, while the other half comprised fresh reinforcements from Australia. Like the 9th Battalion, the majority of the 49th Battalion's soldiers were recruited from Queensland.

After spending an additional three months in Egypt, the 49th Battalion deployed by sea from Alexandria and headed for France to join the British Expeditionary Force on the Western Front. The battalion arrived in Marseilles in June 1916 and entered the trenches of the Western Front for the first time on the 21st of June. It participated in its first major battle at Mouquet Farm in August 1916.

Mouquet Farm, on a ridge near Pozières, was attacked by three Australian divisions in nine offensives from the 8th of August to the 3rd of September 1916. Though the farm's surface was destroyed, its stone cellars remained key to German defences. Over 11,000 casualties were suffered by the 1st, 2nd, and 4th Australian Divisions. Despite their efforts, the farm was never captured. British forces eventually bypassed it, reducing it to an isolated outpost. It finally fell to the Allies on the 27th of September 1916.

It was during this fighting that Private Reg Latchford was Killed in Action on the 8th of August 1916 at Tara Hill, an observation post on the hill located northeast of Albert. He was 21 years, six months old at the time of his death and had served with the AIF for 10 months. He was laid to rest at Bapaume Post Military Cemetery (Plot I, Row C, Grave No.25), France.

Reginald George Latchford's name is inscribed on The Australian War Memorial Roll of Honour (Panel 148), The Fallen Soldiers' Memorial Toowoomba, The City of Townsville WW1 Honour Roll, and The Townsville Cenotaph. Latchford Street, Pimlico, Townsville is named in his memory.

3823 Private Reginald George Latchford was awarded the British War Medal (no:28483) and the Victory Medal (no:28304). A memorial scroll and a memorial plaque were passed to his father in 1921 and 1922, respectively.

Private Reginald Latchford's death was notified in the combined 218th and 219th list of Australian Casualties released on the 23rd of September 1916. After his death was notified, a letter written by Private Reginald Latchford four days before his death was received by his parents. The letter was widely published and used in the conscription debate for the "Yes" case. The referendum was narrowly defeated with 1,015,159 in favour and 1,181,747 against. The letter is reproduced below:

Townsville Daily Bulletin 17th of October 1916

HE YET SPEAKETH.
LATE PRIVATE R.G. LATCHFORD
A PATHETIC DESPATCH.
WHY REINFORCEMENTS ARE ESSENTIAL.

Extracts from a letter received from Private R. Latchford by his parents—dated August 4, on the eve of going into action in which he was killed four days later:

"Received yours and Ethel's, also Nell's, dated June 4, and am glad to know all are well. This will be my last letter for a while as we are now going into action in the Somme front, where the big British advance is being made, and where other Australians have been. We saw them back from the trenches a few days ago. They had a bad time but did well again. We have seen a good deal of France by now and

marched nearly all the way. We have left our packs behind at our last billet; so now have nothing with us except blanket and oil-sheet. We will look a rough lot in a few days' time. We won't get our packs back again until we come out of the trenches. We expect to be there four to seven days, it is impossible to go any longer without rest or sleep (this is what we require reinforcements so badly for). However, I am quite ready. I have been to Holy Communion nearly every week. We have a very fine priest, who is a great worker for 'his flock,' as he calls us. I am writing this on the lid of a mess tin. Remember me to all friends. With fondest love and XXXXX to all and XXX for Nell. —Reg."

(Private Latchford was the son of Mr and Mrs. Latchford, well-known in Charters Towers and Townsville.)

3567 PRIVATE ALEXANDER WHYTE (1895 – 1916)

9th Infantry Battalion / 49th Infantry Battalion

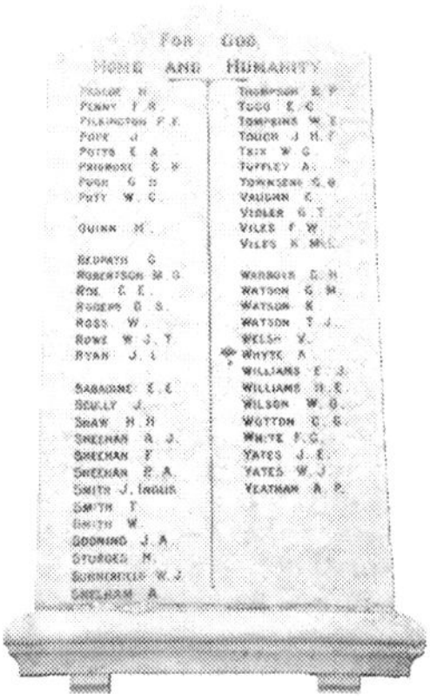

Alexander Whyte was born at Thursday Island, Queensland, on the 3rd of March 1895 to William and Christina Whyte (née Smith). He was the eldest of seven children, with four brothers and two sisters. William Whyte, a labourer, lived with his family on Chester Street, Thursday Island. Alexander grew up on the Island and attended the local school. After completing his education, he worked for the Northern Division of Queensland Government Railways and spent some time in Townsville.

On the 21st of July 1915, Alexander applied to join the Australian Imperial Force (AIF) in Brisbane. He was 20 years and four months old, single, and employed as a Railway Worker. He also had prior military experience with the Naval Reserve in Townsville. His medical report noted that he was 5'8" (173 cm) tall, weighed 143 lbs (65 kg), had a dark complexion, blue eyes, and black hair. He identified as Roman Catholic and listed his father as his next of kin, giving his address as Thursday Island.

Accepted into the AIF, Alexander Whyte was allocated service number 3567, assigned to the Infantry, and posted to the 9th Battalion, 11th reinforcements. The 9th Battalion was deployed to Gallipoli when Alexander Whyte enlisted. The Battalion had been the first ashore for the ANZAC landing on 25 April 1915 at around 4.30 am and was still on the Gallipoli Peninsular when Alexander Whyte enlisted.

Alexander Whyte spent the next three months after enlistment completing basic training in Brisbane. On the 21st of October 1915, Private Whyte and the other reinforcements embarked for Egypt from Brisbane, as part of Convoy No.13. They sailed aboard HMAT A48 *Seang Bee* which carried 1,013 troops on the voyage. The entire convoy of 11 ships transported 11,731 troops and 400 horses to Egypt.

More training followed in Egypt and in January 1916, Private Whyte was admitted to No.4 Auxiliary Hospital in Cairo suffering from mumps. Once discharged from hospital, he re-joined the 9th and continued training.

On the 29th of February 1916, Private Whyte was transferred to the 49th Battalion, which had been formed in Egypt two days earlier as part of the AIF's expansion. Half of the soldiers were Gallipoli veterans from the 9th Battalion, while the other half were new reinforcements from Australia. Like the 9th Battalion, most members of the 49th were from Queensland. The battalion was assigned to the 13th Brigade within the 4th Australian Division.

After forming-up at Tel-el-Kebir, the battalion moved 40 miles (64 km) to the Suez Canal where they undertook a period of training in the desert. On the 5th of June 1916, the 49th embarked from Alexandria, bound for France and the Western Front. Arriving in France on the 12th of June 1916, the 49th moved up to the trenches of the Western Front, around Strazelle, for the first time on the 21st of June.

The 49th Battalion fought in its first major battle at Mouquet Farm in August and suffered heavily. On the 8th of August 1916, the 49th was deployed in the vicinity of Tara Hill, northeast of Albert. The battalion was under constant shelling from the enemy and sustained three casualties. One of those casualties was Private Alexander Whyte who was Killed in Action. He was 231 years five months old and had served in the AIF for one year. He was laid to rest at Bapaume Post Military Cemetery (Plot I, Row C, Grave 24), France and his death was notified in the combined 218th and 219th Lists of Australian Casualties released on the 24th of September 1916.

Alexander Whyte's name is inscribed on The Australian War Memorial Roll of Honour (Panel 149), The City of Townsville WW1 Honour Roll, The Railways Northern Division Honour Board and The Townsville Cenotaph. Whyte Street, Hermit Park, Townsville is named in his memory.

3567 Private Alexander Whyte served in Egypt and on the Western Front during the Great War. He was awarded the 1914/5 Star (no:2659), the British War Medal (no:21661) and the Victory Medal (no:21585). A memorial scroll and a memorial plaque were passed to his father in 1921 and 1922, respectively.

4555 PRIVATE WALTER BERNARD MACKLEY (1889 – 1916)

47th Infantry Battalion

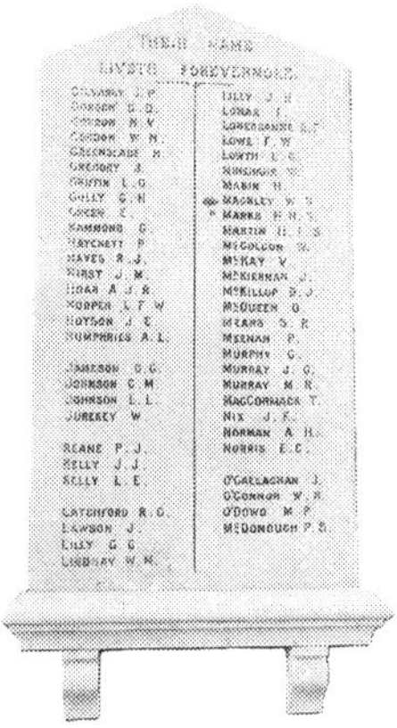

Walter Bernard Mackley (Bernie) was born on the 9th of September 1889 in Charters Towers, Queensland, the first born and only son of Walter and Mary Mackley (née Morgan). He had one younger sister. Walter Mackley Snr worked as a butcher in Charters Towers where Bernie Mackley attended the Queenton State School.

After school was completed, Bernie Mackley worked as a labourer in Mareeba and Townsville where he lived at the Tower Hotel. On the 6th of September 1915, he applied at Townsville to enlist in the Australian Imperial Force (AIF). He declared that he was 26 years old, single, working as a labourer and had no previous military service. His medical examination noted that he was 5'9" (175 cm) tall, weighed 9st 8lbs (60.7kg), had a dark complexion, brown eyes and brown hair. Bernie Mackley identified as Church of England and nominated his mother as his next of kin giving her address as Bluff Rd, Queenton, Charters Towers.

Accepted into the AIF, Bernie Mackley was allocated service number 4555, assigned to the Infantry, and posted to the 15th Battalion, 14th reinforcements. The 15th Battalion was originally raised as part of the all-volunteer AIF in September 1914 and was fighting at Gallipoli when Bernie Mackley enlisted.

After three months of basic training in Australia, Private Bernie Mackley and the other reinforcements embarked for Egypt from Brisbane with Convoy 17 on the 31st of January 1916. They sailed aboard HMAT A69 *Warilda* which carried 1,389 troops on the voyage. The entire convoy of eight ships transported a total of 8,443 troops and 670 horses to Egypt.

In March 1916, Private Mackley arrived at Camp Zeitoun in Egypt. A month later, he was assigned to the newly formed 47th Battalion. This battalion was officially established on March 7, 1916, as part of the AIF's expansion. About half of the 47th Battalion's recruits were seasoned soldiers from the 15th Battalion, who had seen action at Gallipoli, while the other half were new reinforcements from Australia. The battalion's composition mirrored that of the 15th, with most of its members recruited from Queensland and Tasmania. The 47th Battalion was integrated into the 12th Brigade of the 4th Division, alongside the 45th, 46th, and 48th Battalions.

After further training in Egypt, the 47th Battalion departed from Alexandria in June 1916, bound for France to join the British Expeditionary Force. They arrived on the 9th of June and soon experienced their first time in the trenches on the Western Front on July 3, 1916.

The 47th Battalion played a key role in its first major battle at Pozières. Supporting the 2nd Division's assault on 4 August 1916, they provided work parties before being tasked with defending the captured ground.

On 7 August 1916, under heavy bombardment and significant casualties, the 47th relieved the 48th Battalion at the front line. They held their position until 12 August 1916, enduring constant shelling and repelling attacks.

It was during this fighting that Private Bernard Mackley was Killed in Action on the 10th of August 1916. He was one month short of his 27th birthday at the time of his death and had served in the AIF for 11 months. Initially buried near Pozières, his body was later reinterred and buried at the Courcelette British Cemetery in France (Plot IV, Row B, Grave No. 23). His death was notified in the combined 214th and 215th Australian casualty lists issued on 18 September 1916.

Walter Bernard Mackley's name is inscribed on The Australian War Memorial Roll of Honour (Panel 144), The Queenton State School Roll of Honour, The City of Townsville WW1 Honour Roll, and The Townsville Cenotaph. Mackley Street, Garbutt, Townsville was named in his memory.

4555 Private Walter Bernard Mackley was awarded the British War Medal (no:31327) and the Victory Medal (no:31103). A memorial scroll and a memorial plaque were passed to his father in 1921.

The following tribute was published in *The Northern Miner* (Charters Towers, Qld. 1874 - 1954) on 10th August 1917.

ROLL OF HONOR. MACKLEY. —In sad and loving memory of our only son and brother Private Walter Bernard (Bernie) Mackley, of the 47th Battalion, Killed in Action somewhere in France on August 10th. 1916.

> He rose responsive to his country's call,
> And gave for her his life.
> We gave our best, our all.

(Inserted by his sorrowing father, mother and sister).

3791 PRIVATE THOMAS LOMAX (1891 – 1916)

4th Pioneer Battalion

Thomas (Tom) Lomax was born in November 1891 in Clayton, Manchester, England. He was the son of Thomas and Elizabeth Lomax. He had one older sister. Thomas Snr was a master joiner. Tom Lomax followed in his father's footsteps and qualified as a carpenter after attending the National and Board School, England. In 1912 Tom Lomax arrived in Australia when he was 21 years old. He lived in Brisbane until about 1913 and then moved to Townsville where he gained employment as a carpenter with the railways.

On the 20th of September 1915, Tom Lomax applied at Townsville to enlist in the Australian Imperial Force (AIF). He was one month short of his 25th birthday at the time. He declared that he was single, employed as carpenter and that he had no previous military experience. His medical assessment noted that he was 5'8½" (174 cm) tall, weighed 11st (70kg), had a fair complexion, grey eyes and light brown hair. He identified as Church of England and he nominated his father as his next of kin giving his address in England as 20 Green End, Whitechurch, Salop, Shropshire.

Accepted into the AIF, Tom Lomax was allocated service number 3791, initially assigned to the Infantry and was posted to the 15th Battalion/12th Reinforcement. The 15th Battalion was deployed to the Gallipoli Peninsula when he enlisted. After finishing recruit training in Australia, Private Tom Lomax and the reinforcements embarked from Sydney with Convoy No.14, bound for Egypt, on the 30th of November 1915. They sailed aboard HMAT A23 *Suffolk* which carried 973 troops on the voyage. The entire convoy of 14 ships ferried a total of 12,190 troops and 1,561 horses to Egypt.

Fourteen weeks later, in March 1916, they disembarked in Egypt, and Private Lomax joined the 15th Battalion at Tel-el-Kabir on the 6th of March 1916. Ten days after that, he was transferred to the 4th Pioneer Battalion, established on the 16th of March 1916. The battalion was formed as part of the AIF's expansion, a strategic move to relocate from the Middle East to Europe to contribute to the war effort on the Western Front. While these soldiers were primarily involved in infantry operations, many also had trade qualifications from their civilian careers, leading to their assignment as pioneer units. Pioneers were tasked with essential duties such as digging trenches, manual labour, building fortifications, establishing strong points, constructing light railways, and clearing battlefields. These roles required not only basic soldiering skills but also construction and engineering expertise. As a result, soldiers chosen for pioneer units were selected for their prior experience in these fields, which complemented their infantry training.

A bout of influenza saw Private Tom Lomax hospitalized soon after joining the 4th Pioneers, but he recovered in time to sail with the unit from Egypt to France in early June 1916. After arrival and deployment to the trenches in France, the battalion took part in its first major action at Pozières in August 1916. It endured a massive artillery bombardment and defeated a German counterattack on the 7th of August 1916.

On the 11th of August 1916, the 4th Pioneers were engaged in completing engineering works in the vicinity of Becourt Wood. It was during this operation that Private Thomas Lomax was Killed in Action by shell fire. He was 26 years old and had been a member of the AIF for 14 months at the time of his death. He has no known grave. His death was notified in the 211th Australian casualty list released on the 14th of September 1916.

Thomas Lomax's name is inscribed on The Australian War Memorial Roll of Honour (Panel 174), The Australian National Memorial, Villers-Bretonneux, France, The Northern Division Railway Honour Board, The City of Townsville WW1 Honour Roll, and The Townsville Cenotaph. Lomax Street, Mysterton, Townsville is named in his memory.

3791 Private Thomas Lomax was awarded the British War Medal (no:25665) and the Victory Medal (no:25484). A memorial scroll and a memorial plaque were passed to his father in 1922.

4535 PRIVATE CHARLES HENRY WARBOYS (1895 – 1916)

26th Infantry Battalion

Charles Henry Warboys was born on the 12th of December 1895 in Charters Towers, Queensland, to George Warboys and his wife Susan Elizabeth (née Wright). He was the youngest of five children and the only son.

George Warboys was a Saddler, and in 1898, tragedy struck the family when he died at the early age of 38. Charles Warboys was three years of age at the time. Care of the family fell on the shoulders of Susan for the next three years until 1901 when she married Arthur Giles. The Giles family lived in Stubley St and then Pyrites Rd, Charters Towers.

Charles Warboys grew up in the "Towers" and attended the local State School. After completing his schooling, Charles Warboys worked as an assistant amalgamator, a worker in the battery or process plant who used mercury to extract gold from the ore. The mercury was then heated to remove the gold from the amalgam. This work was hot and potentially debilitating, as personal protective equipment was scarce in those early days.

On the 11th of October 1915, Charles Warboys applied in Townsville to enlist in the Australian Imperial Force (AIF). He declared that he was employed as an Assistant Amalgamator, single, and had no previous military experience. He gave his age as 19 years and ten 10 months. His enlistment medical assessment noted that he was 5' 7" (170 cm) tall, weighed 8st 12lbs (56kg), had a fair complexion, grey eyes, and black hair. He identified as Church of England and nominated his mother as his next of kin, giving her address as the corner of Plant & King St, Charters Towers.

Accepted into the AIF, Charles Warboys was allocated service number 4535, initially assigned to the Infantry and was posted to the 26th Battalion, 11th reinforcements. The 26th Battalion was deployed in Gallipoli when he joined the AIF.

After five months of recruit training, Private Charles Warboys and his fellow reinforcements embarked from Brisbane, Queensland, bound for Egypt on board HMAT A16 *Star of Victoria* on the 30th of March 1916. The 26th Battalion spent time resting and rebuilding in Egypt after Gallipoli.

In March 1916, the 26th, as part of the 7th Brigade, proceeded to France. Private Charles Warboys and his fellow reinforcements embarked from Alexandria in May 1916 to join up with the 26th on the Western Front. A few weeks of training at Étaples followed their arrival in France.

It was not until the 28th of July 1916 that the reinforcements were taken on the strength of the 26th Battalion. That was the day the Battalion commenced their first major battle around Pozières in France.

On the 25th of August 1916, the 26th Battalion was deployed in the front line in the vicinity of La Boisselle. It was here on the same day that Private Charles Warboys was wounded by a gunshot to the head. He was immediately evacuated to the 1/1 Midland Field Ambulance, but tragically, he succumbed to his injury soon after arrival and Died of Wounds received in battle.

Private Charles Warboys was 20 years and eight months old at the time of his death. He had served with the AIF for 10 months. He was laid to rest in the Aveluy Communal Cemetery Extension (Row H, Grave 4), France. His death was notified in the 226/227th list of Australian Casualties released on the 2nd of October 1916.

Charles Henry Warboys' name is inscribed on The Australian War Memorial Roll of Honour (Panel 109), The City of Townsville WW1 Honour Roll, and The Townsville Cenotaph. Warboys St, Nelly Bay, Magnetic Island, is named in his memory.

4535 Private Charles Henry Warboys served on the Western Front in The Great War. He was awarded the British War Medal (no: 35913) and the Victory Medal (no: 35644). A memorial scroll and a memorial plaque were also passed to his mother in 1922.

Each year until his mother died in 1949, a tribute was published in the *Townsville Daily Bulletin* The following tributes were published in the *Townsville Daily Bulletin* on 25th August 1917:

ROLL OF HONOR.

WARBOYS. — In loving memory of my dear son and our brother, Private Charles Henry Warboys, who died of wounds in France, August 25th. 1916.

Deeply regretted.

(Inserted by his loving mother, sisters and brothers-in-law.)

WARBOYS. — In loving memory of Private Charles Henry Warboys, who died of wounds received at Pozières on August 25th. 1916.

> The silent grief, but that's in the soul,
> No human eye can trace.
> For many a broken heart is hidden
> Behind a smiling face.

(Inserted by his loving aunts, Mrs. R. Lanskey and Mrs. Asquith.)

4549 SERGEANT GEORGE ALEXANDER TUFFLEY (1894 – 1916)

25th Infantry Battalion

George Alexander Tuffley was born on the 1st of October 1894 in Warwick, Queensland. He was the second son and third of five surviving children born to Joseph and Louisa Tuffley (née Meldrum). He had two brothers and two sisters. Joseph Tuffley was a schoolteacher and taught at schools across Queensland. In 1903, he was School Master at Townsville North State School (renamed Belgian Gardens State School in 1930).

George Tuffley was a student at Townsville Grammar School. After his school years were complete, he started a mechanical engineering apprenticeship with the Railways. He also served with the Citizen Forces and the 2nd Infantry, Kennedy Regiment in Townsville. In August 1914, he was mobilised for war service, and he sailed on the S.S. *Han* to serve as a Sergeant with the Thursday Island Papua group as a member of that expeditionary force on the island.

On the 14th of August 1914, Sergeant Tuffley enlisted for service outside Australia and re-embarked on the 16th of August 1914 on the troopship *Kanowna* to take part in the capture of German New Guinea. He was taken on strength of the Australian Navy and Military Expeditionary Force (A.N.&M.E.F.) on the 6th of September 1914. An incident on board the *Kanowna* involving the firemen forced the return of the troops to Townsville on the 18th of September 1914. On arrival back in Australia, Sergeant Tuffley and all other members of the Force were discharged from service, and he returned to his civilian occupation and service in the 2nd Infantry.

On the 4th of November 1915, George Tuffley applied at Townsville to enlist in the Australian Imperial Force (AIF). He declared that he was single, working as a Mechanical Engineering Cadet with the Railway in Townsville, and currently serving with the 2nd Infantry. His medical examination noted that he was 5'5¾" (167 cm) tall, weighed 8st 2½lbs (52kg), had a fair complexion, grey eyes, and light brown hair. He identified as Church of England. He nominated his father as his next of kin, residing at the State School, North Townsville.

Accepted into the AIF, George Tuffley was allocated service number 4549, assigned to the Infantry and posted to the 25th Infantry Battalion, 11th Reinforcements. The 25th Battalion was raised in March 1915 at Enoggera, Queensland, as a unit of the 7th Brigade. While most of its members were recruited from Queensland, the battalion also comprised a small group of recruits from Darwin. The battalion was fighting at Gallipoli when George Tuffley enlisted.

Following initial training, Private Tuffley and the other 25th Battalion Reinforcements embarked with Convoy No. 19 from Sydney, NSW, bound for Egypt on the 31st of March 1916. They sailed aboard HMAT A16 *Star of*

Victoria (later renamed *Port Melbourne*), which carried 1,552 troops on the voyage. The entire convoy of 10 ships transported a total of 8,099 troops and 1,168 horses to Egypt.

The stay in Egypt was short. Disembarking at Alexandria on the 10th of May 1916, the contingent re-embarked eight days later for movement to England. Once in England, the contingent was transported to the 12th Training Battalion located at Codford in Wiltshire. Further training was completed in England before the contingent was deployed to France, where another two weeks of training was undertaken.

On the 5th of August 1916, Private Tuffley was promoted to Temporary Sergeant. Three days later, on the 8th of August, he and 125 other reinforcements joined the 25th Battalion at Warloy-Baillon (about 24 km northeast of Amiens). Over the next two weeks, the 25th moved to Albert and then near Montrelet. Bivouacking at Brickworks from the 22nd of August 1916, the Battalion was engaged for the next five days in repairing trenches and digging out saps with the Engineers.

On the 27th of August 1916, the 25th relieved the 17th Battalion in Sunken Road. On the 29th of August, 200 men were on duty in the front line, digging a new trench. It was during this activity that Sergeant George Alexander Tuffley was Killed in Action on the 29th of August 1916, three weeks after joining the 25th in the field. He was laid to rest at the Sunken Road Cemetery (Plot I, Row B, Grave No. I), Contalmaison, France.

Sergeant George Tuffley was one month away from his 22nd birthday and had served with the AIF for 11 months at the time of his death. His death was notified in the 228th list of Australian Casualties released on the 6th of October 1916.

George Alexander Tuffley's name is inscribed on The Australian War Memorial Roll of Honour (Panel 106), The Townsville Grammar School Memorial, The Railways Northern Division Honour Roll, The City of Townsville WW1 Honour Roll, and The Townsville Cenotaph. Tuffley Street, West End, Townsville, is named in his memory.

4549 Sergeant George Alexander Tuffley served on the Western Front in The Great War. He was awarded the British War Medal (no: 35844) and the Victory Medal (no: 35576). A memorial scroll and a memorial plaque were passed to his father in 1921 and 1922, respectively.

The following notices were published in the *Townsville Daily Bulletin,* on the 29th of September 1916 and the 10th of October 1916:

The Great War. - Word has been received of the death in action of Sergeant A. Tuffley. He was a son of Mr and Mrs Tuffley, of the Belgian Gardens School, and was a draftsman in the Railway Department. The late Sergeant Tuffley was a member of the *Kanowna* expedition and was educated at the Townsville Grammar School.

TUFFLEY. — On 29th August 1916, George Alexander Tuffley, aged 21, Killed in Action in France. Son of Mr and Mrs J. Tuffley, Townsville North State School."

5064 PRIVATE CHARLES STUART CRAIES (1890 – 1916)

49th Infantry Battalion

PTE. CHARLES STUART CRAIES, Missing since September, 1916. His mother, Mrs. S. Craies, Park-road, Milton, Brisbane, would be glad to hear from any returned soldier who can give information concerning him.

Charles Stuart Craies was born on the 23 of July 1890 in Warwick, Queensland. He was the first born and only son of Stuart and Florence Craies (née Hillcoat). He had one sister. Stuart Craies was an Accountant, and the family lived in Mitchell St, Townsville. Charles Craies grew up in Townsville and attended the Townsville Grammar School.

After his school years were completed, Charles Craies worked as a Clerk with the Adelaide Steamship Company in Townsville. On the 30th of September 1915, he applied at Townsville to enlist in the Australian Imperial Force (AIF). He declared that he had no previous military experience, was 25 years one month old, single, and employed as a clerk. His medical examination noted that he was

5'6" (168 cm) tall, weighed 112 lbs (51 kg), had a dark complexion, grey eyes, black hair and identified as Church of England. Charles nominated his father as his next of kin and gave his address as Park Road Milton Brisbane.

Accepted into the AIF, Charles Craies was allocated service number 5064, assigned to the Infantry and was initially posted to the 9th Battalion/16th reinforcements. The 9th Battalion was the first battalion recruited in Queensland and was later joined by the 10th, 11th, and 12th Battalions to form the 3rd Brigade. The battalion was fighting on the Gallipoli Peninsula when Charles Craies enlisted.

On the 31st of March 1916, after completing basic training in Australia, Private Charles Craies and his fellow reinforcements embarked from Sydney bound for Egypt as part of Convoy No.19. They sailed aboard HMAT A16 *Star of Victoria* (later re-named HMAT *Port Melbourne*) which carried 1,552 troops on the voyage. The 10 ships of Convoy 19 transported a total of 8,075 personnel and 1,168 horses to Egypt.

After a month in Egypt, Private Craies embarked from Alexandria bound for France, arriving in Marseilles in June 1916. In August 1916, he was transferred to the 49th Battalion. The Battalion was formed up in Egypt on the 27th of February 1916, as a unit of the expanded Australian Imperial Force (AIF). Approximately fifty percent of its personnel were seasoned veterans from the 9th Battalion, having previously served at Gallipoli. The remaining half comprised new recruits from Australia. Mirroring the 9th Battalion, the majority of the 49th Battalion's soldiers were recruited from Queensland. The battalion was allocated to the 13th Brigade, in the 4th Australian Division.

Private Craies joined the battalion just as it fought in its first major battle at Mouquet Farm in August 1916. Mouquet Farm, situated on a ridge extending from the village of Pozières, was attacked nine times by three Australian divisions between the 8th of August and the 3rd of September 1916. Despite the obliteration of the farm structures, the robust stone cellars below became strong points of the German defences. These persistent assaults led to more than 11,000 casualties among the 1st, 2nd, and 4th Australian Divisions, yet none succeeded in seizing and maintaining control of the farm. The 49th

Battalion endured significant losses, particularly during the assault launched on the 3rd of September 1916.

It was during this assault that Private Charles Craies was declared Wounded in Action. after a Court of Enquiry was conducted, his fate was changed to Missing believed Killed. An investigation by the Australian Red Cross Society (ARCS), requested by his mother, was unable to find the exact details of Private Charles Craies' death. He was officially recorded as Killed in Action on the 1st of September 1916. He was 26 years and one month old at the time of his death and had served with the AIF for 11 months and two days. He has no known grave.

Private Charles Craies' death was notified in the 235th list of Australian Casualties released on the 17th of October 1916. His mother sought information long after the official declaration of his death. Apart from enlisting the support of the ARCS, Florence Craies placed newspaper advertisements asking any returned veterans to contact her if they had any information in respect to Private Charles Craies.

Charles Craies' name is inscribed on The Australian War Memorial Roll of Honour (Panel 148), The Australian National Memorial, Villers-Bretonneux, France, The City of Townsville WW1 Honour Roll, the Townsville Grammar School Memorial, and The Townsville Cenotaph. Craies Street, Belgian Gardens, Townsville is named in his memory.

5064 Private Charles Stuart Craies was awarded the British War Medal (no:34059) and the Victory Medal (no:33818). A memorial scroll and a memorial plaque were passed to his father in 1921 and 1922, respectively.

The following notice was published in the *Queenslander* (Brisbane, Qld) on the 6th of April 1918:

PRIVATE. CHARLES STUART CRAIES,

Missing since September 1916. His mother, Mrs. S. Craies, Park-road, Milton, Brisbane, would be glad to hear from any returned soldier who can give information concerning him.

4373 PRIVATE JAMES LEONARD BURTON (1894 – 1916)

25th Infantry Battalion

James Leonard Burton was the oldest of three children born to Thomas and Matilda Burton (née Bacchus) on the 18th of September 1894 in Greenmount, near Townsville, Queensland. Thomas Burton was a cattle buyer, and James grew up in the Townsville region, where he attended State School.

After his school years were finished, James started farm work. On the 26th of November 1915, he applied to enlist in the Australian Imperial Force (AIF). He declared that he was single, 21 years and two months old, employed as a farm labourer, and that he had no previous military experience.

His medical examination noted that he was 5'7" (170 cm) tall, weighed 140 lbs (63.5 kg), had a fair complexion, green eyes, and light brown hair, and identified as Church of England. James Burton's father died in February 1915, and he listed his mother as his next of kin, giving her address as Mt Sibley, Greenmount, Queensland.

Accepted into the AIF, James Burton was allocated service number 4373, assigned to the Infantry and posted to the 25th Battalion, 11th reinforcements. The 25th was raised in Enoggera, Queensland, in March 1915 as an Infantry Battalion of the 7th Brigade. While most of its members were recruited from Queensland, a small group of soldiers from Darwin also joined the battalion.

The 25th was fighting on Gallipoli when James Burton enlisted. Six months of training followed enlistment for the reinforcements. On the 31st of March 1916, Private James Burton and his fellow reinforcements embarked from Sydney bound for Egypt as part of Convoy No.19. They sailed aboard HMAT A16 *Star of Victoria* (later re-named HMAT *Port Melbourne*), which carried 1,552 troops on the voyage. The 10 ships of Convoy 19 transported a total of 8,075 personnel and 1,168 horses to Egypt. They disembarked at Alexandria in May 1916.

It was only to be a short stay in Egypt before Private Burton embarked to join the fight in France on the 10th of May 1916. After arrival in France on the 18th of May, further training was undertaken. On the 31st of May 1916, Private James Burton was charged with the offence of breaking out of barracks. He was found guilty and sentenced to three days Field Punishment No.2, which consisted of heavy labouring duties and, possibly, being restrained in handcuffs or fetters.

Punishment completed; Private Burton joined the 25th Battalion on the 16th of July 1916 in the vicinity of Amiens. The 25th Battalion's baptism of fire on the Western Front was the Battle of Pozières from July 25th to August 7th. The Battle of Pozières was a major engagement that endured for more than six weeks. This fiercely contested battle exacted a heavy toll on both sides, marked by intense artillery barrages and steadfast German resistance. The battalion endured 785 casualties during this action.

Following a period in a less active area of the front in Belgium, the 2nd Division moved to the southern part of the front in October 1916 to launch another assault in the Somme Valley. In this offensive, the 25th Battalion participated in two attacks to the east of Flers, both of which failed due to the muddy conditions. It was during this fighting that Private James Burton was Killed in Action on the 13th of October 1916. He was 22 years and one month old. He had served with the AIF for 10 months at the time of his death and was laid to rest at the Railway Dugouts Burial Ground (Transport Farm), Zillebeke, Belgium. His death was notified in the 242nd list of Australian Casualties released on the 10th of November 1916.

James Burton's name is inscribed on The Australian War Memorial Roll of Honour (Panel 104), The City of Townsville WW1 Honour Roll, and The Townsville Cenotaph. Burton Street, Mysterton, Townsville, is named in his memory.

4373 Private James Leonard Burton was awarded the British War Medal (no: 35809) and the Victory Medal (no: 35542). A memorial scroll and a memorial plaque were passed to his mother in 1922.

The following notice was published in the *Townsville Daily Bulletin* on the 24th of January 1917:

Mrs T. Burton, of Hermit Park, has received the sad news of the death of her son Leonard, who was Killed in Action in France on October 13 last.

4801 PRIVATE SYDNEY CHARTERS GIBSON CARSE (1893 – 1916)

26th Infantry Battalion

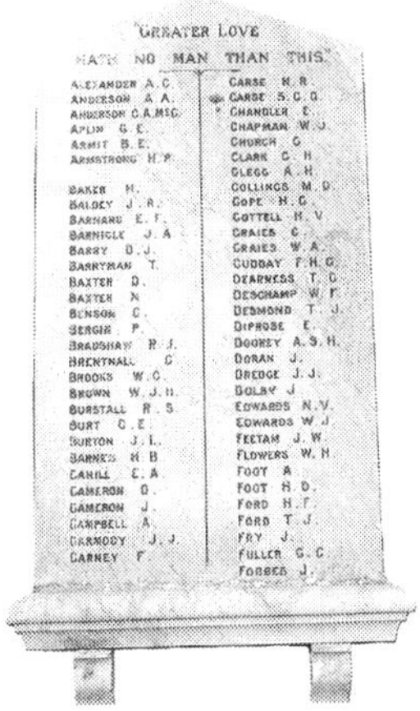

Sydney Charters Gibson Carse was born on the 9th of October 1897 in Belfast, Ireland. He was the son of John and Susan (Susie) Carse (née Gibson). After his arrival in Queensland, John Carse opened a draper's shop on Gill St in Charters Towers. He married twice in his life. In 1894, John's first wife, Hannah, passed away at the age of 34, leaving John as the sole parent to nine children. He married Susan Gibson in 1896. By 1915, John and Susie Carse were residing at premises on the Strand in Townsville and had opened a second draper's shop in Townsville.

Sydney Carse grew up in Charters Towers and Townsville. After his school years were finished, Sydney qualified and worked as a gasfitter. On the 22nd of November 1915, Sydney Carse applied at Townsville to join the AIF. He declared that he had no previous military experience, was 18 years and two months old, single, and employed as a gasfitter. His medical examination noted that he was 5'10" (177 cm) tall, weighed 8st 3lbs (63.5 kg), had a dark complexion, brown eyes, and dark brown hair. He identified as Presbyterian, and he listed his father, residing at The Strand, Townsville, as his next of kin.

Accepted into the AIF, Sydney Carse was allocated service number 4801, assigned to the Infantry and posted to the 26th Battalion/12th Reinforcements. The 26th Battalion was made up of recruits from Queensland and Tasmania and was raised at Enoggera, Queensland, in April 1915. When Sydney Carse enlisted, the 26th Battalion was fighting on the Gallipoli peninsula.

After six months of basic training in Australia, Private Sydney Carse and his fellow reinforcements departed Sydney bound for England on the 12th of April 1916 as part of Convoy No.20. They sailed aboard the RMS Mooltan, which carried 480 troops. The entire convoy consisted of 15 ships and transported a total of 9,249 personnel and 5,590 horses on the voyage.

Further training at the 7th Training Battalion in Rollestone, located on Salisbury Plain in Wiltshire, followed disembarkation in England. Once the training was completed, Private Carse was deployed to France. After 10 days of further training at the 2nd Division Base Depot at Étaples, he joined the 26th Battalion on the 28th of September 1916. The battalion was deployed near Steenvoorde, France.

Prior to Private Carse's arrival, the 26th Battalion had already participated in its first major battle around Pozières, resulting in over 650 casualties. Following this battle, the battalion was assigned to a quieter sector. In November 1916, the 26th moved to the Somme Valley.

During this time, Private Sydney Carse took part in two offensives near a crucial location known as "The Maze," situated east of Flers. Both attacks proved unsuccessful as they became bogged down in the treacherous mud, resulting in over 300 casualties.

On the 5th of November 1916, the 26th Battalion mounted an attack in freezing cold weather from the Cobham Trench in the vicinity of Gueudecourt. The battalion sustained 137 casualties in the attack, of which 30 were killed in action. One of those casualties was Private Sydney Carse, who was listed as Killed in Action at Gueudecourt on the 5th of November 1916. He was 19 years and two months old and had served with the AIF for 11 months. He had been a member of the 26th Battalion for just 38 days at the time of his death. He has no known grave, and his death was notified in the 250th list of Australian Casualties released on the 14th of December 1916.

Sydney Carse's name is inscribed on The Australian War Memorial Roll of Honour (Panel 107), The Australian National Memorial, Villers-Bretonneux,

France, The City of Townsville WW1 Honour Roll, and The Townsville Cenotaph. Carse Street, Hyde Park, Townsville, is named in memory of Sydney Carse and his brother Norman Carse.

4801 Private Sydney Charters Carse was awarded the British War Medal (no: 37402) and the Victory Medal (no: 37100). A memorial scroll and a memorial plaque were passed to his father in 1922.

Sydney Carse's two brothers also joined the AIF, one of whom survived and one who was killed in action. Their details are:

21128 Driver David Andrew Carse, 12th Field Artillery Brigade, returned to Australia on the 1st of August 1919,

5551 Private Norman Reginald Carse, 25th Battalion, Killed in Action on the 4th of October 1917.

The following advice was published in the *Townsville Daily Bulletin* (Qld.) on the 8th of December 1916:

Advice was received last night that Private Sydney Charters Carse had been Killed in Action in France on the 5th of November. He was the youngest of three sons at the front. In consequence, the premises of Messrs Carse and Co. will be closed today.

The following was published in Family Notices in the *Townsville Daily Bulletin* (Qld.) on the 7th of November 1922:

ROLL OF HONOUR.

CARSE. - In loving memory of our beloved son, Private Sydney Charters Carse, who fell at Flers, 6th November 1916.

> Only good night, beloved, nor farewell!
> A little while and all His saints shall dwell,
> In hallowed union, indivisible.
> Good night!

Inserted by his loving father, mother, brothers, and sisters.

CAPTAIN JOHN EDWARD NIX (1891 – 1916)

25th Infantry Battalion

John Edward Nix (Jack) was born on the 23rd of July 1891 in Hillgrove, New South Wales, to George and Elizabeth Nix (née Jarrett). There were no girls born in the family, and Jack was the youngest of four boys. The Nix family moved to Charters Towers, where George Nix worked as a miner.

Jack Nix grew up in Charters Towers, and while at school there, he joined the cadets. After leaving school, he joined the staff of the local newspaper, the *Northern Miner*. In 1914, he joined the staff of the *Townsville Evening Star* as a reporter.

Jack Nix's Senior Cadet service led him to join the Kennedy Regiment. Within days of the outbreak of the war, he was mobilised and sent to Thursday Island in August 1914 as a Sergeant Signaller with the 2nd Infantry Regiment. He stayed on Thursday Island in this role until November 1914.

On the 15th of February 1915, Sergeant Nix applied to transfer to the AIF. He was single and 23 years and six months old. His application for commission noted that he was 5'8½" (174 cm) tall and weighed 146 lbs (66 kg). He identified as Presbyterian, and he nominated his widowed mother as his next of kin, giving her address as Curtis St, Bundaberg, Queensland (later changed to Electra St).

Accepted as an officer in the AIF, Jack Nix was appointed as a 2nd Lieutenant in the 25th Battalion. The 25th Battalion was raised at Enoggera in Queensland in March 1915 as part of the 7th Brigade. Although predominantly composed of men recruited in Queensland, the battalion also included a small contingent of men from Darwin.

For the next four and a half months, 2nd Lieutenant Nix trained with his unit in Brisbane and was promoted to Lieutenant on the 26th of June 1915. Three days later, he embarked with the 25th Battalion from Brisbane bound for

Egypt. They sailed aboard HMAT A60 *Aeneas,* which carried 1,720 troops on the voyage. Their ship was one of four that made up Convoy No.7. The entire convoy transported 5,052 men and 20 horses.

In August 1915, the 25th Battalion underwent training in Egypt. In September, they deployed to Gallipoli as part of the 7th Brigade and provided reinforcement to the diminished New Zealand and Australian Division. The last major Allied offensive had been launched and turned back in August 1915, and consequently, the 25th Battalion had a relatively quiet time.

Lieutenant Nix was mentioned in divisional orders for his *"valuable reconnoitring efforts"* around Apex trenches. He was granted the rank of temporary Captain in December 1915. The 25th Battalion left the peninsula on the 18th of December 1915 as part of the general evacuation. After further training in Egypt, the battalion deployed to France in mid-March 1916 to join the British Expeditionary Force on the Western Front.

Landing in France on the 19th of March 1916, the 25th Battalion was the first AIF battalion to arrive there. The battalion took part in its first major battle at Pozières, fought between July and September 1916, as part of the larger Battle of the Somme. The primary objective was the capture of the strategic village of Pozières in northern France, a crucial point in the German defensive line.

This fiercely contested battle exacted a heavy toll on both sides, marked by intense artillery barrages and steadfast German resistance. The Australians successfully captured Pozières on the 6th of August 1916. However, this triumph came at a significant price, with more than 23,000 Australian casualties. It is remembered as one of the most ferocious battles of the war.

In his book *"Pozières, 1916: Australians on the Somme,"* Peter Charlton makes the following comments about Captain John Nix:

"On the night 28th/29th July, his company managed to make its way through the wire in front of OG1 (a German trench) and discovered the trench line was virtually deserted. Pushing forward, Nix waited until the intense barrage on OG2 lifted and rushed the second trench, only to find it crowded with Germans who had withdrawn (from OG1) and were sheltering, waiting to counter-attack. Nix was fired on from 27 meters by a German machine gun. His attack, which had

probably penetrated further than any Australian assault that night, faltered and failed. Most of the assaulting force in the centre lay behind them, in the shallow saucer-shaped depression which the Germans could cover from the high ground on the rim. Despite the efforts of their platoon and company commanders to get them up and moving, the men were down and staying down. Nix withdrew, bringing with him only eight men from the two hundred plus company he took into the attack, of which fifty were left to assault the OG2 position."

In late July 1916, near Mouquet Farm, Captain Nix was wounded by a rifle shot in the hand. The injury was severe, and he was evacuated back to England for treatment and recovery. He returned to duty with the 25th Battalion in late September 1916.

After a spell in a quieter sector of the front in Belgium, the 2nd Division came south in October 1916 to attack again in the Somme Valley. On the 5th of November 1916, the 25th Battalion mounted an attack in freezing cold weather from the Cobham Trench in the vicinity of Gueudecourt. The War Diary of the 25th Battalion on the 5th of November 1916 records the following:

"Capt Nix, OC B Coy pushed his Coy forward and reached jumping off trench just in time to get in position. 9:10 a.m. The attackers jumped off from trench with assistance of ladders and advanced across NO MANS LAND under cover of our barrage – when about halfway across NO MANS LAND the enemy opened on them with MG fire causing many casualties. The objective of the attack was THE MAZE and Gird trench."

The fire was so intense that the surviving men stayed in No Mans Land all day and returned to their trench at night. The battalion sustained 137 casualties in the attack, of which 30 were killed. One of the casualties was Captain Jack Nix, who was Killed in Action by machine gun fire on the 5th of November 1916. He was 25 years and three months old and had served with the AIF for one year and 10 months at the time of his death. His death was notified in the 243rd list of Australian Casualties released on the 20th of November 1916. He has no known grave.

Captain Jack Nix's commanding officer, Lieutenant Colonel James Walker, wrote to Nix's parents, saying:

"Captain Nix was, without exception, the most popular officer in the Battalion. The Company of which he was in command loved him, both for his gallant leading and attention to the welfare of his men."

John Edward Nix's name is inscribed on The Australian War Memorial Roll of Honour (Panel 105), The Australian National Memorial, Villers-Bretonneux, France, The City of Townsville WW1 Honour Roll, The Townsville Cenotaph (incorrectly inscribed as J.F. Nix), and The Queenton State School Roll of Honour. Nix Street, West End, Townsville, is named in his memory.

Captain John Edward Nix was awarded the 1914/15 Star (no: 35809), the British War Medal (no: 16730), and the Victory Medal (no: 16664). A memorial scroll and a memorial plaque were passed to his mother in 1921 and 1922, respectively. Jack Nix's mother was granted a fortnightly war pension of three pounds, 17 shillings and six pence (£3/17/6) per for the loss of her son. This was slightly less than the national average fortnightly earnings of four pounds (£4) in 1916.

Jack Nix's three brothers also enlisted in the AIF; their details are:

Lieutenant Henry Jarrett Nix, 31st Battalion, returned to Australia on the 3rd of July 1919, 5629 Private Francis William Nix, 26th Battalion, returned to Australia on the 5th of September 1919, and 405 Sergeant George Thomas Nix, 5th Light Horse Regiment, returned to Australia on the 15th of January 1918.

The following notice was published in the *Daily Mercury,* Mackay, on the 20th of November 1916: PERSONAL.

The sad news has been received of the death in action of Captain John Nix, who enlisted from Townsville shortly after the outbreak of hostilities. He went through the Gallipoli campaign unscathed and twice was mentioned in despatches and received promotion for his courageous efforts. He had just returned to the firing line, having recovered from a damaged hand. Captain Nix, prior to enlisting, was engaged in journalistic pursuits in Townsville and

Charters Towers. He has two brothers at the front and another on the instructional staff at Enoggera.

The following notice was published in *The Bundaberg Mail* and *Burnett Advertiser* on the 20th of November 1916:

The Roll of Honour

NIX. — Killed in Action in France on the 5th of November 1916, CAPTAIN J. E. NIX (Jack), "C" Company, 25th Battalion, A.I.F., aged 25 years and 3 months, beloved youngest son of Elizabeth Nix and the late Geo. Nix, and brother of Staff Sergeant Major H. J. Nix, Corporal G. T. Nix, 5th Light Horse, and Private F. W. G. Nix, 25th Battalion, A.I.F. "A good son and a gallant officer."

5454 PRIVATE JOHN LAWRENCE RYAN (1896 – 1916)

12th Infantry Battalion

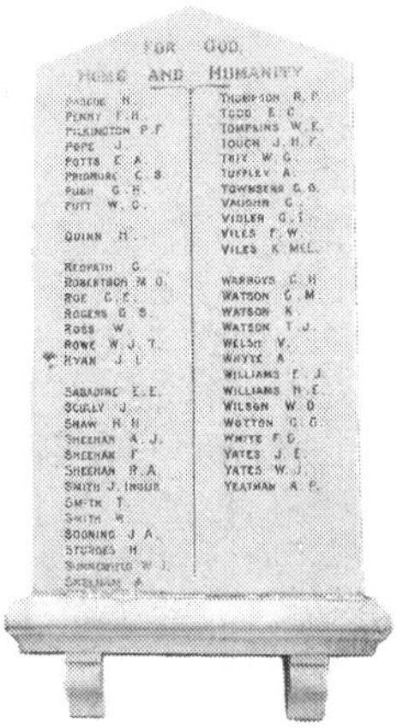

John Lawrence Ryan was born on the 1st of October 1896 in Manton, Queensland, to Thomas and Anne Ryan (née Russell). He was the sixth of 10 children, with seven brothers and one surviving sister. Thomas Ryan worked as a limeburner, and John grew up in Manton, attending the Woodstock State School.

After leaving school, John Ryan worked as a labourer for the Queensland Railways until the 2nd of February 1916, when he applied to enlist in the Australian Imperial Force (AIF). He declared he was 19 years and two months old, single, and had no previous military experience. His enlistment medical noted he was 5' 7" (170 cm) tall, weighed 9st 10 lbs (61 kg), had a fresh complexion, brown eyes, and fair hair. He identified as Roman Catholic and nominated his mother as his next of kin.

Accepted into the AIF, John Ryan was allocated service number 5454, assigned to the Infantry, and posted to the 12th Battalion, 17th reinforcements. The

12th Battalion was consolidating in Egypt after service in Gallipoli when John Ryan joined the AIF. The 12th Battalion sailed from Egypt in March 1916, bound for the Western Front.

Private John Ryan spent February to April 1916 completing Recruit and Basic Training in Australia. On the 20th of April 1916, the 17th reinforcements embarked with Convoy No.20 from Sydney, New South Wales, bound for England. They sailed aboard SS *Hawkes Bay*, which carried 1,000 troops. The 15 ships of Convoy No.20 carried a combined total of 9,249 passengers and 5,590 horses on the voyage.

After disembarking in England, Private Ryan and his fellow reinforcements were transported to the 3rd Training Battalion at Perham Downs on the edge of the Salisbury Plain in Wiltshire, England, to undertake further training. More training followed at the 1st Australian Division Base Depot at Étaples when the reinforcements were transferred to France on the 21st of August 1916.

Fully trained, Private John Ryan was taken on strength of the 12th Battalion on the 8th of September 1916. When he marched into the Unit, the 12th was resting in the Reserve Area after the fighting around Pozières and a stint in the trenches near Vandencourt Wood.

In early October 1916, John Ryan was struck down with mumps and was admitted to the 7th General Hospital, Saint-Omer. He re-joined the 12th Battalion after three weeks in early November. The 12th was located at Bernafay Wood when he re-joined them. This wood was a particularly dangerous part of the Somme battlefield of 1916. It had been heavily fortified by the Germans. Enemy machine-gunners, concealed by the shell-torn tree stumps, inflicted many casualties on the AIF.

On the 12th of November, the Battalion was relieved in place by other units and began moving back to Bernafay Camp. During this move, Private John Lawrence Ryan was Killed in Action at Guedecourt, near Bapaume, France, on the 12th of November 1916. There are no details of his death, but his military file records that he was buried in Pozières, four miles northeast of Albert. Today, he has no known grave.

At the time of his death, Private John Lawrence Ryan was 20 years old, had been a member of the AIF for nine months, and a member of the 12th Battalion for nine weeks. His death was notified in the combined 257th and 258th list of Australian Casualties released on the 30th of December 1916.

Private John Lawrence Ryan served on the Western Front in WW1. His name is inscribed on The Australian War Memorial Roll of Honour (Panel 67), The Australian National Memorial, Villers-Bretonneux, France, The City of Townsville WW1 Honour Roll, and The Townsville Cenotaph. Ryan St, Belgian Gardens, Townsville, is named in his memory.

5454 Private John Lawrence Ryan was awarded the British War Medal (no: 37231) and the Victory Medal (no: 36931). These medals were passed to his father, along with a memorial plaque and a memorial scroll.

862 PRIVATE THOMAS JAMES FORD (1893 – 1916)

25th Infantry Battalion

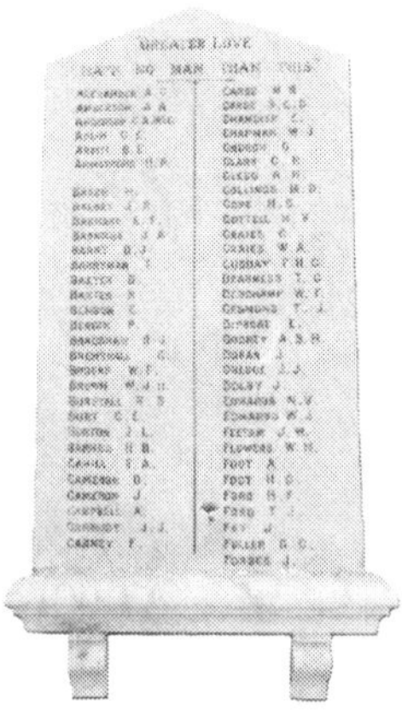

Thomas James Ford was born in Alpha, Queensland, on the 17th of April 1893, to Henry and Christina Ford (née Roche). He was one of five surviving children and had one sister and three brothers. Henry Ford passed away sometime after 1898.

After Henry's death, Christina Ford and the children lived in Townsville at a property on Charters Towers Rd. By 1913, the family had relocated to Mitchell St, North Ward, where Christina Ford worked as a dressmaker. Thomas Ford attended St. Mary's Roman Catholic School in West End, Townsville. Following his schooling, he worked as a labourer.

On the 31st of January 1915, at Ayr, Queensland, Thomas Ford applied to enlist in the Australian Imperial Force (AIF). He declared that he was 22 years and nine months old, single, working as a labourer, and had no previous military experience. His enlistment medical noted that he was 5'7" (170 cm) tall, weighed 150 lbs (68 kg), had a dark complexion, dark eyes, and black hair. He identified as Roman Catholic.

Thomas Ford's brother Henry Ford enlisted two weeks later, and both were assigned to the Infantry and posted to the 25th Battalion, 12 Reinforcements. Both brothers designated their widowed mother as their next of kin and gave her address as Mitchell St, North Ward, Townsville.

The 25th Battalion was raised at Enoggera in Queensland in March 1915, as part of the 7th Brigade. Although primarily consisting of men recruited in Queensland, the battalion also included a small group of men from Darwin. After four months of training in Brisbane, the newly formed battalion prepared for deployment overseas.

On the 29th of June 1915, the 25th Battalion embarked from Brisbane bound for Egypt. They sailed aboard HMAT A60 *Aeneas*, which carried 1,720 troops on the voyage. Their ship was one of four that made up Convoy No.7. The entire convoy transported 5,052 men and 20 horses.

In August 1915, the 25th Battalion underwent training in Egypt. In September, they deployed to Gallipoli as part of the 7th Brigade and provided reinforcement to the diminished New Zealand and Australian Division. The last major Allied offensive had been launched and turned back in August 1915, and consequently, the 25th Battalion had a relatively quiet time.

Private Thomas Ford contracted Typhoid Fever in October 1915. Typhoid is a debilitating bacterial disease transmitted through contaminated food, water, or close contact with infected carriers. He was evacuated to the 21st General Hospital in Alexandria and categorized as dangerously ill. After a couple of weeks, Private Ford recovered sufficiently to be sent to the Enteric Fever Convalescent Camp at Port Said. In November 1915, his medical board deemed him unfit for service for three months and recommended his return to Australia.

Private Thomas Ford arrived back in Australia and was admitted to the 6th Australian General Hospital in Brisbane in January 1916. In April 1916, Thomas Ford re-enlisted by signing new attestation papers. He departed from

Sydney, New South Wales, on the RMS *Mooltan* on the 12th of April 1916, bound for England and France to rejoin the 25th Battalion.

After undergoing further training in England and then France, Thomas rejoined the heavily depleted 25th Battalion in the field on the 12th of August 1916, narrowly missing the battle of Pozières by two weeks. It was during this battle that his brother Henry Ford was initially reported missing and later confirmed as Killed in Action.

Private Thomas Ford continued to serve with the 25th Battalion. After a spell in a quieter sector of the front in Belgium, the 2nd Division (which included the 25th Battalion) came south in October to attack again in the Somme Valley.

The 25th Battalion engaged in two offensives to the east of Flers during the November 1916 attacks named after the village. The harsh conditions on the Somme battlefield, saturated with rain, rendered the assaults extremely challenging. The persistent cloying mud sucked down the advancing troops, causing them to lag behind the creeping artillery barrage. Consequently, they became vulnerable targets for German machine-gunners and riflemen.

It was during this battle that Private Thomas Ford was reported as missing on the 25th of November 1916. It was not until the 28th of July 1917 that he was officially declared Killed in Action on the 14th of November 1916. Private Thomas Ford was 23 years and six months old and had served in the AIF for one year and nine months at the time of his death. He has no known grave, and his death was notified in the 331st list of Australian Casualties released on the 18th of August 1917.

Thomas Ford's name is inscribed on The Australian War Memorial Roll of Honour (Panel 104), The City of Townsville WW1 Honour Roll, The Townsville Cenotaph, and The Villers-Bretonneux Memorial (Australian National Memorial-France). Ford Street, Hermit Park, Townsville, is named in his and his brother Henry's honour.

862 Private Thomas James Ford served in WW1 in Gallipoli and on the Western Front. He was awarded the 1914/15 Star (no:20798), the British War Medal (no:16623), and the Victory Medal (no:16556). These medals, as

well as a memorial scroll and a memorial plaque, were passed to his mother. Thomas' mother was granted a war pension of two pounds (£2/-/-) per fortnight which was 50% of the average fortnightly wage in 1916.

Two of Thomas Ford's brothers, John and Henry, also served in WW1. Henry was Killed in Action at Pozières on the 29th of July 1916, and John returned to Australia on the 30th of June 1918.

The following notice was published in the *Townsville Daily Bulletin* (Qld) on the 14th of November 1917:

FORD. — In most loving remembrance of PRIVATE Thomas James Ford, who was Killed in Action at Flers, November 14th, 1916. R.I.P.

174 PRIVATE ORWELL McQUEEN (1888 – 1916)

25th Infantry Battalion

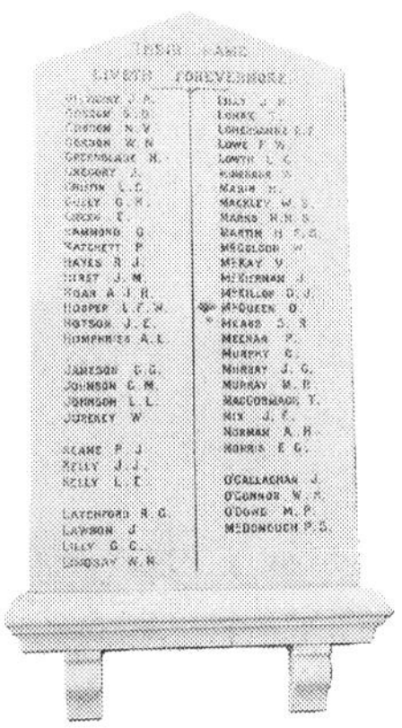

Orwell McQueen was born on the 3rd of May 1888, in Townsville, Queensland, to Peter and Elizabeth McQueen (née Neilsen). He had seven surviving siblings, three brothers and four sisters, while two other children died soon after birth. Orwell's father, Peter McQueen, was a renowned businessman associated with the firm Cruikshank, Love, and McQueen, which specialized in stock and station sales.

During the early 1900s, the McQueen family lived in a property on Mary St, Townsville. Orwell McQueen grew up in Townsville and attended the Townsville West End State School. After completing his school years, he worked as a labourer. In 1915, the McQueen family lived on Stanton Terrace in Townsville.

On the 11th of January 1915, at the age of 26 years and eight months, Orwell McQueen applied to enlist in the Australian Imperial Force (AIF). He was single, without a trade or calling, and declared that he had no prior military training. His enlistment medical noted that he was 5'6" (167 cm) tall, weighed 9st 1lb (57 kg), had a fair complexion, green eyes, and light grey hair. He identified as Presbyterian and nominated his mother as his next of kin.

Accepted into the AIF, Orwell McQueen was allocated service number 174, assigned to the Infantry, and posted as an original member to "A" Company of the 25th Battalion. The 25th Battalion was raised at Enoggera in Queensland in March 1915, as part of the 7th Brigade. Although primarily consisting of men recruited in Queensland, the battalion also included a small group of men from Darwin.

After four months of training in Brisbane, the newly formed battalion prepared for deployment overseas. On the 29th of June 1915, the 25th Battalion embarked from Brisbane bound for Egypt. They sailed aboard HMAT A60 *Aeneas*, which carried 1,720 troops on the voyage. Their ship was one of four that made up Convoy No.7. The entire convoy transported 5,052 men and 20 horses.

In August 1915, the 25th Battalion trained in Egypt. In September, they were deployed to Gallipoli as part of the 7th Brigade, providing reinforcement to the diminished New Zealand and Australian Division. Following the failure of the last major Allied offensive in August 1915, the 25th Battalion experienced a comparatively tranquil period. The battalion was evacuated from the peninsula on the 18th of December 1915. A period of further training was undertaken in Egypt.

In March 1916, the 25th Battalion embarked for France, arriving on the 19th of March 1916, and becoming the first AIF battalion to reach French soil. Operating under the command of the 2nd Division, the battalion faced its baptism of fire on the Western Front at Pozières from the 25th of July to the 7th of August 1916.

On the 28th of July 1916, during a night time assault on the heights, the battalion suffered heavy losses, with 12 officers and 350 other ranks killed or wounded. Many soldiers became entangled in dense wire obstacles that had remained intact despite the artillery bombardment. A subsequent attack on the 4th of August resulted in additional casualties, bringing the total loss for the 25th Battalion in its first battle on the Western Front to 785 personnel. It is remembered as one of the most ferocious battles of the war.

After the Battle of Pozières, Private Orwell McQueen was admitted to the hospital suffering from scabies. He returned to the Battalion in early October 1916. About three weeks later, he was admitted to the South African General Hospital suffering from an abscess. He was treated and re-joined the 25th on the 9th of November 1916, by which time, the Battalion was in trenches near Flers in the Somme Valley.

The 25th Battalion took part in two attacks east of Flers. On 13th of November, they were ordered to capture German trenches at 6:45 a.m. the next day. Heavy rainfall before the attack created poor conditions. The mud slowed the troops, preventing them from keeping pace with the artillery barrage, leaving them exposed to German machine-gunners and riflemen. Despite the weather and difficult terrain, the attack proceeded on schedule but failed.

It was during this attack that Private Orwell McQueen was Killed in Action by a shell burst at approximately 1:00 a.m. on the 14th of November 1916. He was 28 years and six months old and had served in the AIF for one year and 10 months. He was buried in the field on the same day but was his body was later reinterred and laid to rest at Warlencourt British Cemetery (Plot VI, Row G, Grave No.13), France. His death was notified in the 254th list of Australian Casualties released on the 20th of December 1916.

Orwell McQueen's name is inscribed on The Australian War Memorial Roll of Honour (Panel 105), The City of Townsville WW1 Honour Roll, and The Townsville Cenotaph. McQueen Street, Rosslea, Townsville, is named in his memory.

174 Private Orwell McQueen served in WW1 in Gallipoli and on the Western Front. He was awarded the 1914/15 Star (no: 20462), the British War Medal (no: 16722), and the Victory Medal (no: 16656). These medals, as well as a memorial scroll and a memorial plaque, were passed to his father.

The following notice was published in the *Townsville Daily Bulletin* (Qld) on the 6th of February 1917:

Mr. McQueen, of Stanton Terrace, Melton Hill, has received a communication from Signal Sergeant R. Towns concerning the death of his son, Orwell. The letter reads — *'I regret to have to report to you the death of your son Private O. McQueen. A shell burst amongst several of us and, unfortunately, he was killed instantly. His effects were handed to battalion headquarters to be handed on to you. As a soldier and a signaller, he was courageous and one of the best of men. His loss is mourned by the remainder of us and we all send you our deepest condolences. He was killed at 1 a.m. on the 14th of November and was buried the same day.'*

690 PRIVATE CHARLES JAMIESON MURPHY (1889 – 1916)

25th Infantry Battalion

Charles Jamieson Murphy was born on the 3rd of September 1889 in Cooktown, Queensland, to James and Bridget Murphy. After moving to Townsville, the Murphy The family lived at a property on Philp St, West End. Charles Murphy attended the Townsville West State School and after completing his education, worked as a labourer. Sometime before 1909, James Murphy passed away, leaving Bridget widowed. She supported herself and the family by working as a laundress.

On the 22nd of February 1915, Charles Murphy applied in Townsville to enlist in the Australian Imperial Force (AIF). He declared that he was 26 years and five months old, single, employed as a labourer, and without prior military experience. His medical examination noted he was 5'6" (167 cm) tall, weighed 8 stone (51 kg), had a fresh complexion, blue eyes, and light brown hair. He identified as Roman Catholic and nominated his mother as his next of kin, providing her address as Philp St, West End.

Accepted into the AIF, Charles Murphy was allocated service number 690, assigned to the Infantry and posted to "C" Company of the 25th Battalion. The 25th Battalion was raised at Enoggera in Queensland in March 1915 as part of the 7th Brigade. Although primarily consisting of men recruited in Queensland, the battalion also included a small group of men from Darwin.

After four months of training in Brisbane, the newly formed battalion prepared for deployment overseas. On the 29th of June 1915, the 25th Battalion embarked from Brisbane bound for Egypt. They sailed aboard HMAT A60 *Aeneas*, which carried 1,720 troops on the voyage. The four-ship convoy transported 5,052 men and 20 horses.

In August 1915, the 25th Battalion trained in Egypt. In September, they were deployed to Gallipoli as part of the 7th Brigade, providing reinforcement to the diminished New Zealand and Australian Division. Following the failure of the last major Allied offensive in August 1915, the 25th Battalion experienced a comparatively tranquil period.

Approximately three weeks after their arrival at Gallipoli, Private Charles Murphy fell ill and was eventually evacuated to Malta. He underwent treatment for 60 days and was discharged from the hospital in December 1915, after the evacuation of Gallipoli. From Malta, he was sent back to Cairo.

In July 1916, Private Murphy was transported from Alexandria on board the *Arcadian*, bound for Rollestone, England, where he joined the 14th Training Battalion. After two months of training, he crossed to France and finally rejoined the 25th Battalion in the field on the 14th of September 1916. The 25th Battalion had already participated in its first major battle at Pozières between the 25th of July and the 7th of August, resulting in 785 casualties.

Following a period of relative calm in a less active area of the front in Belgium, the 25th Battalion, as part of the 2nd Division, moved south in October to resume their offensive in the Somme Valley. Engaging in two assaults to the east of Flers, both attacks faltered amidst the treacherous mud.

The 25th Battalion took part in two attacks east of Flers. On 13th of November, they were ordered to capture German trenches at 6:45 a.m. the next day. Heavy rainfall before the attack created poor conditions. The mud slowed the troops, preventing them from keeping pace with the artillery barrage, leaving them exposed to German machine-gunners and riflemen. Despite the weather and difficult terrain, the attack proceeded on schedule but failed.

At the end of this fighting, Private Charles Murphy was reported as missing. It was not until the 20th of October 1917 that a Court of Enquiry officially declared Private Charles Murphy Killed in Action on the 14th of November 1916. In a statement made on the 17th of January 1917, Lieutenant A.I Everett said, *"I fear there is no doubt whatever that Murphy was killed on Nov. 14th, 1916, in the attack we made on Gird Trench to the left of Flers. We went over at 7 a.m., it was not successful, and we withdrew. We never got near enough to have any prisoners taken. Any man not accounted for that day must have been killed. He must lie on No Man's Land."*

Private Charles Murphy was 27 years old and had served in the AIF for one year and nine months at the time of his death. His remains were later located, and he was laid to rest at Warlencourt British Cemetery (Plot III, Row D, Grave No.7) in France.

Charles Murphy's name is inscribed on The Australian War Memorial Roll of Honour (Panel 105), The City of Townsville WW1 Honour Roll, and The Townsville Cenotaph. Murphy Street, Rosslea, Townsville, is named in his memory.

690 Private Charles Murphy served in WW1 in Gallipoli and on the Western Front. He was awarded the 1914/15 Star (no: 22382), the British War Medal (no: 16704), and the Victory Medal (no: 16638). These medals, as well as a memorial scroll and a memorial plaque, were passed to his widowed mother.

CAPTAIN CHRISTIAN MARTIN JOHNSON (1892 – 1916)

25th Infantry Battalion

Christian Martin Johnson (Christie) was born on the 25th of May 1892, in Homestead, Queensland, to Peter and Florence Johnson (née Madden). He was the second-born and oldest son in a large family. His family were extensively involved in the Ayr community and owned the Cosmopolitan Hotel in Cloncurry for some time.

Christie Johnson was a scholarship student who attended Townsville Grammar School as a boarder. Gifted academically, Christie attained Sydney University Entrance Honours but chose to work with the AMP Society upon leaving school. He was also a keen sportsman who represented the Townsville Grammar School Old Boys' Association at cricket.

In 1910, Christie Johnson joined the 1st Australian Garrison Artillery (1st AGA), eventually rising to the rank of Sergeant. He was deployed to Thursday Island aboard S.S. *Kanowna* on the 8th of August 1914, with the AGA, just four days after Britain had declared war on Germany. After six months on Thursday Island, he returned to Australia in January 1915. He lived in premises on Denham Street, Townsville, at this time.

About six weeks after returning, Christie Johnson applied to join the Australian Imperial Force (AIF) on the 8th of March 1915. He was 22 years and nine months old. He declared that he was single, employed as a clerk, and had five years' experience with the Garrison Artillery. His medical examination noted that he was 5'8½" (174 cm) tall, weighed 150 lbs (64.5 kg), had a fair complexion, blue eyes, and light brown hair. He identified as Church of England and listed his father, c/o the Cosmopolitan Hotel, Cloncurry, as his next of kin.

When he joined the AIF, the AMP Society continued to pay Christie Johnson a full salary. He initially applied to join the artillery but was unsuccessful. Instead, upon being accepted into the AIF, Christie Johnson was assigned to the infantry as a 2nd Lieutenant (2Lt) and posted to the newly raised 25th Battalion, which was raised at Enoggera in Queensland in March 1915 as part of the 7th Brigade. Although largely made up of men recruited in Queensland, the battalion also included a small contingent of men from Darwin. 2Lt Johnson was posted to "B" Company in the battalion, and on the 1st of June 1915, he was promoted to Lieutenant.

The 25th completed four months of training in Brisbane in preparation for their deployment overseas. On the 29th of June 1915, the battalion embarked from Brisbane bound for Egypt. They sailed aboard HMAT A60 *Aeneas*, which carried 1,720 troops on the voyage. Their ship was one of four that made up Convoy No. 7. The entire convoy transported 5,052 men and 20 horses.

In August 1915, the 25th Battalion completed further training in Egypt. In September, they deployed to Gallipoli as part of the 7th Brigade and provided reinforcement to the diminished New Zealand and Australian Division. The last major Allied offensive had been launched and turned back in August 1915, and consequently, the 25th Battalion had a relatively quiet time.

Lt Johnson fell ill with jaundice (hepatitis) in November 1915 and required evacuation to Egypt, where he was admitted to the 1st Australian General Hospital at Heliopolis. He recovered, and in late January 1916, re-joined the 25th Battalion at Tel-el-Kebir, where they were located following the general withdrawal from Gallipoli. Lt Johnson was only back with the 25th for a few weeks before he contracted mumps and was admitted to the Field Ambulance for three weeks.

In the middle of March 1916, the 25th Battalion embarked at Alexandria to join the British Expeditionary Force on the Western Front in France. Landing on the 19th of March 1916, it was the first AIF battalion to arrive there. Lt Johnson was promoted to Captain in June 1916.

The 25th took part in its first major battle at Pozières, fought between July and September 1916, as part of the larger Battle of the Somme. The primary objective was the capture of the strategic village of Pozières in northern France, a crucial point in the German defensive line. This fiercely contested battle exacted a heavy toll on both sides, marked by intense artillery barrages and steadfast German resistance. It was a bloody and costly battle, and Captain Christie Johnson was one of the casualties when he was shelled and gassed on the 29th of July 1916. These wounds required evacuation by the medical system back to England. Captain Johnson did not return to the 25th Battalion until the 14th of October 1916.

Following a period of relative calm in a less active area of the front in Belgium, the 25th Battalion, as part of the 2nd Division, moved south in October to resume their offensive in the Somme Valley. Engaging in two assaults to the east of Flers, both attacks faltered amidst the treacherous mud.

On the 14th of November 1916, the 25th Battalion was ordered to attack German trenches. Heavy rainfall had turned the Somme battlefield into deep mud, making conditions atrocious. As the troops advanced, they became bogged down, unable to keep up with the creeping artillery barrage, and were exposed to German machine-gun and rifle fire. It was during this battle that Captain Christie Johnson was Killed in Action on the 15th of November 1916, near the town of Gueudecourt in France. He was 24 years and five months old and had served in the AIF for 20 months at the time of his death. His death was notified in the 247th list of Australian Casualties released on the 15th of December 1916. A report of a burial site was made after his death; however, an intensive search made by the War Graves Commission in 1919 failed to identify and reinter his remains. He has no known grave.

Christian Martin Johnson's name is inscribed on The Australian War Memorial Roll of Honour (Panel 105), The Australian National Memorial, Villers-Bretonneux, France, The City of Townsville WW1 Honour Roll, The Townsville Cenotaph, and The Townsville Grammar School Roll of Honour.

Captain Christian Martin Johnson served in WW1 in Gallipoli and on the Western Front. He was awarded the 1914/15 Star (no: 23598), the British War Medal (no: 16666), and the Victory Medal (no: 16599). These medals, as well as a memorial scroll and a memorial plaque, were passed to his father.

The following notice was published in the *Townsville Daily Bulletin* (Qld) on the 14th of November 1919:

JOHNSON. — In fond and ever-loving remembrance of our eldest and dearly beloved son and brother, Captain C. M. Johnson (Christie), of the 25th Battalion, who was Killed in Action somewhere in France, on the 14th of November 1916. (Inserted by his loving parents, sisters, and brothers, Cosmopolitan Hotel, Cloncurry.)

276 PRIVATE CECIL CARRICK WOTTON (1894 – 1916)

25th and 29th Infantry Battalion

Cecil Carrick Wotton was born in Townsville, Queensland, on the 22nd of November 1894. He was the youngest of four children born to Henry Wotton and his wife Frances (née Robinson). Cecil Wotton had two brothers and one sister. Henry Wotton was a music teacher, and the family originally lived in Ayr, Queensland, in the early 1900s before they settled in Cannan Street, South Townsville. In January 1914, Frances Wotton passed away.

Cecil Wotton attended the West State School in Townsville and later completed a three-year apprenticeship to qualify as a gasfitter. Cecil served in the militia as a member of the Kennedy Regiment prior to the outbreak of World War I on the 4th of August 1914. Ten days after the war was declared, Cecil Wotton was mobilized for war service and enlisted in the Australian Naval and Military Expeditionary Force (AN&MEF).

The AN&MEF was tasked with the Australian takeover and occupation of the Pacific colony of German New Guinea in September-November 1914. Private Cecil Wotton embarked at Cairns on the 8th of August 1914, bound for Thursday Island (War Station) Garrison duty. He re-embarked at Thursday Island on the S.S. *Kanowna* on the 16th of August 1914 to take part in the capture of German New Guinea. As events turned out, Private Wotton did not participate in the capture of German New Guinea on account of "trouble caused by the firemen on the troopship." The S.S. *Kanowna* returned to Townsville on the 18th of September 1914, and Private Wotton was discharged from the AN&MEF on the same day.

Three months later, on the 21st of December 1914, Cecil Wotton applied in Townsville to enlist in the Australian Imperial Force (AIF). He declared he was single, a qualified gasfitter, and 21 years old. His medical examination noted that he was 5' 6" (168 cm) tall, weighed 8st 5lbs (53 kg), had a dark complexion, brown eyes, and dark hair. He identified as Church of England and nominated his brother, Frank, as his next of kin and simply gave his address as Post Office, Townsville.

Accepted into the AIF, Cecil Wotton was allocated service number 276, assigned to the Infantry and posted as an original member to "A" Company of the freshly minted 25th Battalion. The 25th Battalion was raised at Enoggera in Queensland in March 1915 as part of the 7th Brigade. The battalion was predominantly composed of men recruited in Queensland but also included a small contingent of men from Darwin. The newly raised battalion underwent four months of training in Brisbane in preparation for their deployment overseas.

On the 29th of June 1915, the 25th Battalion embarked from Brisbane bound for Egypt. They sailed aboard HMAT A60 *Aeneas*, which carried 1,720 troops on the voyage. Their ship was one of four that made up Convoy No.7. The entire convoy transported 5,052 men and 20 horses. Following further training in Egypt throughout August, the battalion proceeded to occupy trenches at Gallipoli in early September. At Gallipoli, the 7th Brigade provided reinforcements for the New Zealand and Australian Division, which had suffered significant depletion. Although the 25th Battalion was part of this reinforcement effort, they experienced a relatively calm period as the last major Allied offensive had been launched and repulsed the previous month.

In late September 1915, Private Cecil Wotton was returned to Australia for medical reasons. Following treatment, he was allocated to the 4th reinforcements of the 29th Infantry Battalion and embarked from Melbourne on the 14th of March 1916 aboard HMAT A68 *Anchises*, bound for Suez. The *Anchises* was part of Convoy No.18 and had 1,319 personnel on board. The total convoy of 13 ships carried 10,826 personnel and 1,920 horses.

After arriving in Suez on the 15th of April 1916, Private Cecil Wotton was taken on the strength of the 29th Battalion on the 25th of May 1916 at Ferry Post, Suez. On the 16th of June 1916, Private Wotton embarked with his unit, bound for France, to join the British Expeditionary Force. He disembarked in Marseilles on the 23rd of June 1916 and afterward was transported by rail to Hazebrouck.

On the 8th of July 1916, the 5th Division, including the 29th Battalion, was called up to the front to replace the 4th Australian Division. The 29th undertook a difficult two-day, 29-mile (47 km) approach march over cobbled roads with loads of up to 70–75 pounds (32–34 kg) before arriving at the front on the night of the 10th/11th of July 1916. The battalion was deployed in a position between Boutillerie and Condonerrie in the Bois Grenier where they relieved the 13th Battalion.

On the 19th of July 1916, the 29th took part in an attack against the German positions around the "Delangre Farm." The next day, the 29th Battalion repulsed a particularly heavy German counterattack. Despite losing 52 men Killed in Action and 164 men wounded, the 29th Battalion went on to hold the line for another 11 days before they were relieved.

On the 28th of July, Private Wotton was attached for duty with the 18th Company Australian Army Service Corps, 5th Divisional Train, located at Sailly. The Unit Diary for the 5th Train records that *"loading parties (1 per supply wagon) from each unit permanently attached to train for fatigue duties in resupplies and to keep wagon clean."* The basic tasks of train companies and motorized columns were to carry baggage and resupply food, forage, and water for the headquarters and units of the division.

Private Wotton continued to serve with the 18th Company as it worked in the vicinity of Sailly and then Becordel. On the 22nd of November 1916, the train started to relocate to Albert. It was during this activity that Private Cecil Wotton was Killed in Action on the 23rd of November 1916. He was laid to rest near Albert in the Dantzig Alley British Cemetery (Plot IX, Row P, Grave No.3), Metz, France. He had turned 22 years old the day before he died and had served in the AIF for 23 months. His death was notified in the 254th list of Australian Casualties released on the 19th of December 1916.

Cecil Carrick Wotton's name is inscribed on The Australian War Memorial Roll of Honour (Panel 116), The City of Townsville WW1 Honour Roll, and The Townsville Cenotaph. Wotton St, Aitkenvale, Townsville, is named in his memory.

276 Private Cecil Carrick Wotton served in WW1 in Egypt and on the Western Front. He was awarded the 1914/15 Star (no: 26191), the British War Medal (no: 16819), and the Victory Medal (no: 20599). These medals, as well as a memorial scroll and a memorial plaque, were passed to his father in 1921 and 1922, respectively. Cecil Wotton's brothers, Frank and Henry, also enlisted in the AIF. Both survived and returned to Australia.

The following article was published in *The Telegraph* (Brisbane) on the 23rd of November 1917:

WOTTON. — In loving memory of Pte. C.C. Wotton, Townsville, Killed in Action 23/11/16. Though lost to sight, to memory dear. Inserted by his friends, Mr. and Mrs. E.J. Emery and family, Northgate.

The following article was published in the *Brisbane Courier* on the 3rd of January 1917:

WOTTON. — Killed in action, in France, November 23rd, 1916, Private Cecil Carrick Wotton, brother of Private H.F. Wotton, and nephew of Mrs. R.R. Robinson, "Seaton Delaval," Lock St., West End, Brisbane.

13135 PRIVATE GEORGE VAUGHN (1897 – 1916)

47th Battalion

George Vaughn was born on the 18th of July 1897 in Townsville, Queensland. He was the second son and third child of eight, born to Samuel and Sarah Vaughn (née McCarty/McCarthy). He had one brother and six sisters. Samuel Vaughn was a dairyman, and the family lived in premises on Stanton St, Townsville in 1903. By 1913, they had moved to Bohle River, north of Townsville.

George Vaughn grew up in Townsville, where he worked as a stockman. He also served in the Citizen Forces. On the 16th of October 1916, George Vaughn applied in Townsville to enlist in the Australian Imperial Force (AIF). He declared that he was 19 years and three months old, single, and employed as a stockman. He also declared his previous military service in the Citizen Forces.

George Vaughn's enlistment medical examination noted that he was 5'4¾" (164 cm) tall, weighed 8st 10lbs (55 kg), had a ruddy complexion, grey eyes, and dark brown hair. He identified as Roman Catholic, listed his mother as his next of kin, giving her address as Bohle River, Townsville. He also allotted her three shillings per day from his pay.

Accepted into the AIF, George Vaughn was allocated service number 13135, assigned to the Infantry and posted to the 47th Battalion, 6th Reinforcement. On the 21st of October 1916, he marched into the 11th Depot Battalion at Enoggera, Queensland, for basic training. One month after enlistment, Private George Vaughn was admitted to Fort Lytton Hospital in Brisbane, suffering from acute nephritis. Tragically, he Died in Service on the 28th of November 1916. He was laid to rest at Toowong Cemetery, Brisbane QLD, Nl/314 (GRM/4) Grave 15.4.6.

George Vaughn's name is inscribed on the Australian War Memorial Roll of Honour (Panel 144), the City of Townsville WW1 Honour Roll, and the Townsville Cenotaph. Vaughn St, West End, Townsville, is named in his memory. A memorial scroll and a memorial plaque were passed to his father in 1923.

The following notice was published in the *Townsville Daily Bulletin* on the 20th of December 1916:

VAUGHN. — On the 28th of November, at Lytton Hospital, Enoggera, George, youngest son of Samuel and Sarah Vaughn, Bohle River. Age 19 years. R. I. P.

3013A PRIVATE FREDERICK GEORGE WHITE (1893 – 1916)

31st Infantry Battalion

Frederick 'Fred' George White was born on the 1st of February 1893 in Townsville, Queensland, to Alfred White and his wife Mary Sophia 'Sophy' (née Parker). Fred White was the fourth of nine children and the second son. He had two brothers and six sisters. Alfred White worked as a labourer, and the family lived in Aitkenvale, Townsville. Fred White grew up in Townsville and attended the Mundingburra State School. Like his father, he became a labourer after finishing school.

On the 3rd of January 1916, Fred White quit his labouring job and applied in Townsville to enlist in the Australian Imperial Force (AIF). He declared that he was single, 22 years and 11 months old, and had no previous military service. His medical assessment noted that he was 5'7¾" (172 cm) tall, weighed 9st 1lb (58 kg), had a fair complexion, grey eyes, and light brown hair. He identified as Methodist and listed his father as his next of kin. He gave his father's address as simply Aitkenvale, Townsville.

Accepted into the AIF, Fred White was allocated service number 3013A, assigned to the Infantry and posted to the 31st Battalion, 6th Reinforcement. The 31st Battalion was in Egypt when Fred White enlisted. The battalion remained in Egypt until June 1916, when they sailed to join the British Expeditionary Force on the Western Front.

After 15 weeks of basic training in Enoggera, Brisbane, Private Fred White embarked for Egypt on the 14th of April 1916 from Sydney, New South Wales, as part of Convoy No.20. He sailed with his fellow contingent of reinforcements aboard HMAT A40 *Ceramic* which carried 2,554 on the voyage. The complete convoy of 15 ships carried a total of 9,249 troops and 5,590 horses to Egypt. They disembarked in Port Said in the middle of May 1916.

Three weeks later, the contingent embarked from Alexandria on the 6th of June 1916 aboard HMAT *Franconia* bound for England. Once in England, further training was completed at the 8th Training Battalion located at Codford near Salisbury Plain.

In September 1916, Private Fred White proceeded overseas to Étaples in France for another 14 days of training. He was hospitalized on the 19th of September 1916, two days before he was taken on the strength of the 31st Battalion in the field, near Armentieres. Fred was suffering from a growth in his chest which he first noticed three days before his hospitalization.

He was evacuated back to the 1st Australian Casualty Clearing Station and then to 7th General Hospital at St Omer in France. Two weeks later, Fred was evacuated back to England, where he was eventually admitted to the Sidney Hall Hospital in Weymouth, Dorset. He was diagnosed with a large tumour extending over his sternum and eventually over the whole of the middle and left chest.

Private Fred White died of sarcoma of the sternum on the 7th of December 1916. He was 23 years and ten months old at the time of his death. He had served with the AIF for 11 months and was laid to rest at the Melcombe Regis Cemetery, Dorset, England, Grave 111.C.2767.

Frederick George White's name is inscribed on The Australian War Memorial Roll of Honour (Panel 119), The No.37 Magnetic Lodge P.A.F.S.O.A. Honor Board, The City of Townsville WW1 Honour Roll, and The Townsville Cenotaph.

3013A Private Frederick George White was awarded the British War Medal (no: 37465) and the Victory Medal (no: 37163). A memorial scroll and a memorial plaque were passed to his father in 1922.

The following notice was published in the *Townsville Daily Bulletin* (Qld.) on the 23rd of October 1916:

DEATH. WHITE — At the Military Hospital, Weymouth, on December 7th, Private F. G. White, of illness, second son of Mr. and Mrs. A. White, Aitkenvale, Townsville, in his 23rd year. He gave his life for King and Country.

The following tribute was published in the *Townsville Daily Bulletin* (Qld.) on the 7th of December 1917:

WHITE — In loving memory of our dear son, brother, and brother-in-law, Private Frederick George White, 31st Battalion, A.I.F., who died in Sydney Hall Hospital on the 7th of December 1917, after seeing service in France. Laid to rest in Melcombe Regis Cemetery, Weymouth, England.

The midnight stars are gleaming,
O'er a grave I cannot see,
Where sleeping without dreaming,
Lies one so dear to me.
He never shunned his country's call,
But gladly gave his life — his all.
He died the helpless to defend,
An Australian soldier's noble end.

(Inserted by his loving Father, Mother, Brother, Sisters, and Brother-in-law.)

PART 3 - 1917

In 1917, Australia's involvement in World War I reached a significant turning point. The Australian Imperial Force (AIF) had already been active in several major campaigns, including the Gallipoli campaign and the battles on the Western Front, but 1917 marked some of the heaviest and most brutal fighting experienced by Australian forces. This year saw the AIF deeply engaged in battles such as Bullecourt and Passchendaele, which became infamous for their intensity and high casualty rates. These battles profoundly affected the soldiers and their communities back home, with some towns experiencing devastating losses.

Australia's Key Contributions in 1917

Australia played a crucial role on the Western Front in 1917, as the Allied forces launched renewed offensives against the Germans. Australian divisions were heavily involved in both the First and Second Battles of Bullecourt in April and May, where they faced formidable German defences. Despite initial successes, these battles proved costly, with poor planning and lack of support leaving the Australian troops vulnerable. The Australians suffered approximately 10,000 casualties, with many soldiers experiencing horrific injuries, imprisonment, or death. Despite the setbacks, the AIF's resilience and bravery during Bullecourt highlighted their commitment to the Allied cause.

Later in 1917, the Australians were called upon again in the Third Battle of Ypres, specifically in the Battle of Passchendaele, fought in Belgium from July to November. Known for its rain-soaked terrain and thick mud, the battlefield turned into a nightmarish landscape where many soldiers drowned in the mud. Here, Australian forces faced some of the heaviest artillery and machine-gun fire of the war. Over 38,000 Australians were killed or wounded in the battle, a shocking loss that left scars both on the soldiers and their communities back home.

The Losses from Townsville

The impact of these battles on the home front was devastating, as communities across Australia mourned the loss of sons, fathers, and friends. Townsville lost 70 men during 1917. The news of these deaths, often relayed in letters or brief news articles, was a constant reminder of the personal toll the war exacted on small communities. Townsville's residents, like many others in Australia, came to understand the depth of the sacrifice made by its young men on distant battlefields.

The impact to Townsville of losing these 70 men was felt on multiple levels. Families lost breadwinners, communities lost promising young leaders, and businesses lost workers, which strained the local economy. Socially and emotionally, the loss of so many young men created a gap in the town that was difficult to fill.

Australia's Sacrifice and Legacy

By the end of 1917, Australians had gained a reputation as formidable fighters, but the human cost was immense. Of the roughly 416,000 Australians who enlisted, around 60,000 would not return, with many more left physically or emotionally scarred. The contributions of Australian soldiers in 1917 solidified their place in history as resilient and courageous, even as the losses continued to weigh heavily on communities like Townsville.

In 1917, the Australian Imperial Force AIF endured its most challenging year on the Western Front, a period marked by a bitter winter, immense casualties and devastating losses. For Townsville, this year was particularly sorrowful, as battles at Bullecourt, Messines, Menin Road and Passchendaele claimed the lives of local soldiers. In total, approximately 76,000 Australian troops became casualties in these campaigns, and Townsville alone mourned the loss of 70 men ranging in age from 19 to 48 years old.

The 41st Battalion, comprised largely of Queensland volunteers, was particularly hard-hit. Ten volunteers from Townsville were lost in this battalion in 1917, marking it as the battalion with the highest casualty rate from Townsville.

At Bullecourt, Australian troops faced formidable German defences, limited support, and suffered substantial casualties. This battle alone resulted in the loss of seven men from Townsville. Later, in June at Messines, a well-planned but hard-fought assault demanded intense coordination, costing the lives of 12 more Townsville soldiers. The impact extended beyond the battlefield, with eight of these fallen men leaving behind widows and grieving families in Townsville. For these families, the personal cost was immense, adding to the broader grief experienced across Australia.

The battles of Menin Road and Broodseinde in September and October of 1917 were critical engagements for the AIF on the Western Front. Both battles formed part of the Third Battle of Ypres, a series of offensives aimed at breaking German defences around the Ypres Salient. Australian divisions, including soldiers from Townsville, faced harsh conditions with deep mud, shell craters, and relentless artillery fire, all of which made the advancement difficult and dangerous. Thirteen soldiers from Townsville perished during these two battles, a loss keenly felt in their hometown. Three Townsville soldiers also fell in Palestine, while two others died as prisoners of war—one in Turkey and one in Germany.

For the Townsville community, the losses of 1917 were a heavy burden, with families and neighbours coming to terms with the absence of fathers, husbands, and sons.

146 SERGEANT FRASER HAROLD GEORGE CUDDAY (1894 – 1917)

41st Infantry Battalion

Fraser Harold George Cudday (aka Cuddy) was born to Richard and Norah (Honora Bridget) Cudday (née Coleman) on the 11th of October 1894 in Mount Fraser via Port Douglas, Queensland. He was the fourth of seven surviving children, and he had two sisters and four brothers. Richard Cudday was employed as a woodcutter, and the family lived in McIlwraith St in Townsville.

Fraser Cudday attended the Ross Island State School, and after completing his school years, he worked as a cleaner with the Railways. He also served two years in the Senior Cadets and three years in the 2nd Infantry Regiment, Citizen Military Forces, Townsville. He served on Thursday Island and Kissing Point Fort while he was a member of the 2nd Infantry Regiment.

On his 21st birthday, the 11th of October 1915, Fraser Cudday applied at Townsville to enlist in the Australian Imperial Force (AIF). He declared that he was single, a Railway employee, and still serving in the 2nd Infantry. His enlistment medical noted that he was 5'9" (175 cm) tall, weighed 9st (57 kg), had a fair complexion, grey eyes, and light brown hair. He identified as Roman Catholic and nominated his mother as his next of kin, giving her address as Ninth Avenue Railway Estate, Townsville.

Accepted into the AIF, Fraser Cudday was allocated service number 146, assigned to the Infantry and was ultimately posted as an original member to the 41st Battalion, which was raised in Brisbane in February 1916. The Battalion formed part of the 11th Brigade of the 3rd Australian Division.

During pre-deployment training in Australia, Private Fraser Cudday was promoted to Corporal in December 1915. Four months later, in April 1916, Corporal Fraser Cudday was promoted to Sergeant. As part of "A" Company of the 41st Battalion, Sergeant Cudday embarked from Sydney on the 18th of May 1916, bound for France via England. They sailed as part of Convoy

No. 21 aboard HMAT A64 *Demosthenes*, which carried 1,612 troops on the voyage. The entire convoy of 18 ships transported 25,333 troops.

Months of further training followed in England before the 41st Battalion arrived in France on the 25th of November 1916, six months after leaving Australia. Two weeks after arrival in France, the 41st entered the line for the first time on Christmas Eve. They then endured a long hard winter and the first half of 1917, rotating through the frontline near Armentieres in France and across the Lys River in Belgium. Enduring continual rain, flooded trenches, and heavy shelling, many of the battalion's platoons dwindled from 35 men to less than 10.

On the 21st of January 1917, the 41st Battalion relieved the 43rd Battalion in the front line in the vicinity of Armentieres. Acting on Brigade orders, the 41st Battalion reorganized their Company dispositions. The new arrangements reduced the garrisoning of the front line from four to three Companies, with the fourth Company, "D" Company, acting as a Reserve Force. This meant that "A" Company was on the front line. It was during this activity that Sergeant Fraser Cudday was Killed in Action on the 21st of January 1917 at Armentieres. There is no known record of the exact details of his death.

Sergeant Fraser Cudday was 22 years and nine months old. He had served with the AIF for one year and nine months. He was laid to rest at Cite Bonjean Military Cemetery (Plot III, Row D, Grave No. 32), France. His death was notified in the 272nd list of Australian Casualties released on the 15th of February 1917.

Sergeant Fraser Harold George Cudday was the first Townsville soldier killed in 1917. He served in WW1 on the Western Front. He was awarded the British War Medal (no:56124) and the Victory Medal (no:55204). These medals, as well as a memorial scroll and a memorial plaque, were passed to his father. His name is inscribed on The Australian War Memorial Roll of Honour (panel 133), The 41st Battalion Roll of Honour, The City of Townsville WW1 Honour Roll, and The Townsville Cenotaph. Cudday Street, Stuart, Townsville, is named in his memory.

146 Sergeant Fraser Harold George Cudday was awarded the British War Medal (no:39797) and the Victory Medal (no:39413). A memorial scroll and a memorial plaque were passed to his mother in 1922, and she was granted a war pension of two pounds (£2/-/-) per fortnight from the 6th of April 1917 for the loss of her son. This was about half of the fortnightly wage paid to a soldier in the AIF in 1917.

Fraser Cudday had one brother and one stepbrother who also served in WW1. Their details are as follows:

357 Sergeant Lewis Charles Cuddy MM, 41st Battalion, returned to Australia on the 19th of October 1918, and

120 Staff Sergeant Francis Coleman, 4th Division Ammunition Column, returned to Australia on the 27th of September 1917.

The following notice was published in the *Townsville Daily Bulletin* (Qld.) on 22nd January 1918:

CUDDAY. —In loving memory of our dear son and brother, Frazer Harold George Cudday, who gave his life for the Empire in France on January 22nd, 1917.

> "Somebody wept as he marched away,
> Looking so handsome, brave, and grand.
> Somebody's kiss on his forehead lay.
> Somebody clung to his parting hand.
> Somebody is watching and waiting for him,
> Yearning to hold him once more to her heart.
> But there he lay in a foreign land,
> In a brave Australian's grave."

(Inserted by his loving mother, sisters, and brothers.)

1600 PRIVATE JOHN O'CALLAGHAN (1893 – 1917)

9th Infantry Battalion

PRIVATE J. O'CALLAGHAN.

John O'Callaghan (Jack) was born on the 21st of January 1893 in Townsville. He was the youngest of two sons of Crosby and Sarah O'Callaghan (née Connor). Jack attended the Sisters of Mercy convent in his early years and later worked as a warder and a drover.

In 1913, Crosby O'Callaghan was a labourer living on Warburton Street, while Sarah lived on Eyre Street. That year, Jack O'Callaghan is listed on the electoral roll as an agent residing on Gregory Street. He joined the Senior Cadets in 1913, serving for 15 months before enlisting in the 2nd Infantry Battalion of the Citizen Forces, where he served for six weeks before being mobilized for war service following the outbreak of WW1 in August 1914.

On the 8th of August 1914, Private Jack O'Callaghan embarked at Cairns for Garrison Duty (War Station) on Thursday Island. Six days later, on the 14th of August 1914, he enlisted for service outside Australia. He re-embarked at Thursday Island on S.S. *Kanowna* on the 16th of August 1914 to take part in the capture of German New Guinea as part of the Australian Naval and Military Expeditionary Force.

The AN&MEF was tasked with the Australian takeover and occupation of the Pacific colony of German New Guinea in September – November 1914. As events turned out, Private O'Callaghan did not participate in the capture of German New Guinea on account of *"trouble caused by the firemen on the troopship"*. Jack O'Callaghan returned to Townsville on the 18th of September 1914 and was discharged from the AN&MEF on the same day.

Three months later, on the 17th of December 1914, Jack O'Callaghan applied at Townsville to enlist in the Australian Imperial Force (AIF). He declared that he was single, employed as a drover, and 21 years and 11 months old. His medical examination noted that he was 5'7" (170 cm) tall and weighed 140 lbs (63 kg). He had a medium complexion, light brown eyes, and light brown hair. He identified as Roman Catholic and nominated his mother as his next of kin, giving her address as Eyre St North Ward, Townsville. He declared his previous service with the Senior Cadets and the 2nd Infantry Battalion.

Accepted into the AIF, Jack O'Callaghan was allocated service number 1600, assigned to the Infantry and posted to the 9th Battalion, 3rd Reinforcements. The 9th Battalion was among the first infantry units raised for the AIF during the First World War. It was the first battalion recruited in Queensland, and when Jack O'Callaghan enlisted, the 9th was already in Egypt undergoing training. The 9th Battalion was soon to make history by becoming the first battalion ashore at Gallipoli on the 25th of April 1915.

After basic training in Australia, Private Jack O'Callaghan and his draft of reinforcements embarked from Brisbane on the 13th of February 1915, bound for Egypt. They sailed aboard HMAT A49 *Seang Choon*, which carried 998 troops and was part of Convoy No. 3. The entire convoy of 21 ships carried a total of 8,829 troops and 6,027 horses on the voyage.

Following a short period of training in Egypt, Private O'Callaghan was one of 120 reinforcements who joined the 9th Battalion at ANZAC Cove on the 6th of May 1915. He was posted to "C" Company. Later in May, the 9th Battalion successfully held against the intense Ottoman counterattack on the 19th. In that battle, the 9th lost 16 dead and 25 wounded. Two hundred enemy dead were counted in front of the 9th's position.

At the end of June, the battalion lost 37 killed and 62 wounded during a two-company attack on Knife Edge and Sniper's Ridge. Private O'Callaghan's company was repelled by heavy machine gun fire from both flanks and shrapnel. Fifteen soldiers from "C" Company were reported missing after the battle. Jack O'Callaghan was one of the missing, and it was not until the 19th of July 1915 that it was confirmed, through the American Embassy in Constantinople (now Istanbul), that he was a prisoner of war.

He had been wounded during the battle and was hospitalized, receiving treatment for a wound to his left knee and right thigh. After treatment and recovery, he was transferred to a detention camp in Angora. Conditions in Turkish prisoner of war camps were very harsh. Prisoners were put to hard manual labour. In 1917, Angora (Ankara) became the centre of the working groups engaged in laying the narrow-gauge line towards Yozgad. It was here that Private Jack O'Callaghan died of enteritis while a prisoner on the 21st of January 1917. It was his 24th birthday, and he had been a member of the AIF for two years and one month.

A letter from 958 Trooper George A. Roberts sent in 1917 confirms Private O'Callaghan's death and the fact that he was buried in an unmarked grave at the prisoner of war camp hospital graveyard. After the war, a concerted search was made to identify the remains of the 265 Australian soldiers who died in Turkish prisoner of war camps. Unfortunately, this proved impossible, and a decision was made to commemorate these fallen soldiers with a special Kipling Memorial in the cemetery at Baghdad and separate permanent headstones in the memorial plot at Baghdad. Accordingly, Private Jack O'Callaghan's resting place is recorded as Baghdad (North Gate) War Cemetery (Angora Memorial No. 125), Iraq.

Private John O'Callaghan's name is inscribed on The Australian War Memorial Roll of Honour (Panel 57), The City of Townsville WW1 Honour Roll, and The Townsville Cenotaph. O'Callaghan Street, Heatley, Townsville, is named in his memory.

1600 Private John O'Callaghan was awarded the 1914/1915 Star (no: 2617), the British War Medal (no: 8530), and the Victory Medal (no: 8490). A memorial scroll and a memorial plaque were passed to his father in 1921 and 1922, respectively.

Private O'Callaghan's capture by the Ottomans was notified on the 7th of March 1916 by the Secretary of the Information Bureau.

The following was published in the *Townsville Daily Bulletin* on the 16th of February 1916:

THE WAR IN EUROPE.

Mr. O. H. Toll has received a note from Private J. O'Callaghan, a native of Charters Towers, who is now a prisoner of war at Kingre, in Turkey in Asia (says the *'Northern Miner'*). He was well, he says, and still lives in hopes of seeing the old goldfield again. Letters were limited to four lines, so that little could be written. He asked to be remembered to all friends and to all his old mates who were with him at Thursday Island.

The following was published in the *Charters Towers Northern Miner* on the 12th of June 1917:

Mrs. Arnesen, of McConkey's Lane, has received a wire from her son-in-law, Mr. Connie O'Callaghan, notifying of his brother's death in Turkey. Private Jack O'Callaghan enlisted in Townsville on returning from service at Thursday Island. He saw service at Gallipoli, where, unfortunately, he was made a prisoner of war, eventually dying while still a prisoner. He was about 25 years of age. Mr. Connie O'Callaghan is at present on a business visit to the Duchess.

The following was published in *The Queenslander* on the 23rd of June 1917:

PERSONAL NOTES. Within the past few days, advice has been received in Maryborough of the deaths, whilst prisoners of war in Turkey, of Pte. D. B. Greedon and Pte. Jack O'Callaghan. Both were members of the famous original 9th Battalion, and were taken prisoners on Gallipoli, where they have remained until death supervened in February last through fever.

5985 PRIVATE JOSEPH AUGUSTINE BARNICLE (1890 – 1917)

13th Infantry Battalion

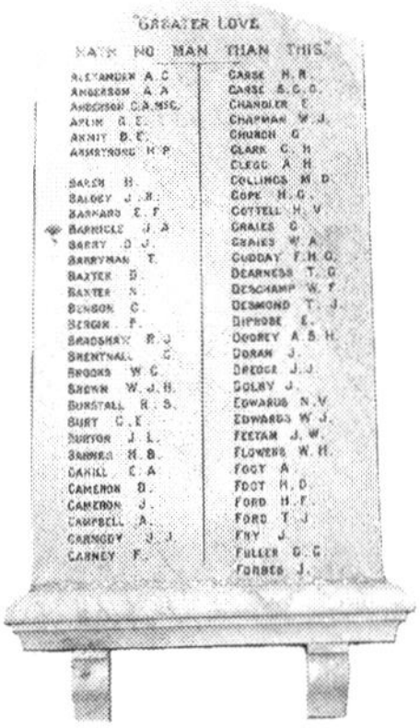

Joseph (Joe) Augustine Barnicle (aka Joe Kelso) was born on the 26th of August 1890 in Townsville, Queensland. He was the oldest son of Joseph Bartholomew Barnicle and Annie Barnicle (née Davis). There were nine children born in the family, and while Joe was the third child born, one of his older sisters did not survive. Joseph Snr. worked as a carpenter, and the family lived in 6th Street, South Townsville.

Joe Barnicle Jnr. grew up in Townsville and attended the Ross Island State School. After his school years were completed, he followed in his father's footsteps and became a carpenter. Joe Barnicle left Townsville and moved to New South Wales. He settled for a time in Lithgow in the Blue Mountains and accepted a job as a letter carrier at the post office. He also assumed the alias of "Joe Kelso" and lived at the Club House Hotel, Main St Lithgow. The *Lithgow Mercury* reported that he was *"well known and highly respected"*.

It was in Lithgow on the 6th of March 1916 that Joe Barnicle applied to enlist in the Australian Imperial Force (AIF) under his real name. He declared that he was 25 years six months old, single, qualified as a carpenter, and had no previous military service. His enlistment medical noted that he was 5'9" (175 cm) tall, weighed 138 lbs (62.5 kg), had a dark complexion, brown eyes, and dark brown hair. He identified as Church of England and nominated his father residing, at 6th Street South Townsville, as his next of kin.

Accepted into the AIF, Barnicle was allocated service number 1715, assigned to the Infantry and posted to the 13th Battalion, 19th Reinforcements. The 13th Battalion was raised in late September 1914, approximately six weeks after the commencement of the First World War. It was primarily comprised of recruits from New South Wales. Alongside the 14th, 15th, and 16th Battalions, it was an Infantry Battalion of the 4th Brigade commanded by Colonel John Monash.

When Joe Barnicle enlisted, the 13th was stationed in Egypt, preparing for deployment to the Western Front. Private Barnicle and the rest of the Reinforcements completed basic training in Kiama, NSW, and embarked from Sydney, bound for England, as part of Convoy No. 23 on the 22nd of August 1916. They sailed aboard HMAT A18 *Wiltshire* which carried 1,781 troops. The entire convoy of 12 ships carried a total of 11,343 troops on the voyage. The ship arrived in Plymouth seven weeks later.

More training and a hospital admission occurred over the next few months before Private Joe Barnicle finally joined the 13th Battalion on the 27th of January 1917 on the Somme in France. During the night of the 4-5 of February 1917, the 13th Battalion was ordered to launch an attack on Stormy Trench. For the next 24 hours, the battalion repelled counterattacks, engaged in brutal close-quarters combat, and endured relentless shelling. Approximately 230 soldiers from the battalion lost their lives in the fierce battle. It was during this action that Private Joe Barnicle was Killed in Action on the 4th of February 1917. He was 25 years six months old at the time of his death and had been a member of the AIF for 11 months and a member of the 13th Battalion for only eight days.

An Australian Red Cross Investigation into Private Barnicle's death records the statements of five witnesses. While the accounts have some inconsistencies, Private Joe Barnicle died at about 10:30 p.m. from either a sniper's bullet or shell fire near Gueudecourt. His death was notified in the 277th list of Australian Casualties released on the 8th of March 1917. Private Joseph Barnicle was laid to rest at Bancourt British Cemetery, France.

Joseph Augustine Barnicle's name is inscribed on The Australian War Memorial Roll of Honour (Panel 68), The City of Townsville WW1 Honour Roll, and The Townsville Cenotaph. Barnicle Street, Railway Estate in Townsville is named in his honour.

5985 Private Joseph Augustine Barnicle served in WW1 on the Western Front. He was awarded the British War Medal (no:45846) and the Victory Medal (no:45320). These medals, as well as a memorial scroll and a memorial plaque, were passed to his father.

The following tribute was published in the *Lithgow Mercury* (NSW) on the 16th of March 1917:

DEATH OF A SOLDIER

PRIVATE J. BARNICLE

Word has been received by Mr. A. C. Kelly, Club House Hotel, of the death at the front of Pte J. Barnicle, better known in Lithgow as "Joe" Kelso. It has not been stated what was the cause of death. The deceased soldier enlisted in Lithgow. Pte. Barnicle was well known and highly respected. He was a member of the Bluebell F.C. He was a carpenter by trade, but for some time was a letter carrier at the post-office. His parents reside in Queensland.

The following tribute was published in the *Townsville Daily Bulletin* (Qld) on the 4th of February 1918:

ROLL OF HONOR

BARNICLE. —In loving memory of our dear son and brother, Pte J. A. Barnicle, who was Killed in Action in France, on February 4th, 1917.

A loving son and brother, true and kind,
Loved by those he left behind,
Forget him, no, we never will.
As time goes on, we miss him still.
We often sit and think of him,
And think of how he died,
But, oh, it was so hard to think,
We could not say good-bye.
R.I.P.

(Inserted by his sorrowing parents, brothers, and sisters.)

2300 PRIVATE THOMAS JOSEPH DESMOND (1887 – 1917)

41st Infantry Battalion

Thomas Joseph Desmond was born on the 30th of March 1887 in Townsville, Queensland. He was the youngest child of Humphrey and Bridget Desmond (née O'Dea) of Eyre Street, North Ward. He had three brothers and one sister.

Thomas Desmond grew up in Townsville and attended the Convent High School. After his school years were completed, he became a clerk and then a customs officer. Humphrey Desmond died in December 1900 when Thomas Desmond was 13 years old. Bridget Desmond was left as a widow to raise the family although John Desmond, the oldest, was 23 years old.

On the 10th of July 1916, Thomas Desmond applied to enlist in the Australian Imperial Force (AIF) at Townsville. He declared that he had no previous military experience, was single, qualified as a clerk, and was 29 years and three months old. His enlistment medical examination noted that he was 5'6½" (169 cm) tall, weighed 8st 7lbs (54 kg), had a fresh complexion, grey eyes, and dark brown hair. He identified as Roman Catholic, and he nominated his mother as his next of kin.

Accepted into the AIF, Thomas Desmond was allocated service number 2300, assigned to the Infantry and posted to the 41st Battalion, 4th Reinforcements. The 41st Battalion had been formed in February 1916 at Bell's Paddock near the current Gallipoli Barracks, Enoggera, in Brisbane, Queensland, with men drawn from Queensland and northern New South Wales. It was a unit of the 11th Brigade, 3rd Division. The 41st Battalion was training in England when Thomas Desmond enlisted in July 1916.

For the next 14 weeks, Private Desmond and the rest of the reinforcements completed basic training in Australia before they embarked from Brisbane on the 21st of October 1916 bound for England. They travelled as part of Convoy No. 25 aboard HMAT A48 *Boonah* which carried 1,237 passengers. The entire convoy of 12 ships carried a total of 12,513 passengers and 5,600 horses on the voyage. Three months later, they marched into the 11th Training Battalion located at Durrington, Wiltshire, England.

It was January and mid-winter when Private Thomas Desmond arrived at the training battalion. Winter temperatures hovered around zero at this time of the year in Durrington. The cold may have affected Private Desmond because about three weeks after his arrival he developed pneumonia and was admitted to the Military Hospital at Fargo. Tragically, Private Thomas Desmond succumbed to his illness and died on the 7th of February 1917. He was 29 years and 10 months old at the time of his death. He had served in the AIF for nearly seven months. He was laid to rest in the Durrington Cemetery (Grave No. 187), Wiltshire, and his death was notified in the 274th Casualty list released on the 24th of February 1917.

Thomas Desmond's name is inscribed on The Australian War Memorial Roll of Honour (Panel 133), The 41st Battalion Roll of Honour, The City of Townsville WW1 Honour Roll, and The Townsville Cenotaph. Desmond Street, Railway Estate, Townsville, is named in his memory.

2300 Private Thomas Joseph Desmond was awarded the British War Medal (no:50292). A memorial scroll and a memorial plaque were passed to his mother in 1922.

The following Notice was published in the *Brisbane Courier*, on the 19th of February 1917:

PERSONAL NOTES

Advice has been received of the death of Private T. J. Desmond at Fargo Military Hospital, England on 7th instant, the cause of death being pneumonia. The late Private Desmond, who enlisted at Townsville in June, 1916 was a brother of Mrs A. T. Jackson, Auchenflower.

338 PRIVATE THOMAS LESLIE BARRYMAN (1889 – 1917)

41st Infantry Battalion

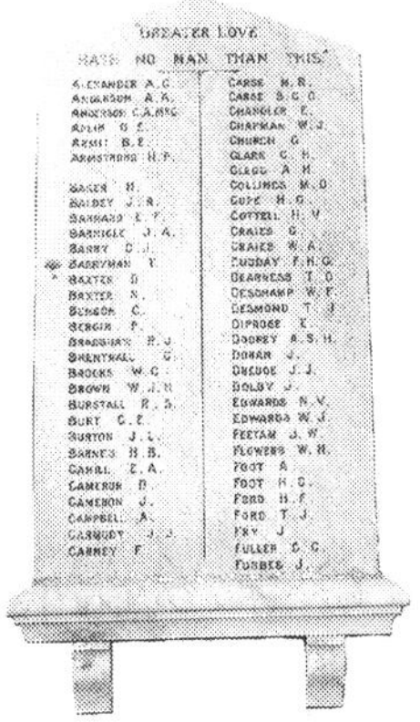

Thomas Leslie Barryman was born on the 10th of June 1889 in Sydney, NSW, and admitted to the care of the State in March 1891. He was born out of wedlock to Thomas Kelly and Elizabeth Barryman. After Thomas Barryman was put into the care of the State, nothing more is known of the birth parents. In his enlistment papers, he declared his foster mother, Mrs Grace Fanny Mills of Indooroopilly, as his next of kin. Mrs Mills cared for Thomas Barryman from May 1891 to June 1901. Thomas attended the Indooroopilly State School while he was in the care of Mrs Mills.

In 1901, when Thomas was 10 years old, he was placed in the care of Mr W Casey in Clifton, Queensland, for the next five years. After his school years were completed, Thomas became a farmer and a labourer.

On the 6th of March 1916, Thomas Barryman applied in Townsville to enlist in the Australian Imperial Force (AIF). He declared that he had no previous military experience, was 26 years nine months old, single, and working as a labourer. He also declared that he had been rejected as unfit for military service previously because of "hydrocile" (sic), which was probably a hydrocele, a collection of fluid in the scrotum.

Thomas Barryman's enlistment medical examination noted that he was 5'2½" (159 cm) tall, weighed 8st 3lbs (52kg), had a fresh complexion, blue eyes, and dark hair. There was no evidence of the previous condition, and he was considered fit for active service. He identified as Church of England and listed Mrs Mills, his foster mother, as his next of kin and gave her address as care of Wilton St Post Office, Indooroopilly.

Accepted into the AIF, he was allocated service number 338, assigned to the Infantry and posted to the newly raised 41st Battalion. The 41st was raised at Bell's Paddock Camp in Brisbane in February 1916 with recruits from Brisbane, northern Queensland, and the northern rivers district of New South Wales. It formed part of the 11th Brigade of the 3rd Australian Division.

Following basic training, Private Barryman with the rest of 41st Battalion moved by train to Sydney on 16th May 1916 to embark for England as part of Convoy No. 21. The battalion sailed aboard HMAT A64 *Demosthenes* on the 16th of May 1916. Their ship carried 1,610 troops, and the entire convoy of 18 vessels carried a total of 25,287 troops on the voyage. Their journey took them through Cape Town and the Cape Verde Islands before finally reaching Plymouth in July 1916.

The Battalion moved with the 3rd Division to training grounds on Salisbury Plain at Larkhill, site of the current Royal School of Artillery. In November 1916, Private Thomas Barryman sailed as part of the 41st Battalion for the Western Front from Portsmouth, England, to Le Havre, France, and moved towards the border area with Belgium near Armentières.

The Battalion entered the frontline for the first time on Christmas Eve, 1916, at the Mushroom Salient, just east of La Chapelle-d'Armentières. For most of the first half of 1917, the 41st Battalion rotated through the frontline around Armentieres in France and across the Lys River in Belgium.

On the 11th of February 1917, the 41st Battalion was in the line and was ordered to deploy a raiding party to attack enemy trenches. The raiding party was tasked to secure prisoners, inflict casualties on the enemy, capture weapons, and destroy enemy trenches. The raiding party of 3 Officers and 38 Other Ranks left their positions at 7:50 pm and laid up in no man's land while artillery pounded the enemy trenches. Despite the barrage, the wire was not effectively cut by the artillery, and the raid was repulsed by the Germans.

The retreating raiding party was illuminated by searchlights as they withdrew and were subjected to machine gun fire. 10 casualties were sustained in the raid. One officer was severely wounded, one soldier was killed, and eight soldiers were wounded. One of these was Private Thomas Barryman, who was wounded by a gunshot to his back and arm. He was evacuated to the 1st Canadian Casualty Clearing Station; however, he succumbed to his wounds two days later and Died of Wounds on the 13th of February 1917.

Private Thomas Barryman was 27 years eight months old, had served in the AIF for 11 months. He had been a member of the 41st Battalion for nine months at the time of his death. He was laid to rest at the Bailleul Communal Cemetery Extension (Plot III, Row A, Grave No. 55), France. Private Thomas Barryman's death was notified in the 276th list of Australian Casualties released on the 3rd of March 1917.

Thomas Barryman's name is inscribed on The Australian War Memorial Roll of Honour (Panel 133), The City of Townsville WW1 Honour Roll, and The Townsville Cenotaph. Barryman Street, Pimlico in Townsville, is named in his honour.

338 Private Thomas Leslie Barryman served in WW1 on the Western Front. He was awarded the British War Medal (no:39384) and the Victory Medal (no:39768). These medals, as well as a memorial scroll and a memorial plaque, were passed to his foster mother.

4720 PRIVATE JOHN WILLIAM FEETAM (1887 – 1917)

15th Infantry Battalion

John William Feetam (Jack) was born in July 1887 in Hull, Yorkshire, the third son of John and Hannah Feetam (née Welburn). Jack Feetam grew up in England and joined the merchant navy when he was about 19 years old. He completed several voyages to Australia and assumed residence before or during 1912. He moved to Townsville, worked as a labourer, and married Ellen Jane Oliver on the 12th of November 1912. They set up home in Bell St, South Townsville, and later moved to Old Rifle Range Rd, South Townsville.

On the 27th of September 1915, Jack Feetam applied at Townsville to enlist in the Australian Imperial Force (AIF). He declared that he was married, employed as a labourer, had no previous military experience and was 27 years and one month old. His medical examination noted that he was 5'6" (167 cm) tall, weighed 9st 9½ lbs (61 kg), had a fresh complexion, brown eyes, and dark brown hair. He identified as Church of England and nominated his wife, Ellen, as his next of kin and gave her address as Old Rifle Range Road, South Townsville.

Accepted into the AIF, Jack Feetam was allocated service number 4720, assigned to the Infantry and posted to the 15th Battalion, 15th Reinforcements. The 15th Battalion was originally raised as part of the all-volunteer AIF in September 1914. Drawing personnel from volunteers from Queensland and Tasmania, it formed part of the 4th Brigade, along with the 13th, 14th, and 16th Battalions.

The 15th Battalion was fighting on the Gallipoli Peninsula when Jack Feetam enlisted. After six months spent completing basic training, Private Jack Feetam and his fellow reinforcements joined Convoy No.18 and embarked from Brisbane on the 28th of March 1916, bound for Egypt. They sailed aboard

HMAT A73 *Commonwealth*, which carried 956 troops. The entire convoy of 13 ships carried a total of 10,826 passengers and 3,840 horses on the voyage.

Private Jack Feetam disembarked in Alexandria on the 6th of May 1916 and spent the next three months training in Egypt. On the 6th of August 1916, he embarked on the S.S. *Megantic* to proceed to England to join the British Expeditionary Force (BEF). Another month of training was undertaken at the 4th Training Battalion located at Codford in England.

From England, Private Jack Feetam and the rest of the reinforcements crossed to the 4th Australian Divisional Base Depot in Étaples, France. Upon reaching France, both reinforcements and "casuals" (soldiers returning from the hospital) would proceed to the Base Depots prior to their allocation to designated units. Although previously certified as fully trained in England, all draft cohorts underwent supplementary assessments in 1916, spanning at least 10 days of additional training. As well, a thorough medical evaluation was conducted.

Private Jack Feetam finally joined the 15th Battalion in the field on the 7th of October 1916. The 15th was deployed in the Ypres salient, around St Eloi and Boorlartbeek, where they spent the rest of 1916 before winter fell on the Western Front. It was the worst winter in Europe in 40 years, and the men suffered heavily from sickness and the cold as they rotated through the line, conducting defensive duties and labouring before moving to Gueudecourt.

During the evening of the 1st of February 1916, the 15th Battalion attacked Stormy Trench. The nearest part of the trench was only 100 yd (91m) from the Australian front line, but on the right of the attack, the Australians were held up by uncut wire. The left-hand section reached the trench and bombed down it to take the rest. German counterattacks began at 0155 hr (1:55 a.m.) on the 2nd of February, but during an attack at 0400 hr (4:00 a.m.), the Australian SOS signals were not seen for 50 minutes. By the time the artillery began to fire, the 15th Battalion had been repulsed with 144 casualties.

It was during these operations around Gueudecourt that Private John William Feetam was critically wounded with multiple gunshot wounds on the 3rd of February 1917. He was evacuated via the Field Ambulance and the Casualty

Clearing Station, finally ending up at the 12th General Hospital in Rouen. Unfortunately, his wounds proved fatal, and Private John William Feetam Died of Wounds on the 15th of February 1917. He was 28 years and five months old at the time of his death. He had been a member of the AIF for 16 months, and a member of the 15th Battalion for 13 weeks.

Private John William Feetam was laid to rest at St. Sever Cemetery Extension (Block O, Plot V, Row E, Grave No. 9), Rouen, France. His death was notified in the 27th list of Australian Casualties released on the 6th of March 1917.

John William Feetam's name is inscribed on The Australian War Memorial Roll of Honour (Panel 75), The City of Townsville WW1 Honour Roll, and The Townsville Cenotaph. Feetam Street, Stuart in Townsville, was named in his memory however, the street name was cancelled on 18th June 1975.

4785 Private John William Feetam was awarded the British War Medal (no:34392) and the Victory Medal (no:34149). A memorial plaque was passed to his widow, Jane, in 1923.

The following notice was published in the *Townsville Daily Bulletin* (Qld.) on the 10th March1917:

FEETAM. —Died of Wounds in France on 15th February 1917, John William Feetam, aged 28 years, third eldest son of Mr and Mrs J. W. Feetam, Hull, Yorkshire, England, and the husband of Mrs Jane Feetam, South Townsville. R.I.P.

The following notice was published in the *Townsville Daily Bulletin* (Qld.) on the 15th of February 1921:

FEETAM. —In loving memory of my dear husband, PRIVATE John William Feetam, died of wounds, in France, 15th February 1917, 15th Battalion.

> Far and often my thoughts do wander,
> To the grave so far away,
> Where they laid my darling Jack,
> Just four years ago today.
> No matter how I pray dear Jack,
> No matter how I call,

There is nothing left to answer,
But your picture on the wall.

Inserted by his loving wife, Jane Feetam.

2425 PRIVATE HENRY FRANCIS SAMUEL MARTIN (1888 – 1917)

4th Australian Pioneer Battalion

Henry Francis Samuel Martin (aka Marten) was born on the 22nd of June 1888 in Normanton, Queensland to Henry and Jane Martin (née Rees). Henry Martin's mother died in 1915, and his father died in 1917. Henry Martin was the fifth of six children. He had two brothers and three sisters, all of whom died before 1897 except for his older sister, Alice.

Henry Martin attended State School as a child and qualified as an Ironmonger upon completing his school years. He lived with his sister, Alice Hayes, and her husband in a property on Sturt Street, Townsville, and worked as an assistant in the hardware department of Burns Philp and Co. He also joined the Salvation Army Band and was a bandsman for about 10 years.

On the 18th of January 1916, Henry Martin applied at Townsville to enlist in the Australian Imperial Force (AIF). He declared that he was single, qualified as an Ironmonger, had no previous military experience and was 27 years and seven months old. His medical examination noted that he was 5'6 ¾" (170 cm) tall, weighed 8st 5lbs (53kg), had a ruddy complexion, blue eyes, and golden hair. He identified as Salvation Army and nominated his sister, Alice Hayes, as his next of kin giving her address as Sturt St, Townsville.

Accepted into the AIF, Henry Martin was allocated service number 2425, assigned to the Engineers and later posted to the 4th Australian Pioneer Battalion, 4th Reinforcements. The 4th Australian Pioneer Battalion was raised on the 16th of March 1916 at Tel-el-Kebir in Egypt.

Pioneer Battalions were specialised infantry skilled in light combat engineering, often at the frontline. Their goal was to enhance allied defence and mobility while hindering the enemy. They built defensive structures, command posts, shelters, and set up barbed wire. They sometimes breached enemy defences with tools like the Bangalore Torpedo. Versatile in construction, road maintenance, and infrastructure, they stayed combat-ready when needed.

After completing recruit training in Australia, Private Henry Martin and his fellow reinforcements joined Convoy No. 23 and embarked from Brisbane, bound for England. They sailed aboard HMAT A50 *Itonus*, on the 8th of August 1916. Their ship carried 780 troops. The entire convoy, of 12 ships, carried a total of 11,343 passengers, including five prisoners from the German cruiser SMS *Emden*, which had been sunk by HMAS *Sydney* in November 1914.

Ten weeks later, Private Martin and the other reinforcements disembarked at Plymouth, England. They underwent two months of further training at Hurdcott Camp near Fovant in Wiltshire before being deployed to France on the 31st of December 1916. Once in France, they completed more training at Étaples.

It was not until the 7th of February 1917 that Private Henry Martin was taken on the strength of the 4th Pioneer Battalion. At that time, the battalion was located on the Somme near Flers and was heavily engaged in the construction of a rail line and other works. On the 16th of February, nine days after Private Martin's arrival, the 4th Pioneers were working around Longueval near Albert.

Private Henry Martin engaged in the construction of a rail line when he was Killed in Action by shellfire, along with two other soldiers who were working with him. In a statement made to the Australian Red Cross in 1917, Private A. Bridlay stated; *"I saw Martin killed in France on February 16th. I was about 50 yards away. We were working at a place called 'Siege Siding' on the Somme front. Death was instantaneous, the shell killing two other men at the same time. It occurred at 11 a.m. I helped bury him at a small cemetery close by. His grave is marked with a nice cross and his name inscribed."*

Private Henry Martin was 28 years and eight months old at the time of his death. He had been a member of the AIF for one year. He was laid to rest at Thistle Dump Cemetery (Row E, Grave No. 33), High Wood, Longueval, France.

Henry Francis Samuel Martin's name is inscribed on The Australian War Memorial Roll of Honour (Panel 174), The Normanton Carpentaria Shire Roll of Honour, The City of Townsville WW1 Honour Roll, and The Townsville Cenotaph.

2425 Private Henry Francis Samuel Martin was awarded the British War Medal (no: 44976) and the Victory Medal (no: 44463). A memorial scroll and a memorial plaque were passed to his sister in 1922.

The following tribute was published in The *Townsville Daily Bulletin* (Qld.) on the 16th of February 1918:

MARTIN. — In loving memory of my dear brother, Private Henry Francis Samuel Martin, who was killed in France on the 16th of February 1917.

> "Somewhere in France" tonight he lies,
> Paying the bitter price.
> "Somewhere in France" tonight he sleeps,
> With the sacrifice.
> There, 'mid the mist and murk of war,
> On the blood-red fields of pain,
> Can he hear the stirring banana palms,
> With their patter like drifting rain?
> Does he dream again how the daylight dips
> Over the blue hills' crest,
> And catch the note of a curlew's call
> To the moon with her silver breast?
> The tide is high on the outer reef,
> Where the breakers surge and swell,
> Have the live shells dulled their memory,
> Or, perchance, he sleeps too well?

Inserted by his loving sister and brother-in-law, Mrs. A. M. J. and Mr. F. J. Hayes.

4429 PRIVATE HAROLD GREENSLADE (1886 – 1917)

17th Infantry Battalion

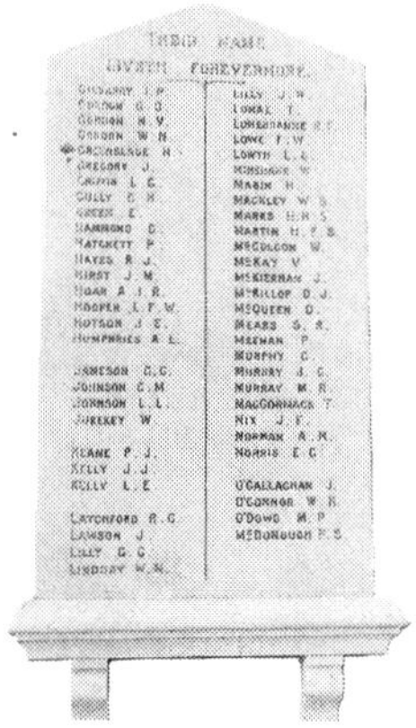

Harold Greenslade was born on the 10th of April 1886 in Townsville, Queensland. He was the sixth of seven children born to George and Johanna Greenslade (née Imbergh). He had three brothers and four sisters. The family lived in a property on Macrossan St, West End, Townsville, and Harold Greenslade attended Mundingburra State School as a child. After his school years were finished, he trained as a boot and harness repairer.

Harold Greenslade spent time in New Zealand and served with the No.1 Garrison Artillery Auckland. He also served with the Fijian volunteer forces for two years until he took his discharge on the 2nd of February 1916. Returning to Australia, Harold was living in Sydney on the 8th of March 1916 when he applied to enlist in the Australian Imperial Force (AIF). He gave his postal address as 199 Hargreaves Street, Paddington. He declared that he was single, qualified as a boot and harness repairer, had some previous military service, and was 29 years and 11 months old. His enlistment medical examination noted that he was 5'4½" (164 cm) tall, weighed 134 lbs (61 kg), had a sallow complexion, blue eyes, and brown hair. He identified as Church of England. Harold Greenslade's father had died in 1908, and he nominated his mother as his next of kin and gave her address as Macrossan Street, West End, Townsville.

Accepted into the AIF, Harold Greenslade was allocated service number 4429, assigned to the Infantry and posted to the 17th Battalion, 11th Reinforcements. The 17th Battalion was raised at Liverpool in New South Wales in March 1915 as part of the 5th Brigade. The 17th Battalion was on its way to France from Egypt when Harold Greenslade enlisted. Five weeks of basic training followed enlistment, before Private Greenslade and the rest of the 11th Reinforcements embarked from Sydney bound for Egypt on the 9th of April 1916. They sailed as part of Convoy No. 19 aboard HMAT A71 *Nestor.*

Their ship carried 1,554 troops, and the entire convoy of 10 ships carried a total of 8,099 passengers and 1,168 horses on the voyage.

After a short stay in Alexandria, Private Greenslade and the other reinforcements re-embarked for England, arriving in Plymouth on the 4th of June 1916. From there, they marched into the 5th Training Battalion located at Fovant, near Salisbury, in Wiltshire. Three months of training followed until they crossed to France in early September 1916. Another week of training at Étaples followed before Private Harold Greenslade joined the 17th Battalion in the field on the 16th of September 1916. At that time, the 17th Battalion was quartered in a quieter sector of the front in Belgium after having fought in the battle of Pozières in July and August.

Not long after joining the 17th Battalion, Private Greenslade was treated for scabies and then trench foot in November 1916. His feet were severe enough for him to be evacuated back to England at the end of November, where he was admitted to the 1st Southern General Hospital. He remained in the hospital until the end of January 1917 when he was discharged from the hospital and granted furlough. Two weeks later, in mid-February, Private Greenslade returned from furlough and marched into Perham Downs, located north of Salisbury, to prepare for his return to duty. While he was in Perham Downs, he became ill in March 1917 and was admitted to the Isolation Hospital, Salisbury.

Private Harold Greenslade did not recover after admission, and he Died in Service on the 20th of March 1917 from cerebrospinal meningitis. He was laid to rest three days after his death at the Stratford-Sub-Castle (St. Lawrence) Churchyard (Grave No. 113), Wiltshire, England. He was 30 years and 11 months old at the time of his death. He had been a member of the AIF for one year and two weeks. Private Harold Greenslade's death was notified in the 285th list of Australian Casualties released on the 13th of April 1917.

Harold Greenslade's name is inscribed on The Australian War Memorial Roll of Honour (Panel 82), The City of Townsville WWI Honour Roll, and The Townsville Cenotaph. Greenslade Street, West End, Townsville, is named in his honour.

4429 Private Harold Greenslade served in WW1 on the Western Front. He was awarded the British War Medal (no:36340) and the Victory Medal (no:36063). A memorial scroll and a memorial plaque were passed to his mother in 1922.

The following notice was published in the *Townsville Daily Bulletin* on 30th March 1917:

GREENSLADE. —Died of meningitis, at Salisbury Plain, England, March 20th, 1917, Private Harold Greenslade, aged 31 years, second youngest son of Mrs Greenslade, Macrossan-street, West End, and brother to Mrs W. Oliver, Morris-street, West End. The best that Australia could offer now sleeps in a soldier's grave.

5669 PRIVATE WILLIAM SMITH (1890 – 1917)

26th Infantry Battalion

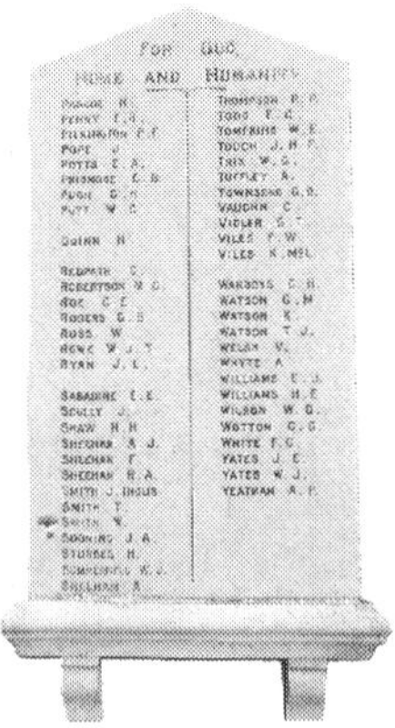

William Smith was born on the 28th of March 1890 in Townsville, Queensland. He was the fourth of six children and the second son born to Thomas and Elizabeth Smith (née Hogan). He had three brothers and two sisters. Thomas Smith worked as a plumber, and the family lived at Morris St and then Flinders Street West in Townsville.

William Smith grew up in Townsville and worked as a stockman once his school years were finished. In 1915, William's older brother, Thomas, had enlisted in the AIF and was Killed in Action on the Gallipoli Peninsula in October 1915, just nine months after he enlisted. Despite the loss of his brother (or perhaps because of it), William Smith decided to join the AIF as well.

He was in Cairns on the 29th of April 1916 when he applied to enlist in the Australian Imperial Force (AIF). He declared that he was single, 25 years and 11 months old, and had no previous military experience. His enlistment medical examination noted that he was 5'7" (170 cm) tall, weighed 130 lbs (59 kg), had a fair complexion, blue eyes, and dark brown hair. William identified as Roman Catholic and listed his father as his next of kin, giving his address as West End, Townsville.

Accepted into the AIF, William Smith was allocated service number 5669, assigned to the Infantry, and posted to the 26th Battalion, 15th Reinforcements. The 26th Battalion was raised at Enoggera, Queensland, in April 1915 from recruits enlisted in Queensland and Tasmania and formed part of the 7th Brigade. The period from May to the end of August 1916 was spent completing basic training.

On the 7th of September 1916, Private William Smith and the other reinforcements embarked from Brisbane with Convoy No. 24 bound for England. They sailed aboard HMAT A46 *Clan Macgillivray*, which carried 1,084 troops. The entire convoy of 13 ships transported 12,600 troops on the voyage. Eight weeks later, in November 1916, they disembarked at Plymouth in England. From there, Private William Smith and the reinforcements were transported to the 7th Training Battalion at Rollestone on the Salisbury Plain for further training. The next move was to France, and Private William Smith embarked from Folkestone, England, on the 13th of December 1916 and marched into the 2nd Division Base Depot at Étaples on the 14th of December.

Three days later, Private Smith was admitted to the hospital for treatment of benign growths and was released back to the depot in early January 1917. He was taken on the strength of the 26th Battalion on the 5th of February 1917.

The 26th was deployed on the Somme at Contalmaison, about six kilometres northeast of Albert. The battalion joined the follow-up of the German withdrawal to the Hindenburg Line and attacked at Warlencourt on the 1st-2nd of March 1917.

On the 26th of March 1917, the 26th Battalion attacked and took the village of Lagincourt. It was during this battle that Private William Smith was Killed in Action. He was 26 years and 10 months old at the time of his death. He had served with the AIF for 11 months. His body was not recovered, and he has no known grave.

Elizabeth Smith received a pension for her son's sacrifice of 14 shillings per fortnight. This was approximately one fifth of the fortnightly wage paid in the AIF in 1917.

William Smith's name is inscribed on the Australian War Memorial Roll of Honour (Panel 109), the Villers-Bretonneux Memorial (Australian National Memorial - France), the City of Townsville WW1 Honour Roll, and the Townsville Cenotaph.

5669 Private William Henry Smith was awarded the British War Medal (no. 47231) and the Victory Medal (no. 46680). A memorial scroll and a memorial plaque were passed to his father in 1921 and 1922, respectively.

1200 COMPANY QUARTER MASTER SERGEANT DUNCAN BAXTER (1872 - 1917)

42nd Infantry Battalion

Duncan Baxter was born in 1872 in Glasgow, Scotland. He was the youngest of five children born to Neil and Anne Baxter (née Brown). Duncan Baxter and his brother Neil came to Townsville from Scotland in 1903, before bringing out their families a year later. Together, they ran the Baxter Furniture Warehouse until Neil Baxter's death in 1908, when Duncan became the sole proprietor.

Duncan Baxter married his wife Annie (née McDonald) in June 1854 in Glasgow. Together, they had seven children. Duncan and Annie Baxter's oldest son, Neil Baxter, joined the Australian Imperial Force (AIF) in November 1914 and was Killed in Action with the 15th Battalion at Gallipoli on the 26th of April 1915, at the age of 21 years. He has no known grave.

After Neil Baxter's death, there was some confusion as to whether he had been killed, or was missing, or was captured. Neil's letters were still being returned to his parents marked 'Missing'. Duncan and Annie Baxter hoped that 'Missing' meant Neil was in hospital somewhere recovering or had been taken prisoner of war. It was not until April 1916 that the Army 'officially' declared Corporal Neil Baxter to have been killed in action.

Duncan Baxter enlisted (with his second eldest son Angus Baxter) in the AIF at Townsville on the 17th of January 1916. Duncan Baxter declared that he was 44 years and four months old and had previously served in the Lanark Royal Engineers from 1890 to 1903 before leaving Scotland for Australia. He was 47 years and four months old. Duncan Baxter's medical examination noted that he was 5'7" (170 cm) tall and weighed 14st (89kg). He had a fair complexion, blue eyes, and grey hair. He identified as Presbyterian and listed his wife as his next of kin, giving her address as Paxton St, Townsville.

Accepted into the AIF, Duncan Baxter was allocated service number 1200, assigned to the Infantry and posted to the 42nd Battalion. His son Angus was also posted to the newly raised 42nd Battalion. Duncan Baxter's previous service placed him in good stead, and he soon found himself promoted to Corporal and then Company Quarter Master Sergeant (CQMS) on the 5th of June 1916. This was the same day that he and the rest of his Unit joined Convoy 21 at Sydney bound for England. They sailed aboard HMAT A30 *Borda* which carried 1,444 troops. The entire convoy of 18 ships carried a total 25,287 troops on the voyage. The contingent disembarked at Southampton at the end of June 1916 and completed further training in England.

The 42nd Battalion sailed for the Western Front from Southampton on the 26th of November 1916. Once landed, the soldiers of the 42nd occupied positions on the frontline in December and endured the winter of 1916 – 17, as well as completing further training and labouring tasks in the rear areas when not in the line.

It was during this phase of the operations that CQMS Duncan Baxter was severely wounded in both feet during an action that occurred on the 17th of March 1917. His son Angus Baxter had been with him in the same dugout when they had been bombarded by high explosive shells.

CQMS Duncan Baxter was evacuated to a clearing station, then hospital, but did not survive the severity of his wounds and he Died of Wounds on the 31st of March 1917, aged 48 years. He had served with the AIF for 14 months. He was laid to rest at the Bologne Eastern Cemetery, France and his death was notified in the 287th list of Australian Casualties released on the 18th of April 1917.

CQMS Duncan Baxter's name is inscribed on The Australian War Memorial Roll of Honour (Panel 135), The City of Townsville WW1 Honour Roll, and The Townsville Cenotaph. Baxter Street, West End, in Townsville, is named in the memory of both Duncan and Neil Baxter.

Company Quarter Master Sergeant Duncan Baxter was awarded the British War Medal (no: 41590) and the Victory Medal (no: 41161). These medals, as well as a memorial scroll and a memorial plaque, were passed to his wife.

The following was published in the *Townsville Daily Bulletin* on 20th June 1917:

"Mrs Duncan Baxter, 'Easdale,' Paxton Street, has received several letters from those who were with her late husband, Company Quartermaster-Sergeant Baxter, before and after he was wounded, and at the time of his death, also one from Mr Baxter himself, describing how he received the wounds which cost him his life. He writes from the hospital two days after being wounded and says:

'Long ere you receive this you will know that one of Fritz's shells got me in both feet. It sliced off a big piece of the right one and the left was full of shrapnel. Thank God, I will not lose my feet, but they will take a long time to heal, and the doctor says I am finished with the firing line. I got hit about 10 p.m. on St. Patrick's Day. It was exactly 14 years since Neil and I left to go to Townsville, and I was thinking about it all day. Angus had charge of the support and had been in my dugout and at the sound of the first shell, he was out with another fellow, very like Neil. I was trying to get my gum boots on, but my feet were a bit swollen, hence the reason for my not being out sooner. But thinking of it now possibly I should have been killed going out of the door. I started to crawl out when I heard Angus calling "Are you there Dad?" I told him I was all right but had no boots on. I then put down my hands and found blood. The socks were blown off. Angus bandaged me up. He had

a miraculous escape, as the place where he had been sitting was struck and three men about 20 yards away were wounded. I did not lose my head, kept as cool as a cucumber, and was complimented by the doctors at the different stations for my efforts to keep the other fellows cheerful. Fritz had had a go at me on the 15th about the same time. They were firing gas shells, and we were wearing our respirators. I was leading a party of men when suddenly an explosive shell burst four yards from me. In the morning, we found that it had dug 3 feet into the ground. The boys got the base of it for me: it was 9 Inches in diameter. I was keeping it for my glory box, but everything has gone now."

The following tributes were published in the *Townsville Daily Bulletin* on 26th April 1917 and 29th March 1918, respectively:

BAXTER. —Died of Wounds, 31st March 1917, at 12th Stationary Hospital, England, Company-Quarter-Master-Sergeant Duncan Baxter, aged 48 years. Dearly beloved husband and father of Annie Baxter and Family, "Easdale," Paxton Street. Through thorny way leads to a joyful end.

BAXTER—In loving memory of our dear husband and father, the late Q M Sergeant Duncan Baxter, who was gassed two days before being wounded on 17th March, at Messines, from which he died on 31st March 1917, in the Boulogne Military Hospital. The late Q M S. was laid to rest in the Boulogne Cemetery, with military honours.

Thy will be done.

1019 2nd LIEUTENANT CHARLES CHALMERS JAMESON (1895 – 1917)

42nd Infantry Battalion

Charles Chalmers Jameson was born on the 6th of December 1895 in Kew, Victoria, to Charles and Maud Jameson (née Macarthur). He was the oldest of three surviving boys. Another child, Joan, passed away in October 1897 at five months of age. Charles Jameson attended Toowoomba and Townsville Grammar Schools in Queensland. After completing school, he followed in the footsteps of his father, Judge Charles Jameson, and started a career as a Judge's associate.

On the 16th of September 1915, Charles Jameson applied at Brisbane to enlist in the Australian Imperial Force (AIF). He declared that he was 19 years and nine months old, single, and employed as a Judge's Associate. He also declared that he had previous military service with the 7th Infantry, served as a Cadet at School, and was a member of the Rifle Club. His enlistment medical noted that he was 5'7" (170 cm) tall, weighed 9st 8lbs (61 kg), had a fresh complexion, brown eyes, and fair hair. He identified as Church of England and nominated his father, Judge Charles Jameson of Brisbane, as his next of kin.

Accepted into the AIF, Charles Jameson was allocated service number 1019, assigned to the Infantry, and posted to "B" Coy of the 42nd Battalion. He spent the time between enlistment and June 1916 completing recruit and basic training in Australia.

The 42nd Battalion joined Convoy 21 at Sydney bound for England on the 5th of June 1916. They sailed aboard HMAT A30 *Borda*, which carried 1,444 troops. The entire convoy of eighteen ships carried a total of 25,287 troops on the voyage. They disembarked at Southampton 48 days later at the end of June 1916.

Training in England followed, and three weeks after his arrival, Charles Jameson was promoted to Warrant Officer Class 2 and appointed as the

Acting Company Sergeant Major. This was not his last promotion, for three weeks later, on the 6th of September 1916, he was commissioned as a Second Lieutenant (2Lt).

After training was completed in England, the 42nd deployed to France on the 25th of November 1916. The Battalion entered the frontline for the first time on the 23rd of December. The winter of 1916-17 was horrendous, and the 42nd spent much of it in the front line, the remainder being spent alternating between training and labouring in the rear areas. 2Lt Charles Jameson contracted mumps and was admitted to hospital in January 1917. Once recovered, he re-joined the 42nd after about three weeks.

In April 1917, the 42nd Battalion was fighting in Belgium in the vicinity of Ploegsteert. It was during this operation that 2Lt Charles was Killed in Action on the 4th of April 1917. A Red Cross Investigation into his death provided the following statement by Private P. Robb:

"I was in a supporting patrol to a bombing party under Lieut. Jameson on the night of April 4th, 1917. They got up to Fritz's wire and two bombs were thrown over by our patrol into their trenches when a revolver shot from the Germans hit him in the forehead and he was carried back by a man from the fighting patrol and died in our trenches. I heard the revolver shot that killed him and saw him brought back. He was my own Platoon Officer and one of the very best."

2Lt Charles Jameson was laid to rest at Berks Cemetery Extension (Row E, Grave No. 24), Ploegsteert, Belgium. He was 21 years and four months old at the time of his death and had been a member of the AIF for 18 months. His death was notified in the 289th list of Australian Casualties released on the 24th of April 1917.

Charles Chalmers Jameson served on the Western Front in WW1. His name is inscribed on The Australian War Memorial Roll of Honour (Panel 135), The City of Townsville WW1 Honour Roll, The Townsville Cenotaph, The 42nd Infantry Battalion AIF Roll of Honour, St Lucia Emmanuel College Dining Room Memorial Plaque, Toowoomba Grammar School WW1 Honour Board, Toowoomba War Memorial (Mothers' Memorial), Townsville Grammar School Roll of Honour, and the University of Queensland WW1 Roll of Honour. Jameson Street, Hermit Park, Townsville, was named in his honour.

2nd Lieutenant Charles Chalmers Jameson was awarded the British War Medal (no: 41685) and the Victory Medal (no: 41253). These medals were passed to his father, as well as a memorial plaque and a memorial scroll, which were also passed to him in 1922.

The following tribute was published in *The Brisbane Courier* 28th April 1917:

JAMESON. — Killed in action, on the 4th inst., in France, 2nd Lieutenant Charles Chalmers Jameson, eldest son of Judge and Mrs. Jameson, Brisbane, age 21 years.

Son of Judge Charles and Maude JAMESON, District Court, Brisbane.

The following tribute was published in *The Telegraph* (Brisbane) on 4th April 1918:

JAMESON. — In loving memory of Lieutenant Charles Chalmers Jameson, killed in action, France, 4th April 1917.

Inserted by D. E. and F. North.

2428 PRIVATE VICTOR RONALD INGHAM McKAY (1896 – 1917)

49th Infantry Battalion

Victor Ronald Ingham McKay was born on the 5th of April 1896 in Ingham, Queensland, to William and Augusta (née Panke/Ponke). He was the sixth of eight children, and he had three brothers and four sisters. Victor McKay's mother was born in Germany, and his father was born in Scotland.

In 1905, the family lived at the Police Station in Ross Street, Townsville, where William McKay was the Acting Sergeant of Police. Victor McKay attended Townsville South State School and later became a qualified gas fitter. He was also a member of the Garrison Artillery, and a newspaper report states that he served on Thursday Island before enlisting in the AIF.

After spending nine months on Thursday Island, Victor McKay attempted to enlist in the Australian Imperial Force (AIF) in Townsville but was rejected because he was a member of the Garrison Artillery. Undeterred, he paid for his own fare to Brisbane, where he successfully enlisted in the AIF on the 27th of June 1915. He claimed to be 21 years old, although he was two years younger. He stated that he worked as a labourer and was single. Surprisingly, despite his previous rejection, Victor McKay mentioned his previous military experience with the Garrison Artillery in Townsville. His medical assessment noted that he was 5'10½" (179 cm) tall, weighed 151 lbs (68 kg), with a fair complexion, blue eyes, and brown hair. He identified as Presbyterian and listed his father, residing at *"Kirrie"* Oxley St, Townsville, as his next of kin.

Accepted into the AIF, Victor McKay was allocated service number 2428, assigned to the Infantry and posted to the 9th Battalion, 7th Reinforcements. Following seven weeks of basic training, Private Victor McKay and the other reinforcements embarked from Sydney on the 20th of August 1915, bound for Lemnos. They sailed as part of Convoy No. 9 aboard HMAT A9 *Shropshire*, which carried 820 troops on the voyage. The entire convoy of eight ships transported 5,732 troops and 320 horses. The *Shropshire* docked in Lemnos on the 18th of November 1915, and Private McKay was taken on strength of the 9th Battalion.

The 9th Battalion was evacuated from Gallipoli on the 19th of December 1915. Although there is a newspaper report stating that Private Victor McKay served on Gallipoli and arrived on the same day his brother was evacuated, his service records do not record movement to the Gallipoli Peninsula.

In January 1916, Private Victor McKay was transferred from Lemnos to Alexandria in Egypt, where he joined the 49th Battalion. This battalion was raised on the 27th of February 1916 as part of the "doubling" of the AIF. About half of the soldiers who joined the battalion were veterans of the Gallipoli campaign from the 9th Battalion, while the rest were fresh reinforcements from Australia. Reflecting the composition of the 9th Battalion, the 49th Battalion primarily consisted of men from Queensland. It was part of the 13th Brigade within the 4th Australian Division.

The battalion completed three-months of training in Egypt before embarking for France on the 5th of June 1916 to reinforce the British Expeditionary Force. Arriving on the 12th of June 1916, the 49th Battalion deployed to positions in the trenches of the Western Front by the 21st of June. Their first major encounter with the enemy occurred during the battle of Mouquet Farm in August, resulting in significant casualties, particularly during the assault launched on the 3rd of September 1916. Subsequently, the battalion undertook a variety of frontline duties while also participating in training and labouring tasks behind the front lines for the remainder of the year. This routine persisted throughout the harsh winter of 1916-17.

In October 1916, Private McKay suffered a high explosive wound to the back of the neck and was evacuated. He soon returned to duty, but a few weeks later was evacuated again with a septic foot, a more serious condition. He was away for six weeks. After recovering, he had three weeks' leave in England, returning to duty in February 1917.

Private McKay was with the 49th Battalion during the advance after the German retreat to the Hindenburg Line, supporting the 13th Brigade's attack at Noreuil on 2 April. During this attack, Private McKay was Killed in Action on the 6th of April 1917 *"while rescuing wounded under heavy machine gun fire."* His body was never recovered, and he has no known grave. He had turned 21 the day before his death and had served in the AIF for one year and nine months. Private Victor McKay's death was notified in the 294th Australian casualty list, released on the 9th of May 1917.

Victor McKay's name is inscribed on The Australian War Memorial Roll of Honour (Panel 148), The Australian National Memorial, Villers-Bretonneux, France, The City of Townsville WW1 Honour Roll, and The Townsville Cenotaph. McKay Street, Hermit Park, Townsville is named in his memory.

2428 Private Victor Ronald Ingham McKay was awarded the 1914/15 Star (no: 2707), the British War Medal (no: 17588), and the Victory Medal (no: 17518). A memorial scroll and a memorial plaque were passed to his father in 1921 and 1922, respectively. Victor McKay's mother was granted a war pension of two pounds (£2/-/-) per fortnight for the loss of her son. This was about half of the fortnightly wage paid in the AIF in 1917.

The following notice was published in the *Townsville Daily Bulletin* on the 2nd of May 1917:

The Great War.

On Tuesday, Sergeant McKay, of the Police Inspector's office, received the sad news of the death of his son, Victor Ronald McKay, killed on service in France on 6th April. The deceased soldier was one of four brothers who are on service, one brother being in France, one in Mesopotamia, and the third at Thursday Island with the A.G.A. Victor Ronald McKay, who was 21 years of age, was the third son of the family, and at the time or his death was attached to the 49th, late 9th Battalion, of the Australian Forces. He had been on service since the commencement of the war, as he was a member of the A.G.A., and was called up to Thursday Island with the contingent which left Townsville on the *Kanowna* immediately after the declaration of war. After nine months service on the island, he came to Townsville and tried for foreign service, but was refused permission as a member of the A.G.A. He then paid his passage to Brisbane, and volunteered there, forgetting to mention his connection with the artillery arm. He joined the 7th Reinforcements of the 9th Battalion and left on foreign service a fortnight later, and at the time of his death had seen about two years' service abroad. He was at Gallipoli (landing there on the same day as his brother left wounded), and remained until the evacuation, going then to France. There he was twice wounded, but returned to duty, and on two occasions received the high commendation or Major-General Cox for his gallant conduct at Pozières and Mouquet Farm. In civil life, prior to the war, the young soldier was an employee of the Townsville Gas Co., where he was learning his business.

The following notice was published in *Townsville Daily Bulletin* on 7th June 1917:

Family Notices

McKay. — Victor Ronald Ingham McKay, killed in action, France, 6th April 1917, aged 21 years and 1 day.

"Greater love hath no man than to lay down his life for his country."

The following notice was published in *Townsville Daily Bulletin* on 6th April 1918:

McKay. —In loving memory of our dear son and brother, Victor Ronald Ingham McKay, who was killed in France on 6th April 1917, while rescuing wounded under heavy machine gun fire.

> But few that have heard their death knell roll
> From the cannon's lips where they faced the foe.
> Have fallen as stout and steady of soul
> As that dead man gone where we all must go.

5187 PRIVATE CORNELIUS EDWIN ROE (1896 – 1917)

49th Infantry Battalion

Cornelius (Con) Roe was born in 1897 at Miriam Vale in Queensland. He was one of four children of Charles and Margaret (Maggie) Coe (née Little). Maggie was one of the Gooreng Gooreng people. Charles Roe was a grazier, and he died in 1904 when Con Roe was seven years old. Maggie Roe was left to raise the four children.

Young Con Roe attended Miriam Vale School. He later worked as a labourer and moved to Townsville to live with his sister Ethel Anderson and her husband Walter at Oonoomba Station.

On the 10th of January 1916, Con Roe applied in Townsville to enlist in the Australian Imperial Force (AIF). He stated that he was 19 years old, and since he was under 21, his sister Ethel signed the consent form, noting that *"Father dead, Mother absent."* Con Roe declared he was single and a labourer.

His medical assessment noted that he was 5'4¾" (164 cm) tall, weighed 8st. (51 kg), had a dark complexion, grey eyes, and brown hair. He identified as Church of England. Initially, he listed his mother as his next of kin, but this was later changed to his sister Ethel, who was also designated as his legatee.

Accepted into the AIF, Con Roe was allocated service number 5187, assigned to the Infantry and posted to the 9th Battalion/16th Reinforcements. After completing 12 weeks basic training, Private Con Roe and the other reinforcements joined Convoy No. 19 on the 31st of March 1916. They sailed aboard HMAT A16 *Star of Victoria* (later renamed *Port Melbourne*), from Sydney bound for Egypt. Their ship carried 1,550 troops, and the entire convoy of 10 ships carried a total of 8,075 troops and 1,168 horses on the voyage.

The reinforcements ship disembarked at Port Said on the 5th of May 1916. One month later, they embarked at Alexandria in Egypt bound for France, where they disembarked at Marseilles on the 14th of June 1916. Prior to arrival in France, Private Con Roe was transferred to the newly formed 49th Battalion. Once in France, it was off to the 4th Australian Division Base Depot at Étaples for further training.

Private Roe joined the 49th Battalion in the field, near Pozières, on the 13th of August 1916. He participated in the battalion's first major battle at Mouquet Farm in August, suffering heavy losses, especially during the assault on the 3rd of September 1916. During this assault, Private Roe was severely wounded in the left thigh and, after initial treatment in France, was evacuated to England. He was admitted to the 5th Southern General Hospital in Portsmouth on the 7th of September 1916, where he remained until the 16th of December 1916.

Private Roe was granted leave in England before returning to France on the 16th of January 1917 and rejoined the 49th Battalion in late January 1917. He fell ill in February 1917 and spent five weeks in a hospital in France, rejoining the 49th on the 6th of March 1917.

The 49th Battalion took part in the advance following the German retreat to the Hindenburg Line, supporting the 13th Brigade's attack at Noreuil. Initially placed in brigade reserve, the battalion launched an attack that captured a railway cutting on the Cambrai–Arras line. During this attack, Private

Con Roe was Killed in Action on the 6th of April 1917. His body was never recovered, and he has no known grave.

At the time of his death, Private Con Roe was 20 years old and had been a member of the AIF for 15 months. His death was reported in the 294th Australian casualty list released on the 9th of May 1917.

Cornelius Edwin Roe's name is inscribed on The Australian War Memorial Roll of Honour (Panel 149), The Australian National Memorial, Villers-Bretonneux, France, The City of Townsville WW1 Honour Roll, and The Townsville Cenotaph. Roe Street, Stuart, Townsville, is named in his memory.

5187 Private Cornelius Edwin Roe was awarded the 1914/15 Star (no: 34128), the British War Medal (no: 17588), and the Victory Medal (no: 33886). A memorial scroll and a memorial plaque were passed to his mother in 1922.

1601 PRIVATE THOMAS McCORMACK (1879 – 1917)

47th Infantry Battalion

Thomas McCormack was born on the 15th of February 1879 in Townsville, Queensland, the youngest son of Thomas and Johanna McCormack (née Luby). The family lived in premises on Ahearne Street, Hermit Park, Townsville. Thomas McCormack grew up in Townsville and attended St Joseph's Catholic School. When his school years were finished, he worked as a Lorry Driver.

On the 23rd of August 1915, Thomas McCormack was single and 36 years old when he applied to enlist in the Australian Imperial Force (AIF). He declared that he had six years previous military service with the 15th Light Horse. His enlistment medical examination noted that he was 5'7" (170cm) tall and weighed 11st 4lbs (72kg). Unusually, his complexion, eye colour, and hair colour were not recorded. Thomas McCormack identified as Roman Catholic and he nominated his brother Michael as his next of kin and gave his address as Grosvenor Street, Hyde Park, Townsville.

Accepted into the AIF, Thomas McCormack was allocated service number 1601, assigned to the Infantry and posted to the 5th Light Horse Regiment, 12th Reinforcements. The 5th Light Horse, which was a mounted Infantry Unit, was operating in the Sinai as part of the ANZAC Mounted Division when Thomas McCormack enlisted.

After three months of basic training in Australia, Trooper Thomas McCormack and his unit embarked for Egypt from Sydney on the 30th of November 1915 as part of Convoy No. 14. They sailed aboard HMAT A23 *Suffolk* which carried 926 troops. The fourteen-ship convoy carried a total of 12,039 troops and 1,561 horses on the voyage.

Arriving in Egypt in the first week of January 1916, Trooper McCormack soon joined the 5th Light Horse at their Training Camp at Maadi near Cairo. In February 1916, Trooper McCormack was admitted to hospital at Abbassia on the outskirts of Cairo. He underwent treatment for a total of 56 days, and in mid-March 1916, he was discharged to duty. Returning to Maadi, he was taken on the strength of the 2nd Reserve Regiment.

At the end of April 1916, Trooper McCormack re-joined the 5th Light Horse in the field near Romani. The 5th Light Horse's main activity in the Sinai was long-range patrolling. Trooper McCormack was engaged in the desert patrols and operations of the 5th Light Horse until early July when he was admitted to hospital suffering from a fever which resulted in him being invalided to England.

He was in hospital for two months before being discharged to the 12th Training Battalion at Rollestone on the Salisbury Plain in England. Here Trooper McCormack underwent training for service in France, and he was transferred to the 47th Battalion in October 1916 and changed from a Trooper to a Private soldier.

After a couple of weeks of further training at Étaples in France, Private Thomas McCormack joined the 47th Battalion in the field on the 13th of October 1916. The 47th was located near the village of Voormezele in Belgium when

Private McCormack was taken on strength. The 47th had endured two stints in the heavily contested trenches of Pozières, as well as a period in reserve in August 1916.

After Pozières, the battalion spent the period up until March 1917 alternating between duty in the trenches and training and rest behind the lines. On the 11th of April 1917, the 47th Battalion took part in the attack mounted against the heavily defended village of Bullecourt. Devoid of surprise, and dependent upon the support of unreliable tanks, the attack failed. It was during this attack that Private Thomas McCormack was Killed in Action on the 11th of April 1917.

In a statement made in October 1917, 1944 Private M.H. Robinson said, *"He was a Lewis Gunner and was killed by MG bullets at about 11.30 in the morning during a counterattack by the Germans at Bullecourt. He was my friend and much liked."*

Private Thomas McCormack was 38 years and one month old and had served with the AIF for 19 months at the time of his death. His body was not recovered, and he has no known grave. Thomas McCormack's name is inscribed on The Australian War Memorial Roll of Honour (Panel 105), The City of Townsville WW1 Honour Roll, The Townsville Cenotaph, and The Australian National Memorial, Villers-Bretonneux, France.

1601 Private Thomas McCormack was awarded the British War Medal (no: 22879) and the Victory Medal (no: 22710). A memorial scroll and a memorial plaque were passed to his brother in 1921 and 1922, respectively.

2135 PRIVATE WILLIAM JAMES HENRY BROWN (1890 – 1917)

47th Infantry Battalion

William James Henry Brown was born at Hughenden, Queensland on the 10th of October 1890. He was the first-born child of James and Emma Brown (née Walker). James Brown worked as a carrier and a horse driver. The Brown family moved to Townsville, and in 1903 lived in Arthur Street. By 1913, they lived in Charlotte Street, Aitkenvale.

William Brown was educated in Townsville, and after his school years were finished, he became a stockman. On the 17th of January 1916, he was working on Longton Station near Charters Towers when he applied at Townsville to enlist in the Australian Imperial Force (AIF). He declared that he was 25 years old, single, employed as a stockman, and that he had no previous military experience. William Brown's enlistment medical noted that he was 5'6" (168 cm) tall, weighed 147 lbs (67 kg), had a dark complexion, brown eyes, and dark brown hair. He identified as Church of England and nominated his father as his next of kin, giving his address as Stamford, Queensland. He also allotted a portion of his pay to his mother who was living in Aitkenvale, Townsville at that time.

Accepted into the AIF, William Brown was allocated service number 2135, assigned to the Infantry and posted to the 47th Battalion 4th Reinforcements. The 47th Battalion was raised in Egypt on the 24th of February 1916 as part of the "doubling" of the AIF. Approximately half of its new recruits were Gallipoli veterans from the 15th Battalion, and the other half were fresh Reinforcements from Australia. The Battalion was composed mostly of men recruited in Queensland and Tasmania. For Private William Brown, the next seven months were spent in Australia preparing for overseas service.

On the 16th of August 1916, Private Brown and the other reinforcements joined Convoy No. 23 and embarked from Brisbane, bound for England.

They sailed aboard HMAT A42 *Boorara* which carried 1,707 troops. The 12-ship convoy carried a total of 11,246 troops on the voyage. Disembarking at Plymouth, England in October 1916, the draft of reinforcements was transported to the 12th Training Battalion at Codford, Wiltshire, near the Salisbury Plain. They underwent further training until December 1916 when they crossed to France.

Australian soldiers arriving in France, whether reinforcements or those returning from hospitals, proceeded to Base Depots prior to deployment to the front. All contingents, despite having completed full training in England, underwent additional assessments, rigorous medical examinations, and a minimum of 10 days of supplementary training. Private Brown and the rest of the draft were marched into the 4th Australian Division Base Depot, France on the 22nd of December 1916 where they stayed until the 5th of February 1917. On the 6th of February, the draft marched into the 47th Battalion in the field.

After fighting at Pozières, the 47th Battalion spent the period up until March 1917 alternating between duty in the trenches and training and resting behind the lines. On the 11th of April 1917, the Battalion took part in the attack mounted against the heavily defended village of Bullecourt. It was during this phase of the operations that Private William Brown was accidentally Killed on Duty on the 12th of April 1917 near Bullecourt when a Mills grenade exploded beneath a nearby fire pit. He was 26 years old and had served with the AIF for 14 months at the time of his death. He has no known grave, and his death was notified in the 342nd list of Australian Casualties released on the 16th of October 1917.

William James Henry Brown's name is inscribed on The Australian War Memorial Roll of Honour (Panel 143), The City of Townsville WW1 Honour Roll, The Townsville Cenotaph, and The Australian National Memorial, Villers-Bretonneux, France.

2135 Private William James Henry Brown was awarded the British War Medal (no: 45135) and the Victory Medal (no: 44619). A memorial scroll and a memorial plaque were passed to his father in 1922.

The following tributes were published in the *Townsville Daily Bulletin* on 11th April 1918 and the 12th of April 1919, respectively:

BROWN. —In loving memory of our dear son and brother, Pte W. J.H. Brown, who was accidentally killed somewhere in France, on the 11th of April 1918.

He answered the call of his country,
But the voice of the cable tells,
That a dauntless lad in khaki,
Was killed somewhere in France.

Inserted by his loving parents, sisters and brothers, and brother-in-law.

BROWN. —In sad and loving memory of our dear brother, Pte W. J. H. Brown, who was accidentally killed on April 12th, 1917.

I could not speak that last good-bye,
But I kissed him o'er and o'er.
And I watched him go,
With a heart of woe,
Go to come back no more.

(Inserted by his loving Sister and Brother-in-law, May and Bert James.)

1941 PRIVATE HARRY PASCOE (1896 – 1917)

42nd Infantry Battalion

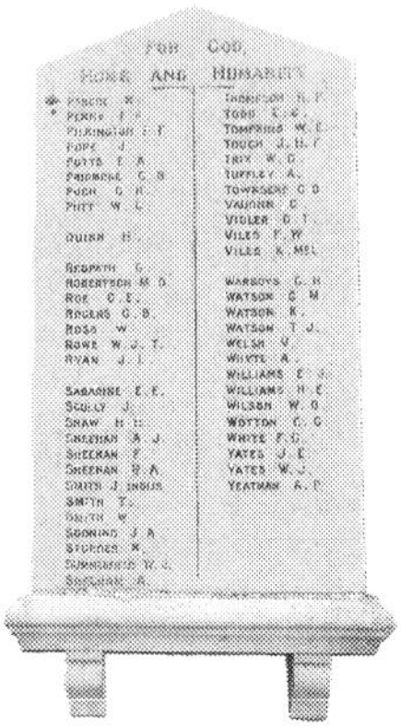

Harry Pascoe was born Harry Norris at Georgetown, Queensland, on the 24th of February 1896. He was the first-born child of William Pascoe and Mary Anne Norris. William and Mary Anne married in May 1896, after which Harry took his father's surname. William Pascoe was a miner, and he died in 1904 when Harry Pascoe was eight years old. Mary Anne Pascoe remarried Arthur Roberts in the same year as William Pascoe died.

Harry Pascoe attended Georgetown State School and, after finishing, worked as a stockman. By the early 1900s, the Roberts family had moved to Townsville, and in 1912, they were living in Cook St, where Arthur Roberts worked as a labourer.

Harry Pascoe was one month short of his 21st birthday when he applied, at Townsville, to enlist in the Australian Imperial Force (AIF) on the 17th of January 1916. He declared that he was single, working as a stockman, and that he had no previous military experience. His enlistment medical noted that he was 5'11" (180cm) tall, weighed 9st 13lbs (63kg), had a fair complexion, blue eyes, and brown hair. He identified as Church of England and nominated his mother as his next of kin, giving her address as West End, Townsville.

Accepted into the AIF, Harry Pascoe was allocated service number 1941 and assigned to the Infantry. He was initially posted to the 47th Battalion 3rd Reinforcements. This posting was later changed to the 42nd Battalion. Harry spent the time between enlistment and May 1916 completing initial training in Australia.

On the 1st of May 1916, Private Harry Pascoe embarked with his fellow draft of reinforcements from Brisbane bound for England via Egypt. They sailed aboard HMAT A46 *Clan MacGillivray* which carried 1,039 troops. Their ship was one of 13 ships that formed Convoy No.22 The entire convoy of transported 8,403 troops on the voyage. The reinforcements arrived in England via Alexandria on the 23rd of September 1916.

Training in England followed, and it was not until the 15th of February 1917 that Private Harry Pascoe crossed to France to join the 42nd Battalion in the field. After three weeks more training in Étaples, Private Pascoe joined the 42nd, near Armentieres, on the 9th of March 1917.

On the 14th of April 1917, the 42nd Battalion was manning trenches near Ploegsteert in Belgium. The Battalion was deployed repairing existing trenches and digging new communication trenches. While completing this task, the soldiers of the 42nd Battalion were subject to heavy bombardment from the German lines and one of the casualties was Private Harry Pascoe who was Killed in Action on the 14th of April 1917. He was 21years old at the time of his death. He had served with the AIF for 15 months and had only been in the field with the 42nd Battalion for three weeks.

Private Harry Pascoe's death was notified in the 294th list of Australian Casualties released on the 9th of May 1917. He was laid to rest at Berks Cemetery Extension (Row G, Grave No. 24), Ploegsteert, Belgium.

Harry Pascoe's name is inscribed on The Australian War Memorial Roll of Honour (Panel 136), The City of Townsville WW1 Honour Roll, and The Townsville Cenotaph. Pascoe St, Railway Estate, Townsville, is named in his memory.

1941 Private Harry Pascoe was awarded the British War Medal (no:37641) and the Victory Medal (no: 37337). A memorial scroll and a memorial plaque were passed to his mother in 1923.

776 PRIVATE WILLIAM JAMES YATES (1889 – 1917)

41st Infantry Battalion

William (Bill) James Yates was born in January 1889 in Marrickville, New South Wales, to his father James and his mother Ellen (née Robinson). He was the oldest of five children and had one brother and three sisters. By 1891 the Yates family had moved to Queensland. Eight years later, on the 5th of November 1899, James Yates died in Townsville at the age of 46. Ellen Yates (Nellie) was left as a widow with five young children, all under the age of nine.

In 1903, Nellie Yates and the family lived in housing on Elizabeth Street, Aitkenvale, Townsville. Bill Yates grew up in Townsville and attended the Mundingburra State School. In 1905 Nellie Yates married Thomas Tobin, a horse team driver who worked the road between Townsville and Georgetown. After his school years were completed, Bill Yates followed his stepfather into the Horse Driver profession.

On the 18th of October 1915, Bill Yates applied at Townsville to enlist in the Australian Imperial Force (AIF). He declared that he was 26 years old, single and working as a Horse Driver. His medical examination noted that he was 5'9¼" (176 cm) tall, weighed 11st 8lbs (73kg), had a dark complexion, blue eyes, and black hair. He identified as Church of England and nominated his mother as his next of kin giving her address as Aitkenvale Townsville.

Accepted into the AIF, Bill Yates was allocated service number 776, assigned to the Infantry and posted as an original member to "C" Company of the 41st Battalion. The 41st Battalion was raised in February 1916 with recruits from Brisbane, northern Queensland, and the northern rivers district of NSW. The 41st Battalion completed the basic training requirements at Enoggera in Brisbane.

After basic training was finished, Private Bill Yates embarked with the rest of the 41st Battalion from Sydney on the 18th of May 1916, bound for France via England. The unit sailed as part of Convoy No. 21 aboard HMAT A64 *Demosthenes*, which carried 1,612 troops on the voyage. The entire convoy of 18 ships transported 25,333 troops. They travelled via Cape Town and the Cape Verde Islands before arriving at Plymouth in July 1916.

Once disembarked, the battalion moved with the 3rd Division to training grounds on Salisbury Plain at Larkhill, the site of the current Royal School of Artillery. Following three months of training, the 41st Battalion sailed for the Western Front from Portsmouth, England in November 1916. They disembarked at Le Havre, France, and moved towards the border area with Belgium near Armentières.

The 41st entered the frontline for the first time on Christmas Eve, 1916, at the Mushroom Salient, just east of La Chapelle-d'Armentières, and spent the bleak winter of 1916-17 alternating between service in the front line, and training and labouring in the rear areas. On the 26th of January 1917, Private Bill Yates was attached for duty to the 2nd Australian Tunnelling Company, re-joining the 41st on the 4th of February 1917.

In April 1917 the 41st Battalion was deployed in Belgium and on the 12th of April, they relieved the 42nd Battalion in the front line near Ploegsteert. They stayed in this location until they were relieved by the 43rd Battalion. On the 15th of April the 41st came under attack as the enemy tried to raid gaps in the line. At 1945 hours (7:45 p.m.) the position was heavily bombarded by enemy artillery. Front line barricades were breached in two locations, killing three Australian soldiers and wounding 11. One of those casualties was Private William James Yates who was Killed in Action by shell fire on the 15th of April 1917.

Private Bill Yates was 28 years and had served with the AIF for 18 months at the time of his death. His body was laid to rest in the Berks Cemetery Extension (Row G, Grave No. 23), Ploegsteert, Belgium and his death was notified in the 294th list of Australian Casualties released on the 9th of May 1917.

William James Yates' name is inscribed on The Australian War Memorial Roll of Honour (Panel 134), The City of Townsville WW1 Honour Roll and The Townsville Cenotaph. Yates St Nelly Bay, Magnetic Island, and Yates St

Railway Estate, Townsville, are named in his memory and in memory of his brother James Ernest Yates who was also Killed in Action in WW1.

776 Private William James Yates served on the Western Front in The Great War. He was awarded the British War Medal (no:39975) and the Victory Medal (no:39589). A memorial scroll and a memorial plaque was also passed to his mother in 1922. His mother received a pension of one pound (£1/-/-) per fortnight for the loss of her son. This was about a quarter of the fortnightly wage paid to a soldier in the AIF in 1917.

The following tributes were published in the *Townsville Daily Bulletin* on the 15th of April 1918:

YATES. — In loving memory of our dear son and brother, William James Yates, Killed in Action in France on April 15th. 1917. Aged 23 years.

> 'Mid the roaring of the battle
> And the rain of shot and shell.
> Fighting for home and country,
> Our darling hero fell.
> No loved ones stood around him
> To bid a fond farewell.
> No word of comfort could we give
> To darling Willie we all loved so well.

(Inserted by his sorrowing mother, stepfather, and sister, E. T. Tobin and Ally Yates.)

433 SERGEANT ROBERT PERCY THOMSON (1874 – 1917)

2nd Light Horse Field Ambulance

Robert (Bob) Percy Thomson was born on the 20th of June 1874, in Brisbane, Queensland, to John Thomson and his wife Annie (née Monks). He was the eldest son and first born of four children. He had two brothers and one sister.

After leaving Brisbane, Robert Thomson lived in Charters Towers, where he was a well-known football player in the local Rugby Union Team. On the 20th of January 1898, Robert Thomson married Laura Margaret Halbert at Queenton (a suburb of Charters Towers). After living in Charters Towers, they moved to Townsville.

In 1913, Robert Thomson was employed as a forwarding clerk and Laura as a housewife, living together on Walter Street in Townsville. On the 5th of October 1914, two months after the declaration of war made on the 4th of August, Robert Thomson applied to enlist in the Australian Imperial Force (AIF). He declared that he was 42 years old (he was 40), married without children, qualified as a clerk, and that he had no previous military experience. His medical examination noted that he was 5'7¾" (171 cm) tall, weighed 157 lbs (71 kg), had a medium complexion, blue eyes, and fair hair. He identified as Church of England, and he listed his wife as his next of kin, giving her address as Wills St, Townsville.

Robert Thomson was accepted into the AIF, allocated service number 433, and posted as a cook to the 2nd Light Horse Field Ambulance (2nd LHFA). The Light Horse units were mounted infantry with efficient mobile medical support. The Field Ambulance ensured the removal of the wounded from the front line to the advanced Dressing Station, where there were surgical and resting tents. The 2nd Light Horse Regiment was raised at Enoggera in Queensland on the 18th of August 1914, with most of the unit recruited from Queensland.

From October to December 1914, Robert Thomson underwent initial training in Australia at Enoggera in Brisbane. He was promoted to Corporal soon after he enlisted. On the 15th of December 1914, the 2nd LHFA contingent joined Convoy No. 2 from Brisbane, bound for Egypt. They sailed aboard HMAT A30 *Borda*, which transported 493 troops and 106 horses. The entire convoy of 17 ships carried 11,127 troops and 4,643 horses on the voyage. They disembarked in Egypt in February 1915 and moved to Maadi Camp to commence training in desert conditions.

In August 1915, Corporal Thomson was deployed to Gallipoli along with other members of the unit. Once landed on the Peninsula at 0400 hours (4:00 a.m.) on the 7th of August, the 2nd LHFA was tasked with assisting the 3rd Field Ambulance in evacuating and treating the wounded from the Lone Pine battle. The conditions were extremely arduous, and there were many casualties to be treated over the coming days. Corporal Thomson was deployed as a cook and did an outstanding job. He was recommended for an award after the evacuation from Gallipoli.

The citation reads: *"For consistent courage, cooking under heavy shell fire on several occasions with practically no cover. On the 7th and 8th Aug 1915 during the 'Lone Pine' engagement, he cooked for wounded and stretcher bearers near 'Browns Dip' with utter disregard for personal safety. He was finally wounded while cooking, some five weeks later. His courage and devotion to duty have been a splendid example to all ranks in his corps."*

On the 10th of September 1915, Corporal Thomson was wounded in the left wrist by shrapnel. The wound was severe enough for him to be evacuated back to Heliopolis in Egypt for treatment. He was released from hospital on the 20th of September and re-joined his unit at ANZAC on the 25th of October 1915. On the 21st of November 1915, Corporal Thomson was promoted to Temporary Sergeant. After the evacuation of Gallipoli in December 1915, the 2nd LHFA returned to Egypt and joined the ANZAC Mounted Division. On the 1st of January 1916, Temporary Sergeant Thomson was promoted to substantive Sergeant.

Between January and May 1916, the 2nd Light Horse Regiment was deployed to protect the Nile valley from bands of pro-Turkish Senussi Arabs. On the

18th of May, as part of its parent brigade, it joined the forces defending the Suez Canal. The 1st Light Horse Brigade played a significant role in turning back the Turkish advance on the canal at the Battle of Romani on the 4th of August 1916.

In November 1916, Sergeant Thomson was admitted to the 31st General Hospital Cairo, Egypt, to have an ulcer removed from his nose. While he was in the hospital, he wrote a long letter to Alderman Frederick Johnson of Charters Towers, which was published in *The Evening Telegraph* (Charters Towers, Qld.) on the 28th of December 1916. In the letter, Thomson outlined the need for more recruits to help the war effort.

About six weeks after he was admitted to the hospital, Sergeant Thomson re-joined the 2nd LHFA on the 10th of January 1917. The 2nd Light Horse Regiment engaged in action at Rafa on the 9th of January 1917. A stint of protective duty along the line of communications through the Sinai followed. The 2nd's next major engagement was the abortive second battle of Gaza on the 19th of April 1917.

On the 17th of April at Shellal, two days before the Gaza action, Sergeant Robert Percy Thomson was Killed in Action when enemy aircraft dropped three bombs amongst the Ambulance. Five other soldiers were wounded, and 10 horses were killed or wounded. Equipment was also destroyed in the attack. Sergeant Thomson was 42 years and 10 months old at the time of his death. He had served in the AIF for two years and six months.

Initially buried in the field on the banks of Waddy Guzzeh, Shellal, his body was reinterred in 1920 and laid to rest at Beersheba War Cemetery, Grave Q.66, Israel and Palestine (including Gaza). Robert Percy Thomson's name is inscribed on The Australian War Memorial Roll of Honour (Panel 184), The City of Townsville WW1 Honour Roll, and The Townsville Cenotaph. Thomson St Aitkenvale, Townsville, is named in his memory.

433 Sergeant Robert Percy Thomson served at Gallipoli, Egypt, and Palestine in The Great War. He was awarded the 1914/15 Star (no: 10642), the British War Medal (no: 5170), and the Victory Medal (no: 5168). His widow, Laura, received a war pension of two pounds and 13 shillings (£2/13/-) per fortnight

for her husband's sacrifice. This was half of the fortnightly wage paid to a soldier in the AIF in 1917. A memorial scroll and a memorial plaque was passed to Laura Thomson in 1922.

The following notice was published in the *Townsville Daily Bulletin* (Qld) on the 5th of June 1917:

Killed in Action Thomson. - On the 17th April Robert Percy Thomson, Q.M.S. Second Light Horse. F. A., beloved husband of Laura Thomson.

"The Supreme Sacrifice."

324 TROOPER PETER MATHEW O'DOWD (1886 – 1917)

11th Light Horse Regiment

Peter Mathew O'Dowd was born in Prairie, Queensland, on the 16th of July 1886, to James and Janet O'Dowd (née Fraser). He was the youngest of 12 children, although only eight were surviving when Peter O'Dowd was born. He had six surviving sisters and two brothers. James O'Dowd died in 1888, and Janet O'Dowd passed away in 1902, leaving the family to survive on their own. Peter O'Dowd's oldest sister was thirty-two, and his oldest brother was thirty-one at the time of their mother's death.

Peter O'Dowd attended Normanton State School, and by 1908, he was a labourer living with his siblings in Normanton, Queensland. He left Normanton and in 1913, he was a storeman residing in premises on Cowley St, Townsville.

On the 2nd of January 1915, Peter O'Dowd applied to enlist in the Australian Imperial Force (AIF) in Townsville. He declared he had no previous military experience, was 27 years and five months old, single, and a baker's assistant. His medical examination noted he was 5'9½" (177 cm) tall, weighed 10 stone

7 lbs (67 kg), had a dark complexion, grey eyes, and brown hair. He identified as Roman Catholic and nominated his brother, John, as his next of kin, giving his address as Friezhand, Prairie, Queensland.

Accepted into the AIF, Peter O'Dowd was allocated service number 324, assigned to the Infantry and posted as an original member to "B" Squadron of the 11th Light Horse Regiment.

On the 11th of February 1915, the formation of the 4th Light Horse Brigade, including the 11th Light Horse Regiment (11th LHR), was announced. Queensland contributed two squadrons to the 11th LHR, while South Australia provided the third. The regiment held its inaugural assembly at Fraser's Paddock Camp, on the outskirts of Brisbane, on the 2nd of May 1915.

The 11th LHR sailed from Australia in two contingents in June 1915. Trooper O'Dowd's contingent joined Convoy No. 6 in Brisbane, bound for Egypt. They sailed aboard A30 HMAT *Borda*, which carried 568 troops. The 15-ship convoy transported 13,949 troops and 2,594 horses.

The first contingent landed at Aden on the 12th of July 1915 to reinforce the British garrison there against a predicted enemy attack; they re-embarked on the 18th of July without having seen action. The regiment was reunited in Egypt on the 23rd of July 1915 and began training as infantry, having been ordered to leave its horses in Australia.

A month later, the 11th LHR deployed to Gallipoli. The regiment was again split up to reinforce three light horse regiments already ashore. A Squadron went to the 2nd Light Horse Regiment, B Squadron to the 5th, and C Squadron to the 9th. Trooper O'Dowd was transferred to the 5th LHR, and in September 1915, he was severely wounded by a gunshot wound to the head. He was evacuated back to Egypt to the 2nd Australian General Hospital where he stayed until November before being discharged as fit for duty.

Trooper O'Dowd was reunited with the 5th Light Horse in Maadi, Egypt, in January 1916 and was eventually posted back to the reunited 11th Light Horse in late February. His return to the unit was short as he was soon admitted to the hospital for three weeks in early March, suffering from mumps.

The period from late April to the end of December 1916 was particularly notable in Trooper O'Dowd's service for the three times he was charged for disciplinary breaches: two for being absent without leave and one for "obscene language and creating a disturbance during church parade."

In April 1917, the 11th Light Horse moved into Palestine to join the main British and dominion advance. It joined its first major battle on the 19th of April when it attacked, dismounted, as part of the ill-fated second battle of Gaza. It was during this battle that Trooper Peter O'Dowd was Killed in Action on the 19th of April 1917. He was unable to be immediately buried due to a *"hurried retirement."* However, he was eventually laid to rest at the Gaza War Cemetery (Plot XVIII, Row E, Grave No. 12), Palestine.

At the time of his death, Trooper Peter O'Dowd was 30 years and nine months old. He had served in the AIF for two years and three months. His death was notified in the 297th Casualty List published on the 4th of May 1917.

Peter Mathew O'Dowd's name is inscribed on The Australian War Memorial Roll of Honour (Panel No.8), The City of Townsville WW1 Honour Roll, and The Townsville Cenotaph. O'Dowd St, Mundingburra, Townsville, is named in his memory.

324 Trooper Peter Mathew O'Dowd served in WW1 in Gallipoli and on the Western Front. He was awarded the 1914/15 Star (no:23262), the British War Medal (no:13020), and the Victory Medal (no:12962). These medals, as well as a memorial scroll and a memorial plaque, were passed to his brother John in 1921 and 1922, respectively.

The following was published in the *Croydon Mining News* - from the *Charters Towers Evening Telegraph* on the 21st of May 1917:

"Mr. J. O'Dowd, of the Croydon railway, has received news that his brother Private Peter O'Dowd was Killed in Action in France on the 19th of April. The deceased had been in the firing line for the last two years. He had previously been wounded, and on recovery returned to the western front. He was a native of Prairie but had spent many years in Normanton, where the family lived in the good days."

The following was published in the *Townsville Daily Bulletin* on the 19th of April 1918:

O'DOWD. —In loving memory of our dear nephew Private P. M. O'Dowd, 11th Light Horse Regiment, who was Killed in Action in Egypt, April 19th, 1917.

> While at the post of duty,
> What more could he do
> Than die for King and country,
> A soldier brave and true?

(Inserted by his loving aunt and uncle, E. and P. Fraser and family, Townsville.)

O'DOWD. —In loving memory of our dear brother, Private P. M. O'Dowd, 11th Light Horse Regiment, who was Killed in Action in Egypt, April 19th. 1917.

> He died as few men get a chance to die,
> Fighting to save a world's morality.
> He died the noblest death a man may die
> Fighting for God and right and liberty,
> And such a death is immortality.
> He sleeps not in his native land,
> But beneath the foreign skies.
> Far from those who loved him best.
> In a hero's grave he lies.

(Inserted by his loving sister and brother-in-law, L. and N. Schneekloth and family, Normanton.)

4116 PRIVATE JAMES CAMERON (1887 – 1917)

26th Infantry Battalion

James Cameron was born in 1887 in Lanarkshire, Scotland, to Alexander and Margaret Cameron (née Ross). He was the first-born son of a large family of eight children, most of whom were born in Australia. Alexander and Margaret emigrated to Australia and arrived soon after the birth of James Cameron. They settled in Reid River, which is south of Townsville. Alexander Cameron worked as a Lengthman (a person employed to maintain a section of road or railway).

James Cameron grew up in Australia, and by 1913 he was employed as a Teamster, working with bullock drays. On the 27th of September 1915, James Cameron applied at Townsville to enlist in the Australian Imperial Force (AIF). The next day, his brother Donald Cameron also applied to enlist. James declared that he had no previous military experience, was 28 years old, single, and employed as a labourer. His medical assessment noted that he was 5'6" (167 cm) tall, weighed 10st 8lbs (67kg), had a dark complexion, grey eyes, and brown hair. He identified as Presbyterian, and he nominated his mother as his next of kin and allotted two-fifths of his pay to her.

Both James Cameron and his brother Donald Cameron were accepted into the AIF and assigned to the Infantry. James Cameron was allocated service number 4116 and, along with his brother, was posted to the 26th Battalion, 10th Reinforcements. The 26th was raised at Enoggera, Queensland, in April 1915 from recruits enlisted in Queensland and Tasmania and formed part of the 7th Brigade.

After six months of basic training, Private James Cameron embarked with his brother and the rest of the Draft of Reinforcements from Brisbane, as part of Convoy No. 18, bound for England via Egypt. They sailed on the 28th of March 1916 aboard HMAT A73 *Commonwealth*, which carried 956 troops.

The entire convoy of 13 ships transported a total of 10,732 troops and 3,840 horses on the voyage.

After a short stay in Egypt, the Draft embarked at Alexandria and sailed to England to join the British Expeditionary Force on the Western Front. Once disembarked in England, they marched into the 7th Training Battalion located at Rollestone on Salisbury Plain in Wiltshire. While at Rollestone, Private James Cameron was charged with being Absent Without Leave for three days. He was found guilty and Confined to Camp for two days and forfeiture of three days of pay.

On the 9th of September 1916, the Reinforcements crossed to France. Once disembarked, they marched into the 2nd Australian Division Base Depot for a further period of training. A little over two weeks later, they joined the 26th Battalion and were taken on strength on the 28th of September 1916. The 26th was deployed near Steenvoorde, in Northern France.

During August, the 26th Battalion had been engaged in the Pozières battle, suffering more than 650 casualties, and were resting and rebuilding when Private James Cameron and the other Reinforcements marched into the unit. In November, the 26th was deployed to the Somme Valley, where they engaged in two offensives near the strategic location called *"The Maze"*, situated to the east of Flers. Unfortunately, these attacks failed as the soldiers got bogged down in the muddy terrain, resulting in over 300 additional casualties.

One of these casualties was Private James Cameron who was wounded in the hand by shell fire on the 15th of November. He was evacuated for treatment and returned to the 26th seven days later. The harsh winter of 1916–17 was spent occupying defensive positions along the front and training behind the lines. Private James Cameron was attached to the 7th Trench Mortar Battery for a week in December.

Christmas was spent in the trenches around Amiens, and in early January 1917, the 26th Battalion was transferred to the Ancre Valley and then to Le Sars later in the month. During the German withdrawal towards the Hindenburg Line, the 26th Battalion participated in various engagements at Warlencourt and Lagnicourt in March 1917. In May, they were engaged in the

Second Battle of Bullecourt, mainly serving in a supportive capacity during the initial assault and held in reserve as part of the 2nd Division.

It was during this engagement that Private James Cameron was reported as Missing in Action. A Court of Enquiry held in December 1917 found that he had been Killed in Action on the 3rd of May 1917. An Australian Red Cross Service investigation confirmed that Private Cameron had been shot in the head by a sniper. Death was instantaneous. The enemy took the ground where he was killed, and he has no known grave.

Private James Cameron was 29 years and 10 months old at the time of his death and had served with the AIF for one year and eight months. He had been a member of the 26th Battalion for seven months. Initially notified as missing in the 313th list of Australian Casualties released on the 22nd of June 1917, Private James Cameron's death was finally notified in the 378th list of Australian Casualties released on the 11th of January 1918.

James Cameron's name is inscribed on The Australian War Memorial Roll of Honour (Panel 107), The Australian National Memorial, Villers-Bretonneux, France, The City of Townsville WW1 Honour Roll, and The Townsville Cenotaph. Cameron Street, Railway Estate, Townsville, is named in his and his brothers' memory. Private Donald Cameron Died from Wounds on the 17th of July 1918.

4116 Private James Cameron was awarded the British War Medal (no: 34711) and the Victory Medal (no: 34464). A memorial scroll and a memorial plaque were passed to his father in 1922.

The following tribute was published in the *Townsville Daily Bulletin* (Qld) on the 3rd of May 1918:

CAMERON. - In loving memory of my dear son, James Cameron, who was killed in action, May 3rd, 1917.

> The midnight stars are gleaming,
> Upon a grave we cannot see,
> Where sleeping without dreaming,
> Lies the one so dear to us.

Peaceful be thy rest, dear James,
'Tis sweet to breathe thy name,
In life we loved you very dear,
In death we do the same.
He lies in a soldier's grave,
Though absent ever present.
For King and Country, he did his best,
May God give him eternal rest.

(Inserted by his loving parents, brothers and sisters.)

5615 PRIVATE LESLIE EDWARD KELLY (1887-1917)

26th Infantry Battalion

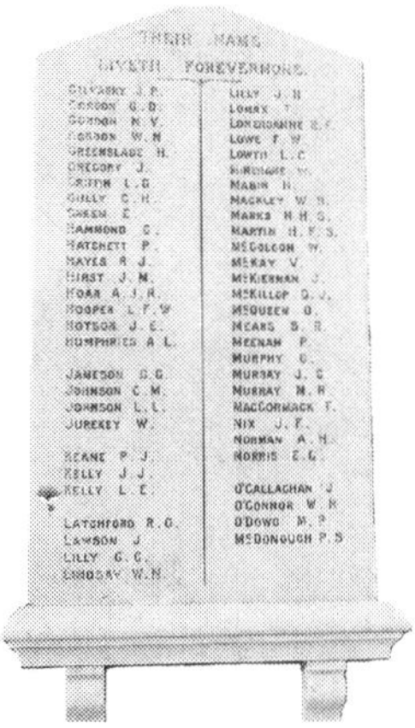

Leslie Edward Kelly was born on the 26th of July 1894 in Charleville, Queensland. He was the second child of John and Ellen Kelly (née McNulty). He had four brothers and one sister. The family moved to Ayr where John Kelly became the licensee of the Club Hotel in Brandon, Queensland, while Ellen was engaged in home duties.

There is no record of where Leslie Kelly attended school, but he did follow in his father's footsteps and became a barman at the Club Hotel. On the 1st of April 1916, Leslie Kelly applied at Townsville to enlist in the Australian Imperial Force (AIF). His older brother, John Kelly, had previously enlisted in August 1915 and was serving overseas with the 47th Battalion. Leslie Kelly declared that he was 21 years old, single, working as a barman, and had no previous military service.

His medical examination noted that he was 5'6" (168 cm) tall, weighed 12st 12lbs (82kg), had a fair complexion, blue eyes, and auburn hair. He identified as Roman Catholic and he nominated his father, c/- the Club Hotel Brandon, as his next of kin. Accepted into the AIF, Leslie Kelly was allocated service number 5615, assigned to the Infantry and posted to the newly raised 26th Battalion, 15th Reinforcements.

The 26th Battalion was raised at Enoggera, Queensland, in April 1915 from recruits enlisted in Queensland and Tasmania and formed part of the 7th Brigade. Twenty-three weeks of preparation and basic training followed enlistment, until the 7th of September 1916, when Private Leslie Kelly and the other Reinforcements joined Convoy No 24 and embarked from Brisbane, bound for England. They sailed aboard HMAT A46 *Clan MacGillivray* which carried 1,083 troops. The entire convoy of 13 ships transported a total of 12,600 troops on the voyage.

The Reinforcements disembarked at Plymouth in November 1916. Further training followed at the 7th Training Battalion located at Rollestone on the Salisbury Plain in Wiltshire, England. In December 1916, Private Kelly crossed to France where he joined the 26th Battalion located on the Somme. He was only in the 26th for a few weeks before he was evacuated to the 25th Stationery Hospital at Rouen in France in early January 1917, suffering from mumps. He re-joined the 26th two months later in early March 1917.

During the German withdrawal towards the Hindenburg Line, the 26th Battalion fought in engagements at Warlencourt and Lagnicourt in March 1917. In May, they were engaged in the Second Battle of Bullecourt, mainly serving in a supportive capacity during the initial assault and held in reserve as part of the 2nd Division. It was during this engagement that Private Leslie Kelly was reported as Missing in Action.

A Court of Enquiry held in December 1917 found that he had been Killed in Action on the 3rd of May 1917. An Australian Red Cross Service investigation confirmed that he had been killed by an artillery round and death was instantaneous. Private Leslie Kelly was 22 years old when he died and had been a member of the AIF for nearly 13 months. He was laid to rest at Queant Road Cemetery (Plot I, Row F, Grave No. 30), Buissy, France. His death was notified in the 313th list of Australian casualties issued on the 21st of June 1917. Leslie's brother, 3048 Private John William Kelly, 47th Battalion, Died of Wounds on the 7th of July 1917 in Belgium.

Leslie Kelly's name is inscribed on The Australian War Memorial Roll of Honour (Panel 108), The Brandon War Memorial, The City of Townsville WW1 Honour Roll, and The Townsville Cenotaph.

Kelly Street, Mundingburra, and Kelly St Nelly Bay were named in Leslie and John Kelly's memory in 1919.

5615 Private Leslie Edward Kelly was awarded the British War Medal (no: 46934) and the Victory Medal (no: 46389). A memorial scroll was passed to his mother in 1921, and a memorial plaque to his father in 1922.

2490 PRIVATE WILLIAM ROSS (1889 – 1917)

9th Infantry Battalion

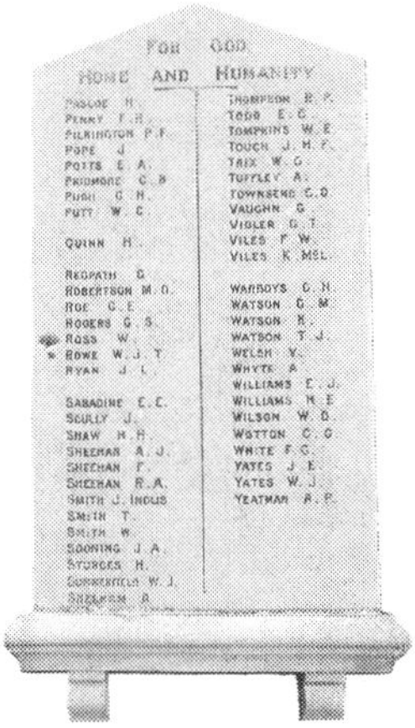

William Ross was born on the 29th of May 1889 in Maryborough, Queensland. He was the first-born child of William and Elizabeth Ross (née Ford). It is not clear when the family left Maryborough and moved to Townsville; however, in 1903 the family lived in a property on Macrossan Street, Townsville. At that time, William Ross Snr. worked as a seaman and William Ross Jnr. attended Ross Island State School. The family later moved to Bell St, South Townsville.

After his school years were finished William Ross Jnr. worked as a labourer until the 26th of August 1915 when he applied to enlist in the Australian Imperial Force (AIF). For reasons unknown, he was at sea on board the Transport A69 HMAT *Warilda* when he applied.

William Ross declared that he was single, 26 years old, and listed his occupation as a jockey. His medical examination noted that he met the minimum height requirements at 5'2⅛" (158 cm) tall. In June 1915, the minimum height requirements for enlistment in the AIF were changed from 5'6" (168 cm) to 5'2" (158 cm). William Ross weighed 117lbs (53kg), had a fair complexion, grey eyes, and fair hair. He identified as Church of England and he listed his mother as his next of kin giving her address as Bell St, South Townsville.

Accepted into the AIF, William Ross was allocated service number 2490, assigned to the Infantry and initially posted to the 25th Battalion. In February 1916, he was re-posted to the 9th Battalion at Habieta in Egypt. Private William Ross was one of several reinforcements that marched into the 9th Battalion. The Unit diaries record that the new arrivals included 28 men that were unfit for service and had to be sent to the hospital and that the draft did not come with the equipment needed for infantry operations.

In March 1916, the 9th Battalion reached full strength, and the battalion sailed for France and the Western Front. The battalion's first major action in France was at Pozières (23rd July – 3rd September 1916) in the Somme valley. The Australian official historian Charles Bean wrote that Pozières ridge "*is more densely sown with Australian sacrifice than any other place on earth.*"

Later in September 1916, the battalion was moved to Belgium where they served in and out of the front lines near Ypres. There, they engaged in training, trench repairs and maintenance, as well as patrols and occasional raids. In early October 1916, they returned to the Somme just in time to spend the bitterly cold winter of 1916-1917 in the trenches.

Private Ross was briefly admitted to Hospital during this period, suffering from fever. Throughout early 1917, the 9th Battalion remained on the Somme as they followed the German retreat to the Hindenburg line, participating in fighting at Lagnicourt in April.

In early May 1917, the Battalion was deployed near Bapaume and Vaulx, about 4 miles (6.4 km) northeast of Bapaume. On the 6th of May, the 9th Battalion received orders to mount an attack the next day on the Hindenburg Line. The attack was launched at 0358 hours (3:58 a.m.). Opposition was met immediately, with the enemy using mortars, hand grenades, flamethrowers, and machine guns.

The 9th pushed on, and by 0425 hours (4:25 a.m.), they had achieved their objectives. The Battalion had taken about 650 yards (approx. 600 meters) of enemy trench, captured 10 prisoners, and seized several weapons and items of equipment. The 9th suffered 160 casualties in this action. One of those casualties was Private William Ross, who was Killed in Action on the 7th of May 1917.

Private William Ross was one month short of his 28th birthday and had served in the AIF for one year and eight months at the time of his death. His body was not recovered, and he has no known grave.

Private William Ross served in Egypt and the Western Front in WW1. His name is inscribed on The Australian War Memorial Roll of Honour (Panel 57), The Australian National Memorial, Villers-Bretonneux, France, The City of Townsville WW1 Honour Roll, and The Townsville Cenotaph.

2490A Private William Ross was awarded the 1914/15 Star (no:20891), the British War Medal (no:19998), and the Victory Medal (no:19923). Elizabeth Ross received a memorial scroll and a memorial plaque in 1921 and 1922. She was also granted a war pension of two pounds (£2/-/-) per fortnight for the loss of her son. This was about half of the fortnightly wage paid to an AIF soldier in 1917.

5582 PRIVATE GEORGE CHARLES FULLER (1884 – 1917)

26th Infantry Battalion

George Charles Fuller was born in January 1884 at Townsville, Queensland. He was the fourth child of Henry and Kathleen Fuller (née Wagner) and had three brothers and two sisters. Growing up in Townsville, George Fuller attended Townsville West State School. After school, he worked as a Storeman and lived in Blackwood Street.

On the 24th of April 1916, George Fuller, who was single, applied to enlist in the Australian Imperial Force (AIF). He was 33 years old. He declared that he was employed as a storeman and that he had no previous military experience. His enlistment medical examination noted that he was 5'7½" (171 cm) tall, weighed 10st 2lbs (64kg), had a dark complexion, grey eyes, and dark brown hair. He identified as Church of England and nominated his mother of Blackwood St, Townsville as his next of kin.

Accepted into the AIF, George Fuller was allocated service number 5582, assigned to the Infantry and posted to the 26th Battalion, 15th Reinforcements. The 26th Battalion was raised at Enoggera, Queensland, in April 1915 from recruits enlisted in Queensland and Tasmania and formed part of the 7th Brigade. The 26th was deployed in France when George Fuller enlisted.

In July 1916, Private Fuller was admitted to the Camp Hospital in Enoggera suffering from pleurisy. Released from the hospital, he spent the next 25 days convalescing. Recovered and finished with basic training, Private Fuller embarked on the 7th of September 1916 with Convoy No 24 from Brisbane, bound for England. He sailed aboard HMAT A46 *Clan MacGillivray*, which carried 1037 troops. The entire Convoy of 13 ships carried a total of 12,600 troops on the voyage.

Private George Fuller and his fellow Reinforcements, disembarked at Plymouth, England, at the beginning of November 1916. From there, they were transported to the 7th Training Battalion at Rollestone on the Salisbury Plain for further training. The next move was to France, and Private Fuller embarked from Folkestone, England, on the 13th of December 1916 and marched into the 2nd Division Base Depot at Étaples on the 14th of December.

After completing more training at Étaples, he joined the 26th Battalion in December 1916. The harsh winter of 1916–17 was spent occupying defensive positions along the front and training behind the lines. Christmas was spent in the trenches around Amiens, and in early January 1917, the 26th Battalion was transferred to the Ancre Valley and then to Le Sars later in the month.

During the German withdrawal towards the Hindenburg Line, the 26th Battalion participated in various engagements at Warlencourt and Lagnicourt in March 1917. In May, they were engaged in the Second Battle of Bullecourt, which was part of the larger Battle of Arras, a significant British offensive in 1917. This offensive aimed to break the deadlock on the Western Front and penetrate the formidable German defences of the Hindenburg Line. The 26th mainly served in a supportive capacity during the initial assault and was held in reserve as part of the 2nd Division. Almost 7,500 Australians and over 8,000 British soldiers were killed or wounded during the Second Battle of Bullecourt.

It was during this battle that Private George Fuller was Killed in Action on the 8th of May 1917. He was 34 years old and had served with the AIF for 13 months at the time of his death. His body was not recovered, and he has no known grave. Private George Fuller's death was notified in the 306th list of Australian Casualties released on the 4th of June 1917.

George Fuller's name is inscribed on the Australian War Memorial Roll of Honour (Panel 107), the City of Townsville WW1 Honour Roll, the Townsville Cenotaph, the Townsville West State School Memorial Gate, and the Villers-Bretonneux Memorial (Australian National Memorial - France).

5582 Private George Charles Fuller was awarded the British War Medal (no:47904) and the Victory Medal (no:47346). A memorial scroll and a memorial plaque were passed to his father in 1921.

The following notice was published in the *Townsville Daily Bulletin* (Qld) on the 3rd of May 1918:

FULLER. - In loving memory of my dear son and our brother, George Charles Fuller, who was Killed in Action at Bullecourt, on May 3rd, 1917.

> Under the flag of the five-starred cross,
> The pride of the southern sea,
> He laid down his life in a deadly strife
> To keep Australia free.
> He sleeps not in his native land,
> But beneath the foreign skies,
> Far from those who loved him best,
> In a hero's grave he lies.

(Inserted by his loving parents, sisters, and brothers.)

2405 GUNNER ROY ALLEN SHEEHAN (1893 – 1917)

5th Light Horse Regiment/ 11th Field Artillery Brigade

Roy Allan Sheehan was born on the 10th of September 1893 in Cunnamulla, Queensland. He was the oldest son and first-born child of William and Mary Sheehan (née Ambrose). He had two brothers and one sister. William Sheehan worked as a miner at Monteagle in Queensland in 1903 and by 1913, he was a contractor living in a property on Charters Towers Road, Townsville. Roy Sheehan attended State School and started work as a labourer later in life.

On the 3rd of January 1916, at Townsville, he applied to enlist in the Australian Imperial Force (AIF). He declared that he had no previous military experience, was single, employed as a labourer, and that he was 22 years old. He identified as Roman Catholic and nominated his mother as his next of kin, giving her address as Short St, Townsville. His enlistment medical examination noted that he was 5'8½" (174 cm) tall, weighed 11st (70kg), had a fair complexion, blue eyes, and fair hair.

Accepted into the AIF, Roy Sheehan was allocated service number 2405, assigned to the Infantry and posted to the 5th Light Horse Regiment, 16th Reinforcements. The Light Horse was deployed as part of the ANZAC Mounted Division defending the Suez Canal from a Turkish advance across the Sinai Desert when Roy Sheehan enlisted.

After three months of basic training in Australia, Trooper Roy Sheehan embarked with other unit members from Sydney, New South Wales, bound for Egypt. They sailed with Convoy No. 19 aboard HMAT A47 *Mashobra* on the 5th of April 1916. Their ship carried 132 troops, and the entire convoy of 10 ships carried 8,075 troops and 1,168 horses on the voyage.

Seven weeks later, in May 1916, they joined the 2nd Light Horse Training Regiment at Tel-el-Kebir and soon after Trooper Sheehan was transferred to the Artillery. This was the time when the AIF was doubled in size, and each battalion was required to provide 100 men suitable for the artillery. It is not recorded whether Trooper Sheehan applied for transfer or was offered up and compulsorily posted. Either way, he was no longer a Trooper, and he became a Gunner.

After spending a further three months in Egypt as part of the Artillery Details, Gunner Sheehan and the rest of his unit left Alexandria aboard H.M.T. *Corsican* on the 28th of May 1916, bound for England. Two weeks later, they disembarked at Plymouth. From there, they were transported to Bulford in Wiltshire, close to the Salisbury Plain, for further training. The training lasted until the 23rd of August 1917 when they crossed to France.

A month of further training occurred at Étaples, and it was not until the 24th of September 1917 that Gunner Roy Sheehan joined the 4th Division Ammunition Column (4th D.A.C.). Each Division had an Ammunition Column to keep ammunition up to the guns by moving it from "Third line" storage up to the Front ("First Line"). It was a dangerous and mammoth task involving motor and horse-drawn transport, heavy and light rail, and tramways.

Three weeks after joining the D.A.C., Gunner Sheehan was involved in an accident on the 6th of October 1916. He dislocated his elbow and shoulder joint while on duty and working. His Commanding Officer recorded that *"he is in no way to blame."* The injury was severe, and Gunner Sheehan was evacuated back to England where he was admitted to Reading War Hospital and then to the 2nd Auxiliary Hospital. His injury took several months to heal, and it was not until the 9th of January 1917 that he arrived back in France at Étaples.

While training in Étaples, Gunner Sheehan contracted mumps which were sweeping through the AIF. He was admitted to the 18th General Hospital at Camieres, near Calais. He was in the hospital for 19 days before being released. Gunner Sheehan re-joined the 4th D.A.C. on the 8th of February 1917, 18 weeks after he had been injured.

Five weeks later, he was taken on the strength of the 11th Field Artillery Brigade and was immediately posted to the 111th Battery. The 111th was a Howitzer Battery equipped with the excellent new British 4.5-inch Howitzer.

Following a disastrous attempt in April 1917, the Allied Forces mounted a renewed attempt to secure the fortified village of Bullecourt in the period 3-17 May 1917. Although the locality was of little or no strategic importance, the actions were nevertheless extremely costly: AIF casualties totalled 7,482 from three Australian Divisions. It was during this battle that Gunner Roy Sheehan was Killed in Action on the 12th of May 1917. He was 23 years and eight months old at the time of his death. He had been a member of the AIF for 16 months.

Gunner Sheehan was laid to rest in the Honourable Artillery Company (H.A.C.) Cemetery (Plot III, Row H, Grave No. 25), Ecoust-St. Mein, France. His death was notified in the 311th list of Australian Casualties released on the 16th of June 1917.

Gunner Roy Allan Sheehan served in Egypt and on the Western Front in WW1. His name is inscribed on The Australian War Memorial Roll of Honour (Panel 17), The City of Townsville WW1 Honour Roll, and The Townsville Cenotaph.

Sheehan Street, Hermit Park, Townsville, is named in his memory as well as the memory of his brother Albert John Sheehan (Died of Wounds on the 22nd of September 1917) and his cousin William Frederick Francis Alexander David Sheehan (Killed in Action on the 8th of August 1917).

2405 Gunner Roy Allan Sheehan was awarded the British War Medal (no: 36991) and the Victory Medal (no: 36697). A memorial scroll and a memorial plaque were passed to his mother in 1922.

Roy Sheehan's mother was granted a pension of fourteen shillings in respect to his death. This was later increased by one pound and 10 shillings (£1/10/-) per fortnight following the death of her second son Albert. This combined amount of two pounds and four shillings (£2/4/-) was about a quarter of the fortnightly wage for two soldiers in the AIF in 1917.

4720 PRIVATE WILLIAM JAMES EDWARDS (1888 – 1917)

26th Infantry Battalion

William James Edwards was born on the 25th of June 1888 in London, England. He was the youngest son of John and Sarah Edwards (née Shepherd). He had four brothers and one sister. John Edwards was employed as a labourer in the building industry. William Edwards grew up in England and attended Latymer School, Hammersmith.

William Edwards qualified as a carpenter and joiner, after his school years were finished. In about 1910, he emigrated to Australia and arrived in Queensland. He moved to Townsville and boarded in the Hermit Park Hotel. William Edwards had some family connections with Townsville as his aunt, Mrs Elizabeth (Lizzie) Hitchcock, lived in Hermit Park.

On the 28th of October 1915, William Edwards applied at Townsville to enlist in the Australian Imperial Force (AIF). He declared that was 27 years old, single, qualified as a carpenter and had no military experience. His enlistment medical examination noted that he was 5'4" (163 cm) tall, weighed 9st 2lbs (58kg), had a fair complexion, brown eyes, and light brown hair. He identified as Church of England and nominated his mother as his next of kin, giving her address as 86 Edinburgh Rd Walthamstow, London. This was changed after his mother's death in 1918 to his father at 51 Queens Park Rise Brighton, England.

Accepted into the AIF, William Edwards was allocated service number 4720, assigned to the Infantry and posted to the 26th Battalion, 12th Reinforcements. The 26th Battalion was raised at Enoggera, Queensland, in April 1915 from recruits enlisted in Queensland and Tasmania and formed part of the 7th Brigade. At the time of William's enlistment, the 26th Battalion was fighting on the Gallipoli Peninsula. Seven months of basic training followed enlistment for Private William Edwards until the 12th of April 1916 when he embarked with Convoy No. 20 from Sydney, bound for England. He sailed aboard RMS *Mooltan*, which carried 480 troops. The entire convoy of 15 ships carried 9,213 troops and 5,590 horses on the voyage.

Once in England, Private Edwards and the other reinforcements were transported to the 7th Training Battalion at Rollestone on the Salisbury Plain for further training. They stayed there until the 5th of September 1916 when they crossed to France and marched into the 2nd Division Base Depot at Étaples. Following a further two weeks of training, Private William Edwards and the rest of the reinforcements marched into the 26th Battalion in the field on 30th September 1916.

About a month prior to Private Edwards arrival, the 26th had been engaged in the battalion's first major battle around Pozières. After this battle, the 26th Battalion was sent to a quieter sector in Belgium, having suffered over 650 casualties. In November 1916, the 26th moved south to the Somme Valley and throughout the month participated in two attacks around a position known as "The Maze," a salient to the east of Flers. Both attacks ultimately proved fruitless, grinding to a halt in the muddy conditions and resulting in over 300 casualties.

In early 1917, the 26th Battalion joined the follow-up of the German withdrawal to the Hindenburg Line and attacked at Warlencourt (1st-2nd of March) and Lagincourt (26th of March). On the 3rd of May 1917, the 26th was involved in the second attempt to breach the Hindenburg Line defences around Bullecourt. It was during this attack that Private William Edwards was Killed in Action at Bullecourt on the 19th of May 1917. He was 28 years and 10 months old at the time of his death. He had served with the AIF for 18 months and had been a member of the 26th Battalion for seven months. He has no known grave, and his death was notified in the *Townsville Daily Bulletin* (Qld.) on the 25th of June 1917.

William James Edwards' name is inscribed on The Australian War Memorial Roll of Honour (Panel 107), The Australian National Memorial, Villers-Bretonneux, France, The City of Townsville WW1 Honour Roll, and The Townsville Cenotaph.

4720 Private William James Edwards was awarded the British War Medal (no: 37405) and the Victory Medal (no: 37103). A memorial plaque and a memorial scroll were passed to his father in England in 1933.

11345 PRIVATE GORDON HENRY PUGH (1894 – 1917)

41st Infantry Battalion

Gordon Henry Pugh was born on the 17th of August 1894 at Glengallen, Darling Downs, Queensland, to Walter and Annie Pugh (née Eberhardt). His parents both died when Gordon Pugh was young; his mother in 1899 and his father in 1901. Before she died, Annie Pugh gave birth to seven children. The youngest child was born prematurely and died a few weeks after birth. Gordon Pugh had five surviving siblings: four brothers and one sister.

After their parents died, the children were put into the care of foster parents and relatives. Gordon Pugh and his older brother, Alfred, were fostered in Townsville by Mrs. Henrietta Witham, who was widowed in 1908. They lived at an address on Flinders Street. The Withams were a prominent Townsville family of bakers and confectioners. After his school years, Gordon Pugh took up a career as a cook.

On the 7th of February 1916, Gordon Pugh applied at Townsville to join the AIF. He declared that he was 21 years old, single, and a qualified cook. He also declared that he had 4½ years' previous military service with the Citizen Military Forces in the Australian Regimental Corps. His enlistment medical examination noted that he was 5'2" (157 cm) tall, weighed 8st 2lbs (52 kg), had a dark complexion, blue eyes, and dark hair. He identified as Church of England and nominated his foster mother, Mrs. Witham, as his next of kin, describing her as his "friend."

Accepted into the AIF, Gordon Pugh was allocated service number 11345, assigned to the Regimental Corps, and posted to the 12th Depot Unit of

Supply. Each depot unit of supply could handle the supplies for about 4,000 men, so five were required per division. The 12th was formed at Enoggera, Queensland, in April 1916. Three months of training in Brisbane followed enlistment.

Private Gordon Pugh and his unit embarked with Convoy No. 21 from Sydney bound for England on the 17th of May 1916. They sailed aboard HMAT A64 *Demosthenes,* which carried 1,610 troops. The 18 ships of the convoy carried 25,287 troops on the voyage. They docked at Plymouth on the 20th of July 1916. Four more months of training in England followed before the unit crossed to France at the end of November 1916 and established at Calais. It was here that the unit was split up and allocated to other supply depots on the western front.

Private Pugh transferred to the 41st Infantry Battalion during this period and was taken on strength of that battalion on the 4th of January 1917. The 41st Battalion had arrived in France on the 25th of November 1916. It entered the front line for the first time on Christmas Eve and spent the bleak winter of 1916-17 alternating between service in the front line, and training and labouring in the rear areas.

In early March 1917, Private Gordon Pugh spent two weeks in the hospital suffering from an abscess near the mouth. He re-joined the battalion in mid-March. Three months of life in the trenches followed.

On the 3rd of June 1917, the 41st Battalion was deployed in the vicinity of Pont de Nieppe, situated by the Belgian border, close to Armentières. They occupied a position just south of Ploegsteert Wood, a sector in Flanders and part of the Ypres Salient. It is located around the Belgian village of Ploegsteert, Wallonia.

The unit diary of the 41st Battalion records that on the 3rd of June 1917, enemy bombardment, in retaliation for an Allied preliminary bombardment, was severe, and the battalion suffered casualties. One of those casualties was Private Gordon Pugh, who was Killed in Action on the 3rd of June 1917. He was 22 years and nine months old at the time of his death. He had served in the AIF for 16 months and was laid to rest at the Strand Military Cemetery

(Plot III, Row D, Grave No. 8), Ploegsteert, Belgium. His death was notified in the 313th list of Australian casualties released on the 21st of June 1917.

Private Gordon Henry Pugh's name is inscribed on the Australian War Memorial Roll of Honour (panel 134), the 41st Battalion Roll of Honour, the City of Townsville WW1 Honour Roll, and the Townsville Cenotaph. Pugh Street, Aitkenvale, Townsville, is named in his memory.

11345 Private Gordon Henry Pugh served in WW1 on the Western Front. He was awarded the British War Medal (no. 40002) and the Victory Medal (no. 39616). These medals, as well as a memorial scroll and a memorial plaque, were passed to his foster mother in 1922.

In her letters to the Army following Private Pugh's death, Mrs. Witham sent the cost of postage to Base Records, Melbourne, asking for photographs of his grave to be sent to his *"brothers and sisters and also to his poor sweetheart who he left behind."* There are several letters to the Army from Miss Ethel Edwards of Darlinghurst, Sydney, requesting details of Private Pugh's whereabouts.

Gordon Pugh's younger brother, Oswald, had also enlisted in the AIF. He lived in Western Australia and had been adopted by his uncle. He enlisted as James Pugh Mann and tragically died in May 1916 of pneumonia while on the voyage to England from Australia.

The following tributes were published in the *Townsville Daily Bulletin* on the 18th of June 1917:

Gordon Henry Pugh was Killed in Action on June 3rd in France. Gordon, or 'Darkie,' as he was familiarly known to all the boys, was a general favourite, possessing an easy temperament in work or play. His loss is keenly felt by his sorrowing brother, Alfred Pugh, as another brother, Oswald, died of pneumonia on a transport ship in May 1916 when going to the front. Gordon Pugh was at Thursday Island and left Townsville for the front early in 1916.

PUGH. — Gordon Henry aged 21 years. Killed in Action, June 3rd, 1917, brother of Agnes Ingram Pugh, and adopted son of Mrs. H. P. Witham, Townsville.

He being dead yet speaketh.

566 CORPORAL JOHN JOSEPH CARMODY (1890 – 1917)

9th Infantry Battalion

John Joseph Carmody (Joe) was born on the 28th of April 1890, in Arrilalah, Queensland. He was the fourth of six children born to Michael and Ellen Carmody (née Burns). He grew up in Brisbane, where he attended school. After completing his schooling, Joe Carmody worked as a Guard with the Railways.

In 1912, Joe Carmody's mother was living in Sturt Street, Townsville, while his father served as a Police Officer stationed in Charters Towers. Joe Carmody lived in Townsville and joined the Garrison Artillery for two years before eventually leaving the district. On the 1st of September 1914, just four weeks after Australia entered World War I, Joe Carmody applied at Maryborough Queensland to enlist in the Australian Imperial Force (AIF). He stated that he was 24 years old, single, and employed as a Railway Guard. He also disclosed his previous service with the Artillery. His enlistment medical noted that he was 5'7½" (172 cm) tall, weighed 150 lbs (68 kg), had a fresh complexion, grey eyes, and fair hair. Joe Carmody identified as Roman Catholic and listed his brother Michael, who managed a hotel in Redbank, as his next of kin.

Accepted into the AIF, Joe Carmody was allocated service number 566, assigned to the Infantry and posted as original member to the newly raised 9th Battalion. After assembling, equipping, and training, Private Joe Carmody and other members of the 9th Battalion embarked on the 24th of September 1914 from Brisbane, Queensland, as part of Convoy No.1, also known as the *Great Convoy*. The convoy gathered at Albany in Western Australia, where escorts were secured before setting sail across the Indian Ocean towards the Middle East. Private Carmody and his unit travelled aboard HMAT *Omrah*, which carried a total of 1,142 troops. In total, the convoy consisted of 29 ships transporting 21,431 troops and 7,882 horses. HMAS *Berrima*, one of the ships in the convoy, towed the Australian submarine *AE2*.

Private Joe Carmody was among the first to land at Gallipoli on the 25th of April 1915. He sustained a knee injury on the campaign's first day and was subsequently hospitalised in Malta. He rejoined the 9th Battalion in October 1915. After the general withdrawal from Gallipoli in December 1915, Private Joe Carmody was transferred to the 49th Infantry Battalion and promoted to the rank of Corporal in March 1916.

The 49th Battalion was established in Egypt on the 27th of February 1916 as part of the "doubling" of the AIF. Approximately half of its enlistees were veterans of Gallipoli from the 9th Battalion, with the remaining half being new recruits from Australia. Reflecting the makeup of the 9th Battalion, a significant number of the 49th Battalion's soldiers originated from Queensland.

In June 1916, the 49th Battalion left Alexandria to join the British Expeditionary Force on the Western Front. Sailing on the transport *Arcadian*, the 49th Battalion landed in Marseilles, France, on the 12th of June 1916 and moved up to the front line around Strazelle, arriving on the 21st of June. The battalion's first significant action came during the Battle of Mouquet Farm, taking part in two efforts in August and early September. During their initial engagement, the 49th Battalion experienced devastating losses with 14 officers and 417 soldiers killed or wounded.

It was in this battle that Corporal Joe Carmody was wounded for the second time on the 14th of August 1916, sustaining a gunshot wound to his abdomen. He was evacuated for medical treatment and rejoined the battalion six weeks after being wounded. As a result of the high casualty rate, the 49th Battalion was not deployed in major offensives for the remainder of 1916. However, they did undertake several rotations along the front lines, initially in the Ypres region and later in the Somme. During these periods, they engaged in patrols and conducted raids, interspersed with intervals of rest, training, and performing manual labour in the rear areas.

In November 1916, the 49th Battalion was deployed at Delvile Wood (often referred to as *"Devil's Wood")*, directly, east of the village of Longueval and 38 km (24 miles) northwest of Amiens. The 49th Battalion was held in Reserve and carrying out labouring duties for the front line. During their time in the support and front line, the 49th lost 15 killed and 27 wounded.

One of these casualties was Corporal Joe Carmody who was wounded for the third time, sustaining a gunshot wound to his right leg. He was then sent to England for treatment, and while on convalescent leave, he was hospitalised with mumps. He re-joined the 49th Battalion five months later, in April 1917.

In June 1917, the 49th Battalion fought in the Battle of Messines. It was during this action, that Corporal Joe Carmody was Killed in Action on the 7th of June 1917. Sergeant Benjamin Hughes provided the following statement in the investigation of Carmody's death conducted by the Australian Red Cross in August 1917: "*J. Carmody was a Cpl. in B.Coy. No. 6 pltn. Shortly before we went over the top at Messines, about 2 p.m. on the 7th June, a shell came over which killed Carmody and wounded four other men. I did not actually see Carmody hit, but I was in the same trench and saw his body, which was badly cut up. We had to go over shortly afterward, so I am unable to say whether Carmody's body was buried, but I think that it would have been. He came from Rockhampton, Queensland.*"

Corporal Joe Carmody was 27 years old at the time of his death. He had served with the AIF for two years and nine months. He has no known grave. His death was notified in the 323rd list of Australian Casualties released on the 16th of July 1917.

John Joseph Carmody's name is inscribed on several memorials, including The Australian War Memorial Roll of Honour (Panel No. 147), The Ypres (Menin Gate) Memorial, The City of Townsville WW1 Honour Roll, and The Townsville Cenotaph. In Townsville, Carmody Street in Hermit Park is named in his memory.

566 Corporal John Joseph Carmody served at Gallipoli and on the Western Front. He was awarded the 1914/15 Star (No: 896), the British War Medal (No: 344), and the Victory Medal (No: 300). A memorial scroll and a memorial plaque were given to his father in 1922. His mother, Ellen, received a war pension of one pound (£1/-/-) per fortnight for the loss of her son. This was about a quarter of the fortnightly wage paid to a soldier in the AIF in 1917

The following article and tribute were published in the newspapers as listed:

The Northern Miner (Charters Towers, Qld) 20 Dec 1916:

WAR NEWS

Mrs. E. Carmody, Sturt Street, Townsville, has been advised that her son, Corporal J. J. Carmody, has been wounded in France (third occasion).

The *Townsville Daily Bulletin* (Qld.) 7th June 1918:

CARMODY. — In Memory of Corporal J. J. Carmody, who was killed on active service, June 7th. 1917.

> Dearest brother, thou has left us.
> Thou doth dwell with angels now:
> And a wreath of glory priceless,
> Sparkles on thy shining brow.
> good and gentle was thy lifetime,
> You have died for your country true.
> Wait a little, dearest brother,
> And we shall soon follow you.

(Inserted by his loving Father, Sister, and Brothers)

3058 SIGNALLER THOMAS GEORGE DEARNESS (1893 – 1917)

47th Infantry Battalion

Thomas George Dearness was born on the 17th of August 1893 in Townsville, Queensland. He was the first son and oldest of nine children born to George and Mary Dearness (née Naismith). George Dearness worked as a compositor, and the family lived in Nelson St, South Townsville. Young Thomas Dearness grew up in Townsville and attended the Ross Island State School. After completing his school years, he found employment as an insurance clerk.

On the 18th of September 1915, at the age of 22 years and one month, Thomas Dearness applied to enlist in the Australian Imperial Force (AIF) in Townsville. He stated that he was single, had two years of service in the military cadets, and worked as an insurance clerk. His medical examination noted that he was 5'6" (167 cm) tall, weighed 7st 12lbs (52kg), had a dark complexion, brown eyes, and brown hair. He identified as Presbyterian and nominated his mother as his next of kin, giving her address as Nelson St, Ross Island, Townsville.

Accepted into the AIF, Thomas Dearness was allocated service number 3058, assigned to the Infantry and posted to the 25th Battalion, 7th Reinforcements. After assembling, equipping, and training Private Thomas Dearness and the other Reinforcements departed from Brisbane on the 30th of December 1915 bound for Egypt. They sailed aboard HMAT A50 *Itonus* as part of Convoy No.15. Their ship carried a total of 803 troops, while the entire convoy of 11 ships transported 9,123 troops on the voyage.

Shortly after arriving at Tel-el-Kebir in March 1916, Private Thomas Dearness was transferred to the 47th Battalion as a signaller. The 47th Battalion had been established in Egypt on the 24th of February 1916 as part of the AIF's reorganization, which aimed to expand the existing forces. The battalion's composition included both seasoned veterans of the Gallipoli campaign from the 15th Battalion and fresh recruits from Australia, with a significant number recruited from Queensland and Tasmania.

The 47th Battalion sailed to France from Egypt, on the 2nd of June 1916, disembarking at Marseilles seven days later. The battalion deployed in the trenches of the Western Front and engaged in its inaugural combat on the 3rd of July, entering the Battle of Pozières. Initially, the battalion supported with labour duties during the 2nd Division's assault on the 4th of August. Following this, it defended the territory that had been seized.

The 47th Battalion endured two rotations amidst the fiercely contested trenches of Pozières. These alternated with periods of reserve duty. During this time, on the 2nd of August 1916, Private Thomas Dearness sustained a gunshot wound to the head, leading to his evacuation to England for medical treatment. After being discharged from the hospital, he was temporarily attached to other units, including a dental section, as he recovered his fitness for active service.

It was not until the 14th of April 1917, after an absence of approximately eight months, that Private Dearness rejoined the 47th Battalion, just in time to participate in the ill-fated first battle of Bullecourt. Following this battle, the battalion was withdrawn from the front line for rest and training. Later in the year, the focus of the AIF's operations shifted to the Ypres sector in Belgium, where the 47th Battalion took part in the battles of Messines and Passchendaele.

It was during the battle of Messines that Private Thomas Dearness was Killed in Action on the 7th of June 1917. He was buried in the field near where he fell, but he has no known grave. He was 23 years and nine months old at the time of his death and had served in the AIF for one year and eight months.

Thomas George Dearness' name is inscribed on The Australian War Memorial Roll of Honour (Panel 143), The Ypres (Menin Gate) Memorial, The City of Townsville WW1 Honour Roll, and The Townsville Cenotaph. Dearness Street in Garbutt, Townsville, is named in his memory.

3058 Signaller Thomas George Dearness was awarded the British War Medal (No:29768) and the Victory Medal (No:29586). In 1921 and 1922, his father received a memorial scroll and plaque, respectively. His mother, Mary, was granted a pension of fourteen shillings per fortnight for the loss of her son. This was about 1/5th of the wage paid to an AIF soldier in 1917.

Thomas Dearness' brother, 6007 Private James Naismith Dearness, also served in France with the AIF as a member of the 9th Battalion. He survived the war and returned to Australia on the 19th of February 1919.

The following notices were published in the *Townsville Daily Bulletin* (Qld) on the 7th of June 1918:

DEARNESS. — In loving Memory of Signaller Thomas George Dearness, Killed in Action in France, June 7th, 1917. Aged 23 years.

> In the bloom of his life God claimed him,
> In the prime of his manhood days;
> None knew him but to trust him.
> None spoke his name but to praise.

There came a day when the roll was called,
That he did not answer "Here!"
He slept with comrades his last long sleep,
And he died without a fear.

(Inserted by his sorrowing Parents, George and Mary Dearness, Sister, and Brothers).

DEARNESS.—In loving Memory of Signaller T. G. Dearness, killed at Messines, June 7, 1917. Aged 28 years.

(Inserted by his Uncle and Aunt W. J. Dearness and Family).

2446 PRIVATE GEORGE HAMMOND (1892 – 1917)

47th Infantry Battalion

George Hammond was born on the 4th of May 1892 in Port Douglas, Queensland, to William and Clara Hammond (née Cook). He was the youngest boy of five surviving children in the family. William Hammond was an engineer and the family lived in Innisfail, Far North Queensland. George Hammond grew up in Innisfail and attended the local State School. After he finished his school years, George Hammond worked as a labourer. On Christmas Eve in 1913, he married May Murray. They had two children, both girls. One of the girls was George Hammond's adopted daughter, and the other was his biological daughter. George Hammond and his family relocated to Townsville and lived in Morris St, West End, and then Piccadilly Street, Hyde Park, Townsville.

On the 17th of April 1916, George Hammond applied to enlist in the Australian Imperial Force (AIF). He declared that he was one month short of his 24th birthday, married and working as a labourer. His medical examination noted that he was 5'6" (168 cm) tall, weighed 8st 8lbs (54.5kg), had a fair complexion, brown eyes, and brown hair. He identified as Church of England and nominated his wife as his next of kin.

Accepted into the AIF, George Hammond was allocated service number 2446, assigned to the Infantry and posted to the 47th Battalion 5th Reinforcements. The 47th Battalion was raised in Egypt on the 24th of February 1916 as part of the AIF's restructuring efforts to double its forces. Comprising veterans of the Gallipoli campaign from the 15th Battalion alongside new recruits from Australia, particularly from Queensland and Tasmania, the battalion was allocated to the 12th Brigade of the 4th Australian Division.

When George Hammond enlisted, the 47th was deployed to Egypt, preparing to join the fight on the Western Front. George Hammond spent the next five months in Enoggera, Brisbane, training and equipping for overseas service. On the 19th of September 1916, Private George Hammond and the contingent of Reinforcements departed from Brisbane aboard HMAT A49 *Seang Choon* as part of Convoy No.24 bound for England. The ship carried a total of 1,702 troops, while the entire convoy of 13 ships transported 12,785 troops during the voyage.

Three months later, in December 1916, the contingent marched into the 12th Training Battalion at Codford on the Salisbury Plain, where training continued. Private Hammond contracted mumps while at Codford and spent two weeks in the hospital, including Christmas and New Years.

In February 1917, Private Hammond and his fellow reinforcements crossed to the 4th Australian Division Base Depot, France. Upon reaching France, Australian soldiers, whether they arrived as Reinforcements or as individuals returning from hospitals, were initially dispatched to Base Depots before being deployed to the frontlines. Despite having been certified as fully trained in England, all contingents were subjected to further evaluations, underwent thorough medical inspections, and participated in a minimum of 10 days of additional training.

Private George Hammond joined the 47th Battalion in the Somme valley on the 11th of February 1917. The 47th Battalion was swiftly deployed in the advance towards the Hindenburg Line. German forces had withdrawn to a new defensive stance behind Bapaume.

On the 11th of April 1917, units from the 4th Division, including the 47th Battalion, initiated an assault on the formidable Hindenburg defences at Bullecourt. Faced with significant challenges such as the absence of tank support and effective artillery cover, they suffered heavy casualties. Initially held in reserve for the initial offensive, the 47th was soon called upon to reinforce the attacking units in response to enemy counterattacks. In this fierce engagement, the battalion endured substantial losses, finding itself nearly surrounded and trapped. The encounter proved to be a costly tragedy.

After Bullecourt, the 4th Division was withdrawn from the front lines to recuperate, receive reinforcements, and undergo further training. Eight weeks later, the 47th Battalion found themselves in action once more, this time at Messines, south of Ypres in Belgium, where they suffered losses exceeding half their strength. Advancing during the afternoon of the battle, the Battalion encountered formidable pillboxes that obstructed their progress. Trenches wrested from the Germans and occupied by the Australians came under inadvertent attack by allied artillery, leading to the abandonment of hard-won objectives.

It was during the battle of Messines that Private George Hammond was Killed in Action on the 7th of June 1917. He was 25 years old and had been a member of the AIF for 13 months. His body was recovered and buried *"750 yards east of Messines at (map reference) Sheet 28SW.O.34.CD.U.4.AB"*. Today he has no known grave. Private George Hammond's death was notified in the 323rd list of Australian Casualties released on the 14th of July 1917.

George Hammond's name is inscribed on The Australian War Memorial Roll of Honour (Panel 143), The Ypres (Menin Gate) Memorial (Panel 27), Belgium, The City of Townsville WW1 Honour Roll, and The Townsville Cenotaph.

2446 Private George Hammond was awarded the British War Medal (no:48175) and the Victory Medal (no:47605). A memorial scroll and a memorial plaque were passed to his widow in 1921 and 1922, respectively. Hammond Street, Mysterton, Townsville, was named in his memory in 1926.

Fortnightly war pensions were paid to Private George Hammond's family. His widow, May, received two pounds (£2/-/-), his adopted daughter, Lily May, was granted one pound (£1/-/-), and his daughter, Heatherbelle, received fifteen shillings. The total amount of three pounds and 15 shillings (£3/15/-), was about ninety percent of the fortnightly wage paid to a soldier in the AIF in 1917.

The following tribute was published in the *Townsville Daily Bulletin* on 7th June 1918:

HAMMOND. — In loving Memory of Private George Hammond, 47th Battalion. Killed in Action on the 7th of June 1917.

(Inserted by his loving Father, Brothers and Sisters, William Hammond.)

5118 PRIVATE LINARD LEWIS JOHNSON (1891 – 1917)

47th Infantry Battalion

Linard Lewis Johnson was born on the 11th of October 1891, in Cairns, Queensland. He was the son of Lewis and Ellen (aka Helen) Johnson (née Floyd). Linard had one younger sister and no other siblings, and in 1913, he lived with both parents at a property on Stanton Crescent in Townsville. Lewis Johnson worked as a seaman, while Ellen Johnson was engaged in home duties at that time. Between 1913 and 1915, Ellen Johnson abandoned the family, although the exact date of her departure is not recorded.

Linard Johnson worked as a salesman until the 22nd of November 1915, when he applied in Townsville to enlist in the Australian Imperial Force (AIF). He declared he was 24 years old, single, working as a salesman, and had served for two years with the Senior Cadets in Townsville. His medical examination noted that he was 5'10½" (179 cm) tall, weighed 8st 11lbs (56kg), had a fair complexion, grey eyes, and light brown hair. His religious denomination was Church of England, and he listed his father as his next of kin, giving his address as Stanton Terrace, Townsville.

Accepted into the AIF, Linard Johnson was allocated service number 5118, assigned to the Infantry, and posted to the 15th Battalion, 16th reinforcements. The 15th Battalion, established in September 1914, was formed by recruiting volunteers from Queensland and Tasmania. It was one of the four infantry battalions within the 4th Brigade, alongside the 13th, 14th, and 16th Battalions. At the time of Linard Johnson's enlistment, the 15th Battalion was committed to the Gallipoli campaign, having landed at ANZAC Cove on the 25th of April 1915.

After completing basic training, Private Linard Johnson and his fellow reinforcements joined Convoy No.18, departing Sydney on the 31st of March 1916, bound for Egypt. They sailed aboard A16 HMAT *Star of Victoria*, which carried 1,466 troops. The entire convoy of 10 ships transported a total of 8,075 troops and 1,168 horses. The reinforcements disembarked in Egypt on the 5th of May 1916 and were immediately transferred to the 47th Battalion, which had been established in Egypt on the 24th of February 1916 as part of the "doubling" of the AIF. The battalion's composition included both seasoned veterans of the Gallipoli campaign from the 15th Battalion and fresh recruits from Australia, with a significant number recruited from Queensland and Tasmania. The 47th Battalion was allocated to the 12th Brigade of the 4th Australian Division.

After commencing training in Egypt, Private Linard Johnson fell ill with pleurisy and pneumonia on the 27th of May 1916, just a week after joining the 47th Battalion. He remained in the hospital until July 1916. In August 1916, Private Johnson embarked from Alexandria for England to rejoin his unit, which had already deployed to France as part of the British Expeditionary Force (BEF). They had entered the trenches for the first time on the 3rd of July 1916. Private Johnson underwent an additional month of training in England before crossing over to France, where he rejoined the 47th Battalion on the 29th of September 1916, from the 4th Division's Base Depot in Étaples.

The 47th was engaged in a cycle of alternating duty between the trenches and training/rest behind the lines. The battalion had endured two deployments to the heavily contested trenches of Pozières, as well as a period in reserve. In November 1916, Private Linard Johnson was hospitalized for exhaustion and shell shock. While at the 5th Australian Field Hospital, he contracted influenza and rejoined the 47th Battalion on the 13th of January 1917, near Pozières during the harsh winter.

On the 11th of April 1917, the 47th Battalion attacked Bullecourt, facing strong enemy defences and suffering heavy losses. A mere eight weeks later, the 47th faced another engagement, this time at Messines, situated south of Ypres in Belgium. Advancing during the afternoon of the battle, the battalion encountered formidable pillboxes that obstructed their progress. Trenches wrested from the Germans and occupied by the Australians came under inadvertent attack by allied artillery, leading to the abandonment of hard-won objectives.

It was during the battle of Messines that Private Linard Johnson was Killed in Action on the 7th of June 1917. He was 25 years and eight months old and had been a member of the AIF for 18 months. His body was recovered and buried *"750 yards east of Messines at (map reference) Sheet 28SW.O.34. CD.U.4.AB."* His body was not reinterred and today he has no known grave. His death was officially recorded in the 323rd list of Australian casualties issued on the 14th of July 1917.

Linard Johnson's name is inscribed on The Australian War Memorial Roll of Honour (Panel 143), The Ypres (Menin Gate) Memorial (Panel 27) in Belgium, The City of Townsville WW1 Honour Roll, and The Townsville Cenotaph.

5118 Private Linard Lewis Johnson was awarded the British War Medal (no:34411) and the Victory Medal (no:34168). Lewis Johnson was granted a war pension of 30 shillings per fortnight for his son's sacrifice.

Following his son's death, Lewis Johnson was admitted to the Goodna Asylum and passed away shortly thereafter. Despite efforts to locate Linard Johnson's mother after the war, she could not be found. Consequently, Linard Johnson's service medals, along with a memorial scroll and memorial plaque, were passed to his older sister, Florence Parsons, who lived in New Westminster, British Columbia, Canada.

3104 PRIVATE KEITH McLEAN VILES (1898 – 1917)

15th Infantry Battalion / 47th Infantry Battalion

K. McL. VILES

Keith McLean Viles was born on the 22nd of January 1898 in Croydon, Queensland. He was the fourth of six children born to Douglas and Minnie Viles (née McLean). He had three brothers and two sisters. Douglas Viles was a miner. Young Keith Viles grew up in Croydon and attended the Croydon State School. By 1913, the Viles family had relocated to Townsville and lived in a property on Melton Hill where Douglas Viles worked as a clerk.

Keith Viles trained as a Draper and joined the Cadets and the Citizen Forces. On the 13th of August 1915, he applied at Townsville to enlist in the Australian Imperial Force (AIF). He declared that he was a Draper, single and 18 years old (he was six months younger). He also declared his previous military experience. His medical examination noted that he was 5'7¼" (171 cm) tall and weighed 114 pounds (52 kg). He had a fresh complexion, brown eyes, and dark brown hair. He identified as Church of England and nominated his father living at Paxton St, Townsville as his next of kin.

Accepted into the AIF, Keith Viles was allocated service number 3104, assigned to the Infantry and posted to the 15th Battalion, 10th Reinforcements. The 15th Battalion was originally raised as part of the all-volunteer AIF in September 1914. Drawing personnel from volunteers from Queensland and Tasmania, it formed part of the 4th Brigade, along with the 13th, 14th, and 16th Battalions.

The 15th was deployed to Gallipoli when Keith Viles enlisted but was soon to return to Egypt as part of the evacuation of the Gallipoli Peninsula in December 1915. After two months spent completing basic training, the Reinforcements joined Convoy No.12 and embarked from Brisbane on the 5th of October 1915, bound for Egypt. They sailed aboard HMAT A69 *Warilda*, which carried 674 troops from Brisbane and 687 from Sydney, a total of 1,575 troops. The entire convoy of five ships carried 5,560 troops to Egypt.

Two months later, the Reinforcements reported in at the Moascar Isolation Camp. The isolation camp screened soldiers arriving in Egypt as reinforcements for two weeks, checking for any illnesses such as measles, which can break out when people are crowded together for long periods.

On the 3rd of March 1916, Private Keith Viles was training with the 15th Battalion at Tel-el-Kebir when he was transferred to the newly raised 47th Battalion. The 47th had been established in Egypt on the 24th of February 1916 as part of the AIF's reorganization, which aimed to expand the existing forces. The battalion's composition included both seasoned veterans of the Gallipoli campaign from the 15th Battalion and fresh recruits from Australia, with a significant number coming from Queensland and Tasmania.

The 47th Battalion was allocated to the 12th Brigade of the 4th Australian Division. Three months later, Private Keith Viles embarked at Alexandria for the one-week voyage to Marseilles in France. Arriving in France on the 9th of June 1916, the 47th entered the trenches of the Western Front for the first time on the 3rd of July 1916. It participated in its first major battle at Pozières. Initially, the battalion provided working parties during the 2nd Division's attack on the 4th of August 1916, and then, with its own division, defended the ground that had been captured.

It was immediately after this battle that Private Keith Viles' brother, Private Frank Viles, marched into the 47th and no doubt spent time with his brother when the situation allowed. The reunion was to be short as on the 1st of September, Private Keith Viles was Wounded in Action by a gunshot wound to the wrist, and he was evacuated back to England for hospitalization. Following treatment and some furlough, Private Keith Viles returned to the 47th in the field on the 19th of January 1917.

With the coming of spring 1917, the 47th was soon deployed in the push towards the Hindenburg Line. The German forces had retreated to a new defensive position behind Bapaume. On the 11th of April 1917, units from the 4th Division, including the 47th, launched an attack on the formidable

Hindenburg defences at Bullecourt. They faced immense challenges as they lacked tank support and effective artillery backing, resulting in heavy casualties.

Initially, the 47th Battalion remained in reserve for the initial assault, but they were soon called upon to reinforce the attacking units due to enemy counter-attacks. During this action, the battalion took heavy casualties, was trapped, and almost surrounded. The battle proved to be a costly disaster. As a result, the 4th Division was withdrawn from the front lines to recuperate, receive reinforcements, and undergo further training.

Eight weeks later, in June 1917, the 47th was in action again. This time at Messines south of Ypres in Belgium. The 47th Battalion advanced east of Messines village in the afternoon of the battle. They were tasked to attack, capture, and hold enemy positions in 'Uncanny trench, Unbearable trench, Owl Trench, and the line up to Huns Walk'. Enemy resistance was strong, and it was during this battle that Private Keith Viles was Killed in Action.

Initially, Private Keith Viles was reported as Missing in Action, however, this was later changed to Killed in Action on the 7th of June 1917. His remains were recovered, and he was eventually laid to rest at Messines Ridge British Cemetery (Plot VI, Row C, Grave No. 38), Belgium. He was 19 years and four months old at the time of his death and had been a member of the AIF for one year and 10 months.

Private Keith Viles' death was notified in the 323rd list of Australian Casualties released on the 21st of July 1917. His older brother, Corporal Frank McLean Viles, was Killed in Action during the same battle, on the day after Private Keith Viles' death.

Keith McLean Viles' name is inscribed on The Australian War Memorial Roll of Honour (panel 144), The Croydon War Memorial, The City of Townsville WW1 Honour Roll, and The Townsville Cenotaph. Viles St Rosslea, Townsville, is named in his and his brother Frank's memory.

3104 Private Keith McLean Viles was awarded the 1914/15 Star (no:15092), The British War Medal (no:23233), and the Victory Medal (no:23064). A memorial plaque and a memorial scroll were passed to his father in 1921 and 1922, respectively. Minnie Viles was granted a fortnightly war pension of 14 shillings for the death of Keith Viles and two pounds (£2/-/-) for Frank Viles' death. The combined amount of two pounds, 14 shillings (£2/14/-) was about one third of the fortnightly wage paid to two soldiers in the AIF in 1917.

The following Notice was published in the *Townsville Daily Bulletin* on the 7th of June 1918:

Family Notices

VILES. — In sad and loving Memory of Corporal Francis Watts Viles, 47th Battalion, and Private Keith McLean Viles, 47th Battalion, Killed in Action at Messines, 7th June, 1917, sons of Mr. and Mrs. D. P. Viles, Paxton street.

> He is not dead who gave his life.
> That other men might live from tyrants free;
> But in our hearts forever enshrined shall be,
> And safe from all that hurts and harms,
> They rest within the "Everlasting Arms."

(Inserted by their loving Comrades, Walter and Arthur Jones, from France.)

VILES-KELLY. — In proud Remembrance of our two beloved sons, Corporal Francis Watts Viles, and Keith McLean Viles; also, their dear mate, Jack W. Kelly, of Brandon, all of the 47th Battalion. Killed in Action at Messines on June 7th, 1917.

> In life they were mates and brothers,
> And in death were not divided.

(Inserted by D. P. Viles and Family.)

5205 CORPORAL FRANCIS WATTS VILES (1891-1917)

47th Infantry Battalion

Francis (Frank) Watts Viles was born on the 12th of March 1891 in Croydon, Queensland. He was the first of six children born to Douglas and Minnie Viles (née McLean). He had three brothers and two sisters. Douglas Viles was a miner. Young Frank Viles grew up in Croydon and attended the Croydon State School. By 1913, the Viles family had relocated to Townsville and Douglas Viles worked as a clerk. The Viles family lived in a property on Melton Hill except for Frank Viles who, lived on Paxton Street. Like his father, he was also employed as a clerk.

On the 13th of August 1915, Frank Viles' younger brother, Keith Viles, enlisted in the AIF. Four months later, on the 20th of November 1915, Frank Viles also applied to enlist in the Australian Imperial Force (AIF) at Townsville. He declared that he was single, 24 years old, without previous military experience, and working as a clerk. His medical examination noted that he was 5'6½" (169 cm) tall and weighed 9st. 9lbs. (61 kg). He had a dark complexion, brown eyes, and brown hair. He identified as Church of England and nominated his mother Minnie as his next of kin, giving her address as Paxton St, Townsville.

Accepted into the AIF, Frank Viles was allocated service number 5205, assigned to the Infantry and posted to the 15th Battalion, 16th Reinforcements. The 15th Battalion was established in September 1914 and was made up of recruits from Queensland and Tasmania. It was one of the four Infantry Battalion allocated to the 4th Brigade, alongside the 13th, 14th, and 16th Battalions. When Frank Viles enlisted, the 15th Battalion was preparing to withdraw from Gallipoli to Egypt, as part of the general evacuation from the Gallipoli Peninsula in December 1915.

After four months spent completing basic training, Private Frank Viles and the other reinforcements joined Convoy No.19 and embarked from Brisbane bound for Egypt. They sailed aboard HMAT A16 *Star of Victoria* on the 31st of March 1916. Their ship carried 1,550 troops while the entire convoy of 10 ships carried 8,075 troops and 1,168 horses on the voyage to Egypt.

Six weeks later, the reinforcements reported in at Tel-el-Kebir, and Private Frank Viles was transferred to the 47th Battalion. The battalion had been established in Egypt on the 24th of February 1916 as part of the AIF's "doubling" of the AIF reorganisation. The 47th's manning included both seasoned veterans of the Gallipoli campaign from the 15th Battalion and fresh recruits from Australia, with a significant number coming from Queensland and Tasmania. The battalion was allocated to the 12th Brigade of the 4th Australian Division.

Frank Viles's brother, Private Keith Viles, had joined the 47th Battalion in March 1916 and was located with the 47th in France when Private Frank Viles arrived in Egypt.

In early June 1916, the 47th Battalion embarked at Alexandria for the one-week voyage to Marseilles in France. Private Frank Viles joined the 47th in the field on the 10th of August 1916, immediately after the battle of Pozières. The battalion was defending the ground that had been captured during that battle. Private Frank Viles would have had the opportunity to catch up with his brother when the situation allowed.

The 47th endured two stints in the heavily contested trenches of Pozières, as well as a period in reserve. In October 1916, Private Frank Viles was promoted to Temporary Corporal and then, in December 1916, Substantive Corporal. The 47th Battalion spent the bitterly cold, miserable winter months in trenches near Pozières.

With the coming of spring 1917, the 47th was soon deployed in the push towards the Hindenburg Line. The German forces had retreated to a new defensive position behind Bapaume. In March 1917, Corporal Viles was hospitalized with mumps and then flu. He re-joined the 47th on the 1st of May 1917. The 47th was resting and training behind the lines after its participation in the disastrous first battle of Bullecourt on the 11th of April 1917.

Later in the year, the focus of the AIF's operations switched to the Ypres sector in Belgium, where the 47th took part in the battle of Messines on the 7th of June 1917. The 47th Battalion advanced east of Messines village in the afternoon of the battle. They were tasked to attack, capture, and hold enemy positions in *'Uncanny trench, Unbearable trench, Owl Trench, and the line up to Huns Walk.'* Enemy resistance was strong, and it was during this battle that Corporal Frank Viles was Killed in Action.

Initially, Corporal Frank Viles was reported as wounded and missing; however, a Court of Enquiry later confirmed that he was "killed instantaneously by the explosion of a shell" on the 8th of June 1917. In testimony to the Court of Enquiry, it was stated that Corporal Viles' remains had been later buried by a party from the 4th Pioneers, and the official burial location was recorded as (map reference) Sheet 28SW.O.34.CD.U.4.AB. He was not reinterred, and today he has no known grave.

Corporal Frank Viles was 26 years old and had been a member of the AIF for 19 months. His death was notified in the 333rd list of Australian Casualties released on the 27th of August 1917.

Frank Viles' younger brother Private Keith McLean Viles was also serving in the 47th Battalion, and he was Killed in Action at the same battle, one day before Corporal Frank Viles.

Francis Watts Viles' name is inscribed on The Australian War Memorial Roll of Honour (Panel 144), The Ypres (Menin Gate) Memorial Belgium, Roll of Honor Burke Shire Council, The Croydon War Memorial, The City of Townsville WW1 Honour Roll, and The Townsville Cenotaph. Viles St Rosslea, Townsville is named in his and his brother Keith's memory.

5205 Corporal Francis Watts Viles was awarded the British War Medal (no:34457) and the Victory Medal (no:34213). A memorial plaque and a memorial scroll were passed to his father in 1921 and 1922, respectively. Minnie Viles was granted a fortnightly war pension of two pounds (£2/-/-) for the loss of Frank Viles and 14 shillings (14/-) for the death of Keith Viles. The combined amount of two pounds and fourteen shillings (£2/14/-) was about one third of the fortnightly wage paid in the AIF for two soldiers in 1917.

The following Notice was published in the *Townsville Daily Bulletin* on the 7th of Jun 1918:

VILES. — In sad and loving Memory of Corporal Francis Watts Viles, 47th Battalion, and Private Keith McLean Viles, 47th Battalion, Killed in Action at Messines, 7th June, 1917, sons of Mr. and Mrs. D. P. Viles, Paxton street.

> He is not dead who gave his life.
> That other men might live from tyrants free;
> But in our hearts forever enshrined shall be,
> And safe from all that hurts and harms,
> They rest within the "Everlasting Arms."

(Inserted by their loving Comrades, Walter and Arthur Jones, from France.)

VILES-KELLY. — In proud Remembrance of our two beloved sons, Corporal Francis Watts Viles, and Keith McLean Viles; also, their dear mate, Jack W. Kelly, of Brandon, all of the 47th Battalion. Killed in Action at Messines on June 7th, 1917. In life they were mates and brothers, And in death were not divided.

(Inserted by D. P. Viles and Family.)

53 SAPPER THOMAS JOHN WATSON (1893 – 1917)

41st Infantry Battalion/ 3rd Div Signal Company

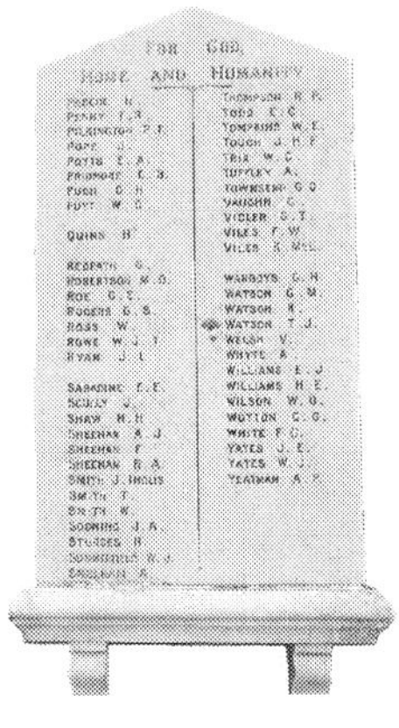

Thomas John Watson was born in May 1893 in Islington, London, England. He was the first-born child of Thomas Watson and his wife Mary Anne (née Tuck). Thomas Watson Snr. was a fireman, and the family lived at the Old Fire Station, 216 Pavilion Rd, London.

Thomas Watson Jnr. grew up in England and attended the Capland St, London County Council School. He trained as a motor mechanic; however, his true interėst was mining. In 1912, he emigrated to Australia, arriving in Sydney, New South Wales, on the 5th of December. He found employment as a wolfram (tungsten) miner at Ollera Creek, north of Townsville in Queensland.

On the 10th of January 1916, Thomas Watson quit mining and applied at Townsville to enlist in the Australian Imperial Force (AIF). He declared that he was 22 years old, single, and that he had previous military service with the British Navy. His enlistment medical examination noted that he was 5'6" (168 cm) tall, weighed 9st 3lbs (59kg), had a dark complexion, brown eyes, and brown hair. He identified as Church of England and listed his mother as his next of kin, giving her address as Old Fire Station, 216 Pavilion Rd, London, England.

Accepted into the AIF, Thomas Watson was allocated service number 53, allocated to the infantry and posted as an original member to the 41st Infantry Battalion. The 41st was raised at Bell's Paddock Camp in Brisbane in February 1916, with recruits from Brisbane, northern Queensland, and the northern rivers district of New South Wales. It formed part of the 11th Brigade of the 3rd Australian Division.

Private Thomas Watson commenced basic training in Brisbane with No. 11 Depot Battalion in the middle of January 1916, soon after enlistment. Three weeks later, he was moved to the Depot Signal Company, and at the end of April 1916, he was transferred to the 3rd Divisional Wireless Sub Section.

Following basic training, Private Thomas Watson moved by train to Sydney on 16th May 1916 to embark for England as part of Convoy No. 21. Travelling with the 41st Battalion, his unit sailed aboard HMAT A64 *Demosthenes.* Their ship carried 1,610 troops, and the entire convoy of 18 vessels carried a total of 25,287 troops on the voyage. Their journey took them through Cape Town and the Cape Verde Islands before finally reaching Plymouth in July 1916.

Once disembarked, they were transported to training grounds on Salisbury Plain at Larkhill. In November 1916, Private Watson crossed to France and the Western Front from Southampton, England. Further training occurred in France, and on the 5th of April 1917, he was placed on a one-week buzzer course.

Upon his return to the 41st Battalion, Private Watson became Sapper Watson because of his transfer to the 3rd Divisional Signalling Company located at Croix-du-Bac near Steenwerck. Divisional Signals Companies belonged to the Royal Australian Engineers during WW1. Later, the Royal Australian Signals Corps was created. The Divisional Signals Company was tasked with providing communications from the Headquarters to those of subordinate formations.

For the next six weeks, the Signals Company was busy laying lines, burying cables, and setting up the equipment to provide communications. The War Diary records the fact that personnel were under constant fire, and casualties were common. The tempo of the Unit was at its height in June 1917 as the Battle of Messines commenced.

On the 10th of June 1917, Sapper Thomas Watson was located at the 11th Brigade Forward Station when he was Killed in Action. The Unit War Diary records; *"11th Brigade. Forward Station blown out. Sappers Simons and Watson killed. Communications re-established within 5 minutes and relief effected."*

Sapper Thomas Watson was 24 years old at the time of his death and had been a member of the AIF for 17 months. He was laid to rest at Messines Ridge British Cemetery, Belgium.

Thomas John Watson's name is inscribed on The Australian War Memorial Roll of Honour (Panel 26), The City of Townsville WW1 Honour Roll, and The Townsville Cenotaph.

53 Sapper Thomas John Watson served in WW1 on the Western Front. He was awarded the British War Medal (no:39951) and the Victory Medal (no:39565). These medals, as well as a memorial scroll and a memorial plaque, were issued in England to his mother.

3048 PRIVATE JOHN WILLIAM KELLY (1892 – 1917)

47th Infantry Battalion

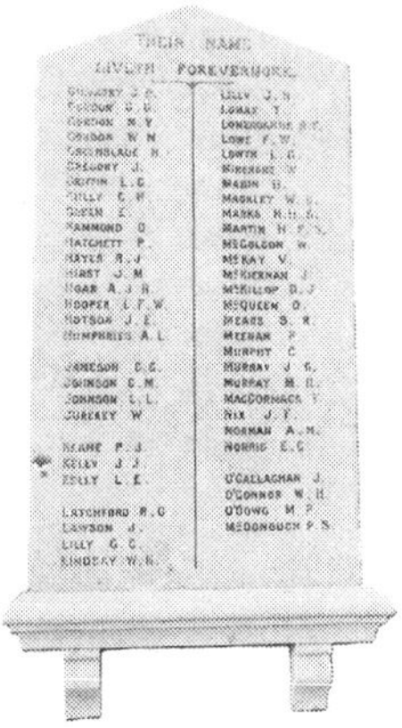

John William Kelly was born on the 12th of December 1892, in Charleville, Queensland. He was the oldest child of John and Ellen Kelly (née McNulty). He had four brothers and one sister. The family moved to Ayr where John Kelly Sr. became the licensee of the Club Hotel in Brandon, Queensland, while Ellen was engaged in home duties. John Kelly Jr. trained as a draper.

On the 13th of August 1915, John Kelly Jr. applied at Brisbane to enlist in the Australian Imperial Force (AIF). He declared that he had no previous military service, was 22 years and eight months old, single, and working as a draper's assistant. John Kelly's enlistment medical examination noted that he was 5'4½" (164 cm) tall, weighed 118 lbs (53.5 kg), had a fair complexion, light brown eyes, and light brown hair. He identified as Roman Catholic and he nominated his father, c/- the Club Hotel Brandon, as his next of kin.

Accepted into the AIF, John Kelly was allocated service number 3048, assigned to the Infantry and posted to the 15th Battalion, 10th Reinforcements. The 15th Battalion was originally raised as part of the all-volunteer AIF in September 1914. Drawing personnel from volunteers from Queensland and Tasmania, it formed part of the 4th Brigade, along with the 13th, 14th, and 16th Battalions. The Battalion was fighting on the Gallipoli Peninsula when John Kelly enlisted.

After seven weeks spent completing basic training, Private John Kelly and the other reinforcements joined Convoy No.12 and embarked from Brisbane on the 5th of October 1915, bound for Egypt. They sailed aboard HMAT A69 *Warilda* which carried 1,361 troops. The entire convoy of five ships carried a total of 5,560 passengers. The reinforcements disembarked in December 1915. After some further training and a bout of illness, Private John Kelly joined the 15th Battalion at Moascar in early February 1916.

Approximately six weeks later, he was transferred to the newly formed 47th Battalion, established as part of the AIF's expansion. The 47th drew its ranks from a mix of Gallipoli veterans of the 15th Battalion and fresh recruits from Australia. Much like its predecessor, the 15th Battalion, the 47th was predominantly composed of men recruited from Queensland and Tasmania.

Assigned to the 4th Division, the 47th Battalion, along with the 45th, 46th, and 48th Battalions, formed the 12th Brigade. The battalion commenced training in Egypt, setting sail from Alexandria bound for France approximately three months later, in June 1916. On the 9th of June 1916, the 47th Battalion arrived in France and was swiftly deployed to the trenches of the Western Front by the 3rd of July 1916.

The 47th Battalion fought its first major engagement at Pozières. Initially, the battalion was tasked with providing support through working parties as the 2nd Division launched its assault on the 4th of August 1916. Following the successful capture of territory, the 47th assumed responsibility for defending the hard-won positions alongside its division. Enduring two rotations in the fiercely contested trenches of Pozières, the battalion also spent a period in reserve during this time.

The end of August 1916 saw Private John Kelly admitted to the hospital for a week suffering from an unspecified illness. After Pozières, the 47th spent the period up until March 1917 alternating between duty in the trenches and training and resting behind the lines. In November 1916, Private John Kelly was granted leave in England. While on leave, he was admitted to Bulford Military Hospital. He was hospitalized until the 8th of March 1917.

Following the mandatory training in England, required after a period of hospitalization, Private Kelly crossed to France on the 29th of April 1917. He rejoined the 47th Battalion from Étaples in France on the 4th of May 1917 along with 52 other reinforcements. A few days later, John would have received the news that his younger brother, Private Leslie Kelly, had been Killed in Action at Bullecourt on the 3rd of May 1917 while serving with the 26th Battalion.

From the 7th to the 14th of June the 47th Battalion was engaged in the pivotal Battle of Messines, serving as a vital prelude to the Battle of Passchendaele. The main goal was the seizure of Messines Ridge. To accomplish this objective, powerful high-explosive mines were strategically placed beneath the German lines, resulting in tremendous detonations that reverberated as far as London, approximately 130 miles (209 km) distant.

During the afternoon of the battle, the 47th Battalion encountered significant resistance in the form of pillboxes, impeding their advance. Subsequently, they gained control of German trenches, only to face a tragic turn of events when they were mistakenly targeted by allied artillery. This resulted in the objectives that had been won being abandoned. The toll on the Australian forces was immense, with over 6000 casualties suffered in this intense battle. Tragically, the 47th Battalion alone lost more than half of its personnel during the conflict.

One of those casualties was Private John Kelly, who was wounded by gunshot in the thigh and neck on the 7th of June 1917. He was quickly evacuated to the 2nd New Zealand Field Ambulance but Died of Wounds on the same day. He was 24 years and five months old and had been a member of the AIF for one year and 10 months. His body was recovered, and he was laid to rest at Kandahar Farm Cemetery (Plot II, Row A, Grave No. 35), Neuve-Eglise, Belgium.

Private John Kelly's death was notified in the 316th list of Australian casualties issued on the 27th of June 1917. The news would have been a crippling tragedy for John's parents as the notification of their younger son's death (Private Leslie Kelly) would have only reached them a couple of weeks before the death of John Kelly.

John Kelly's name is inscribed on The Australian War Memorial Roll of Honour (Panel 143), The Brandon War Memorial, The City of Townsville WW1 Honour Roll, and The Townsville Cenotaph (mistakenly engraved as J.J. Kelly instead of J.W. Kelly). Kelly Street, Mundingburra, and Kelly St Nelly Bay were named in his and his brother's memory in 1919.

3048 Private John William Kelly was awarded the 1914/15 Star (no.15378), the British War Medal (no:20308), and the Victory Medal (no:20232). A memorial scroll and a memorial plaque were passed to his father in 1921 and 1922, respectively. John Kelly's mother was granted a pension of 14 shillings per fortnight for the loss of her son this amount was about one fifth of the fortnightly amount paid to an AIF soldier in 1917.

737 PRIVATE EUGENE EDWARD SABADINE (1892 – 1917)

41st Infantry Battalion

Eugene Edward Sabadine was born on the 17th of June 1892 in Ravenswood, Queensland. He was the fifth of six children born to Joseph and Elizabeth Sabadine (née O'Rourke). Joseph Sabadine was an Austrian immigrant who worked as a labourer. By 1903, the Sabadine family had moved to a property on Hale Street, Townsville. Joseph Sabadine died in 1909 at the age of 59. Growing up, Eugene Sabadine received a catholic education. When his school years were finished, he started a career in Townsville as an ironmonger and a salesperson with Smellies & Co. hardware and machinery merchants.

On the 18th of October 1915, Eugene Sabadine applied to enlist in the Australian Imperial Force (AIF) at Townsville. He declared that he was 23 years old, single and had previous military service with the Australian Garrison Artillery. His enlistment medical examination noted that he was 5'6" (168 cm) tall, weighed 9st 10lbs (62 kg), had a dark complexion, grey eyes, and black curly hair. Eugene Sabadine identified as Roman Catholic and listed his widowed mother as his next of kin giving her address as Hale St Townsville.

Accepted into the AIF, Eugene Sabadine was allocated service number 737, allocated to the infantry and posted as an original member to the 41st Battalion, "C" Company. The 41st was raised at Bell's Paddock Camp in Brisbane in February 1916 with recruits from Brisbane, northern Queensland, and the northern rivers district of New South Wales. It formed part of the 11th Brigade of the 3rd Australian Division.

After initial training in Australia, Private Eugene Sabadine embarked with his battalion from Sydney on the 18th of May 1916, bound for England. They sailed as part of Convoy No. 21 aboard HMAT A64 *Demosthenes*, which carried 1,612 troops on the voyage. The entire convoy of eighteen ships transported 25,333 troops.

The 41st Battalion disembarked at Plymouth in late July 1916 and moved with the 3rd Division to training grounds on Salisbury Plain at Larkhill, site of the current Royal School of Artillery. Months of further training followed in England before the 41st Battalion crossed to France on the 25th of November 1916, six months after leaving Australia.

Private Sabadine was admitted to the hospital soon after arrival in France, suffering from mumps which swept through the AIF. The 41st entered the frontline for the first time on Christmas Eve, 1916, at the Mushroom Salient, just east of La Chapelle-d'Armentières. Four soldiers were killed by artillery fire on the first night. Two days later, Private Sabadine re-joined the battalion from hospital.

The 41st cycled through the frontline near Armentieres in France and across the Lys River in Belgium for the majority of the first half of 1917. The primary offensive of that year unfolded in June during the Battle of Messines. Here, the 41st Battalion maintained the entirety of the 11th Brigade frontage prior to the detonation of the mines and the commencement of the assault.

Following this, Private Sabadine participated in a succession of smaller "Bite and Hold" manoeuvres to capitalise on the gains made during the Battle of Messines. Among these actions, the most significant was the attack on the 'Windmill' position held by the Germans near Warneton, Belgium.

On the 9th of July 1917, Private Sabadine's Company was under heavy shelling and sniper fire as it occupied a portion of the Messines - Warneton Line. It was during this shelling that Private Eugene Sabadine was Killed in Action. In a statement made to the Australian Red Cross in September 1917, 2593 Pte W.J. Sheahan said: *"I knew Pte. Sabadine of C Coy 10 Pltn (or 11th). He was in our front-line trench at Messines on July 9, 1917, opposite Warneton. I was within 30 yards of him when he was killed instantly by a shell about midnight. I saw his body after. We held the ground. The Stretcher Bearers took the body away to bury it next morning. I do not know where the grave is, but I think it would be in the Military Cemetery at Messines."*

Private Eugene Sabadine was 25 years old and had been a member of the AIF for one year and nine months at the time of his death. He was laid to rest at Messines Ridge British Cemetery (Plot I, Row E, Grave No. 7), Belgium and his death was notified in the 328th list of Australian Casualties released on the 1st of August 1917.

Eugene Edward Sabadine's name is inscribed on The Australian War Memorial Roll of Honour (Panel 134), The 41st Battalion Roll of Honour, The City of Townsville WW1 Honour Roll, and The Townsville Cenotaph. Sabadine Street, Aitkenvale Townsville, is named in his memory.

737 Private Eugene Edward Sabadine served in WW1 on the Western Front. He was awarded the British War Medal (no:39915) and the Victory Medal (no:39531). These medals, as well as a memorial scroll and a memorial plaque, were passed to his widowed mother in 1922. His mother received a war pension of two pounds (£2/-/-) per fortnight from the 19th of September 1917 for her son's sacrifice. This was about half of the wage paid to an AIF soldier in 1917.

The following tribute was printed in the *Townsville Daily Bulletin* on 9th April 1918:

SABADINE. — In loving memory of our son and brother, Eugene Edward (Sabo), who was Killed in Action, 9th April 1917.

> He died for his country.
> May God in his blessing,
> Have mercy on his soul.

Inserted by his loving mother, sister, and brothers.

1848 LANCE CORPORAL WILLIAM JUREKY (1887 – 1917)

44th Infantry Battalion

William Jureky was born on the 28th of August 1887 in Hughenden, Queensland. He was the only child of William and Ellen Jureky (née Dulahentey/ Delahunty). William Jureky Snr. died in Hughenden when William Jureky Jnr. was a young child. After her husbands death, Ellen Jureky moved the family to Townsville where she worked as a laundress. William Jureky attended the Sisters of Mercy school and worked as a labourer and lorry driver when his school years were finished.

On the 31st of January 1912, William Jureky married Susan Anne Mary Eden in Townsville, Queensland. Following their marriage, they lived with William's mother at a property on Mitchell Street for a while and then set up their own place on Flinders St West. They had two children during their marriage.

On the 3rd of April 1916, William was 27 years old and working as a lorryman when he applied at Townsville to enlist in the Australian Imperial Force (AIF). His enlistment medical examination noted that he was 5'11½" (182 cm) tall, weighed 10st 7lbs (67 kg), had a dark complexion, blue eyes, and dark hair. He identified as Roman Catholic and nominated his wife as his next of kin, giving her address as Flinders St West, Townsville.

Accepted into the AIF, William Jureky was allocated service number 1848, assigned to the infantry and posted to the 41st Battalion, 2nd Reinforcements. The 41st Battalion was raised at Bell's Paddock Camp in Brisbane in February 1916 with recruits from Brisbane, northern Queensland, and the northern rivers district of New South Wales. It formed part of the 11th Brigade of the 3rd Australian Division.

For Private William Jureky and the other reinforcements, the next five months were spent in Australia preparing for overseas service. On the 16th of August 1916, they joined Convoy No. 23 and embarked from Brisbane, bound for England. They sailed aboard HMAT A42 *Boorara,* which carried 1,707 troops. The entire convoy of 12 ships carried a total of 11,246 troops on the voyage. Private William Jureky disembarked at Plymouth on the 13th of October 1916.

From there, the troops underwent further training in preparation for deployment to the Western Front. Two weeks after arrival in England, Private Jureky was transferred from the 41st Battalion to the 44th Battalion, which had been raised at Claremont, Western Australia, in February 1916. The battalion underwent further training on Salisbury Plain in Wiltshire until November 1916, at which time the 44th crossed to France.

Upon disembarking in France, Australian soldiers, whether reinforcements or returning from hospitals, were directed to Base Depots before their deployment to the frontlines. Despite having completed full training in England, each draft underwent rigorous testing, a thorough medical examination, and at least 10 days of supplementary training.

The 44th Battalion entered the front-line trenches for the first time on the 29th of December 1916. There they spent the bleak winter of 1916-17 alternating between service in the front line, and training and labouring in the rear areas. This routine was broken by only one major raid, an ill-fated effort involving almost half the battalion on the 13th of March 1917.

Private William Jureky must have acquitted himself well in the 44th Battalion because he was promoted to Lance Corporal in April 1917. The battalion fought in its first major battle at Messines, in Belgium, between the 7th and 10th of June 1917. In the months that followed, it was heavily employed in the Ypres sector, taking part in another major battle to capture Broodseinde Ridge on the 7th of June 1917, and participating in costly defensive operations in horrendous conditions.

Of the 992 men from the battalion who were involved in the Ypres operations, only 158 emerged unwounded. Lance Corporal William Jureky was one of the casualties of this fighting when he was severely wounded on the 9th of July 1917. He was evacuated to the 9th Australian Field Ambulance, but he succumbed to his wounds and Died of Wounds on the 10th of July 1917.

William Jureky was 29 years and 10 months old and had served in the AIF for 15 months at the time of his death. He was laid to rest at Trois-Arbres Cemetery (Plot I, Row U, Grave No. 30), Steenwerck, France, and his death was notified in the 328th list of Australian casualties released on the 1st of August 1917.

William Jureky's name is inscribed on The Australian War Memorial Roll of Honour (Panel 138), The City of Townsville WW1 Honour Roll, and The Townsville Cenotaph. Jureky Street, Cluden Townsville, is named in his memory.

1848 Lance Corporal William Jureky served in WW1 on the Western Front. He was awarded the British War Medal (no:45376) and the Victory Medal (no:44855). These medals, as well as a memorial scroll and a memorial plaque, were passed to his widowed wife, Susan, in 1921 and 1923. Susan Jureky was granted a war pension of two pounds (£2/-/-) per fortnight from 19th September 1917. The two children also received a fortnightly pension from the same date, Ivy Jureky received one pound (£1/-/-), and Henry Jureky 15 shillings. The total amount of these pensions was the equivalent of ninety percent of the fortnightly wage paid to a soldier of the AIF in 1917.

The following tribute was published in the *Townsville Daily Bulletin* on the 25th of July 1917:

JUREKY. — In loving remembrance of my darling son (only child), Lance Corporal Jureky, who was Killed in Action on July 10th. 1917.

R.I.P.

Deeply regretted by his affectionate mother, Ellen Jureky.

The following tribute was published in the *Townsville Daily Bulletin* on the 26th of July 1917:

Jureky, - In loving remembrance of my dear husband and father, Lance-Corporal William Jureky, who Died of Wounds on July 10th, 1917.

R.I.P.

(Deeply mourned by his loving wife and children. Susan Jureky)."

The following tribute was published in the *Townsville Daily Bulletin* on the 10th of July 1918:

JUREKY. — In loving memory of our dear husband and father, Lance Corporal William Jureky, 44th Battalion, who Died of Wounds in France on July 10th, 1917.

Only a grave of a hero,
Only a mound of earth.
Far from the land of the wattle,
The place that gave him birth,
Somewhere in France he is lying,
He answered his country's call
He died an Australian hero,
Fighting to save us all.

(Inserted by his loving wife and children, Ivy and Henry, Susan Jureky.)

Rest in Peace.

1899 PRIVATE JOHN GREGORY (1889 - 1917)

52nd Infantry Battalion

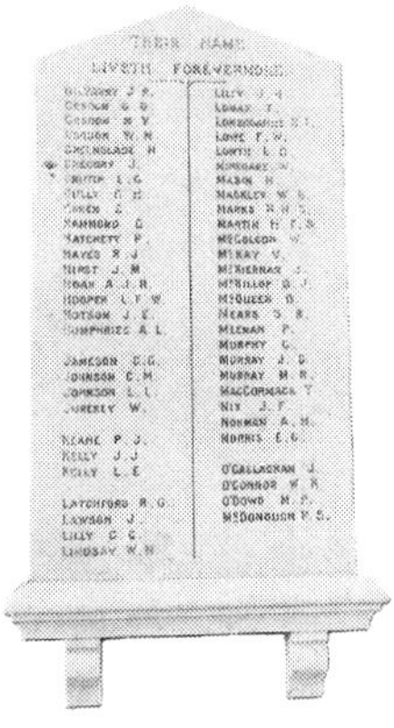

John Gregory (Greg) was born in October 1889 in Parramatta, New South Wales. He was the second son of a large family born to William and Maria Gregory (née Irving). William Gregory was a farmer, and by 1903 the Gregory family lived at Hillgrove Farm, Brandon, near Ayr, North Queensland. Brandon is south of Townsville, and John Gregory grew up in the town and attended the Brandon State School. When his school years were finished, John followed his father's livelihood and took up farming.

On the 27th of December 1915, John Gregory applied at Townsville to enlist in the Australian Imperial Force (AIF). He declared that he was 26 years old, single, and a farmer. He had previously entered the AIF but was rejected as *"med unfit – teeth"* after three days. His enlistment medical examination in December 1915 noted that he was 5'11½" (182 cm) tall, weighed 10st 13lbs (69 kg), had a fair complexion, blue eyes, and brown hair. He identified as Roman Catholic and nominated his father as his next of kin, giving his address as Brandon, Qld.

Accepted into the AIF, John Gregory was allocated service number 1899, assigned to the Infantry and posted to the 52nd Battalion/3rd Reinforcements. The 52nd Battalion was established during the reorganization and expansion of the AIF after the Gallipoli campaign. Its composition consisted of two groups: one comprising veterans from the 12th Battalion recruited from South Australia, Western Australia, and Tasmania, while the other half consisted of Queenslanders.

Private John Gregory and the rest of the reinforcements spent the five months after enlistment training and preparing for overseas service. On the 1st of May 1916, they joined Convoy No. 22 and embarked from Brisbane, bound for Egypt. They sailed aboard HMAT A46 *Clan MacGillivray*, which carried 1,039 troops. The entire convoy of 13 ships carried a total of 8,462 troops on the voyage.

From Egypt, the reinforcements embarked for England on the 6th of August 1916. Ten days later, they disembarked at Plymouth and marched into the 13th Training Battalion, located at Rollestone on Salisbury Plain in Wiltshire. Here, the troops underwent further training in preparation for deployment to the Western Front.

After three months of training, Private John Gregory and the other reinforcements crossed to France in November 1916 and spent two more weeks in training at the 4th Australian Division Base Depot at Étaples. Private Gregory was taken on the strength of the 52nd Battalion on the 18th of November 1916. The 52nd Battalion was positioned on the Somme when he joined them.

After the second attack on Mouquet Farm, the battalion was in a phase of recovery. During the attack, the 52nd had undertaken a crucial role and endured significant casualties, amounting to 50% of its fighting strength. Following this battle, the battalion spent the remainder of the year alternating between front line duty and engaging in training and labouring tasks behind the lines.

This routine endured through the harsh winter of 1916-17, during which time Private John Gregory took up the task of a stretcher bearer. Within the infantry battalions, these bearers carried out the vital duty of retrieving wounded soldiers. Their responsibilities encompassed applying bandages to wounds and conveying the injured to the Regimental Aid Post (RAP) for treatment by the battalion's medical officer. The soldiers held deep respect for the stretcher bearers. Despite lacking arms and relying solely on their 'SB' armbands for identification, they courageously worked even amidst enemy fire, bravely transporting wounded soldiers to safety.

In early 1917, the 52nd Battalion was committed to the offensive of following the German withdrawal to the Hindenburg Line, launching an assault on Noreuil on the 2nd of April 1917. Later that same year, AIF operations shifted to the Ypres sector in Belgium. Here, Private Gregory played a role in the Battle of Messines from the 7th to the 12th of June 1917. He survived that battle, but in later action around Messines, Private John Gregory was Killed in Action by shellfire on the 11th of July 1917.

An Australian Red Cross Investigation provided the following detail from a witness (1881 Private C.C. Collins) to John's death, who stated inter alia; *"Pte Gregory was killed at 'Canteen Corner' by shellfire. It was after Messines position was consolidated. I saw him hit. It was in daylight. His dead body was taken away with two others who were also killed."*

Private John Gregory was 27 years and nine months old and had been a member of the AIF for 18 months at the time of his death. He was laid to rest at the Pont D'Achelles Cemetery (Plot 1, Row C), France. His death was notified in the 328th list of Australian Casualties released on the 1st of August 1917.

John Gregory's name is inscribed on The Australian War Memorial Roll of Honour, the Brandon War Memorial, The City of Townsville WW1 Honour Roll, and The Townsville Cenotaph. 1899 Private John Gregory served in WW1 on the Western Front. He was awarded the British War Medal (no:37810) and the Victory Medal (no:37502). A memorial scroll and a memorial plaque were passed to his father in 1922.

3207 PRIVATE RICHARD FRANCOIS LONERGANNE (1896 – 1917)

49th Infantry Battalion

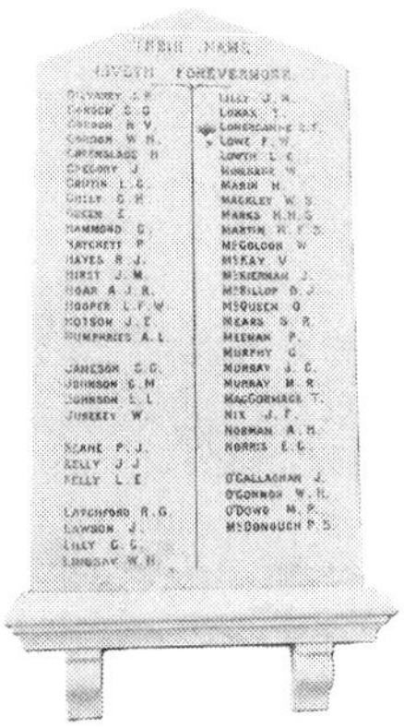

Richard Francois Lonerganne was born on 12th February 1896 in Brisbane, Queensland, to Frederick and Margaret (Marguerite) Lonerganne (née Quenet). He was the youngest of three children and the only son. Frederick Lonerganne was a clerk and in 1916, the family lived in Mitchell St, Townsville.

Richard Lonerganne followed in his father's footsteps and became a clerk while his two sisters worked as shorthand typists. On the 21st of October 1916, Richard Lonerganne applied at Townsville to enlist in the Australian Imperial Force (AIF). He was 20 years old and single. His parents had to provide written consent for him to enlist as he was younger than 21 years.

On enlistment, he declared that he served in the Senior Cadets and attained the rank of Lieutenant. His medical assessment noted that he was 6'1" (185cm) tall, weighed 9st 5lbs (59kg), had a fair complexion, grey eyes, and brown hair. He identified as Church of England and nominated his father as his next of kin, giving his address as Mitchell St Townsville. Later in his service, he nominated his mother at the same address as the only beneficiary in his will. He also allotted three shillings per day (half of his Army pay) to his mother.

Accepted into the AIF, Richard Lonerganne was allocated service number 3207, assigned to the Infantry and was posted to the 49th Battalion, 8th Reinforcements. Two months of basic training in Australia followed enlistment. Private Richard Lonerganne and the other reinforcements embarked from Sydney bound for England on the 23rd of December 1916. They sailed aboard HMAT A64 *Demosthenes* as part of Convoy No. 27, and their ship carried 1,240 troops. The entire convoy of 13 ships carried 15,278 troops and 311 other passengers on the voyage.

Ten weeks later, the Reinforcements disembarked at Plymouth in England and were transported to Codford in Wiltshire, where they completed another three months of training. In June 1917, Private Richard Lonerganne crossed to France and marched into the 4th Division's Base Depot in Harfleur, near Le Havre, France. Three weeks later, Private Lonerganne and the rest of the reinforcements marched into the 49th Battalion in the field on the 4th of July 1917. The 49th was in camp at Desule in the Messines sector in Belgium at this time, having taken part in the Battle of Messines in June.

Private Lonerganne was allocated to the Lewis machine gun section of 'B' Company. On the 14th of July 1917, the 49th moved back into the front line, relieving the 13th Battalion. The 49th was required to remain there until the 19th of July 1917. Once in the front lines, the Battalion conducted patrols in the ground to their front and in the vicinity of the Lys River. They spent time improving trenches and wiring the support line. During this period, six soldiers were killed, and 20 were wounded.

One of those killed was Private Richard Lonerganne, who was Killed in Action on the 15th of July 1917. In a later statement made by Private F.J. Thurston in an Australian Red Cross investigation, it was established that Richard was manning a machine gun forward of the 49th's position when he was struck in the heart by a piece of shrapnel, and that he died instantly. He also reported that Private Lonerganne was buried, near where he fell, on the next day. His body was never located, and today he has no known grave.

Private Richard Lonerganne was 21 years and five months old at the time of his death and had been a member of the AIF for nine months. His death was notified in the 329th Australian casualty list released on the 9th of August 1917.

Richard Lonerganne's name is inscribed on The Australian War Memorial Roll of Honour (panel 148), The Ypres (Menin Gate) Memorial, Belgium, The City of Townsville WW1 Honour Roll, and The Townsville Cenotaph. Lonerganne Street, Garbutt, Townsville, is named in his memory.

3207 Private Richard Francois Lonerganne was awarded the British War Medal (no:54786) and the Victory Medal (no:54024). After their son's death, his mourning parents left Townsville and moved to Maroubra Bay in the southern part of Sydney. They called their new residence, located in Bona Vista Street, 'Messines'. A memorial scroll and a memorial plaque were passed to his father in 1922. Margaret Lonerganne was granted a war pension of two pounds (£2/-/-) per fortnight for the loss of her son. This was about half the fortnightly wage of an AIF soldier in 1917.

21430 DRIVER WILLIAM EDWARD TOMPKINS (1890 – 1917)

107th Howitzer Battery

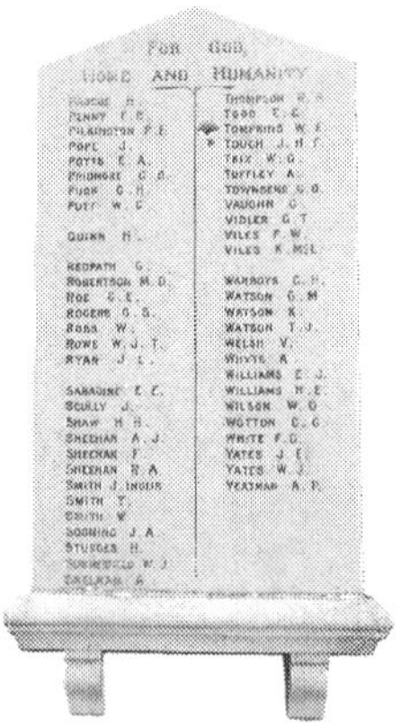

William Edward Tompkins was born on the 27th of July 1890, at Townsville Queensland, to William and Elizabeth Tompkins (née Wagland). He was the first born of five children, and he had two brothers and two sisters. William Tompkins Snr. was a Woodcutter, and the family lived at Bohle River in Townsville. William Tompkins Jnr. grew up in Townsville and attended the West End State School. After completing his school years, he lived in Mundingburra, Townsville, and worked as a bus driver in 1913 and as a teamster in 1915.

On the 6th of September 1915, William Tompkins Jnr. applied to enlist in the Australian Imperial Force (AIF). He declared that he was 25 years old, single, qualified as a teamster, and had no previous military experience. His medical examination noted that he was 5'9½" (177 cm) tall, weighed 10st 7lbs (67 kg), had a fair complexion, brown eyes, and brown hair. He identified as

Church of England and listed his father as his next of kin, giving his address as Bohle River, Ingham Road, Townsville.

Accepted into the AIF, William Tompkins was allotted service number 21430, allocated to the Artillery and posted to the 9th Field Artillery Brigade, which formed in New South Wales in April 1916. From September 1915 to May 1916, Gunner Tompkins underwent initial training in Australia.

On the 11th of May 1916, the 9th Field Artillery Brigade embarked from Sydney, as part of Convoy No. 21. They sailed aboard HMAT A8 *Argyllshire* which carried 1,763 troops for the 10-week journey to Plymouth, England. The 18-ship convoy carried a total of 25,333 troops on the voyage.

Once disembarked, more training followed at Larkhill, Salisbury Plains, England. On the 25th of August 1916, Gunner Tompkins became Driver Tompkins when he was, unsurprisingly, appointed as a Driver. A Driver (Dvr) denoted a military rank within the AIF, equivalent to that of a private. This rank pertained specifically to individuals in the Artillery responsible for manoeuvring teams of horses that pulled both the artillery guns and the wagons carrying ammunition. It was a hazardous job, and Drivers and their teams were high-value targets for enemy artillery.

A war correspondent records a statement of Major Manton, Battery Commander, *"It was looked on almost as a cold-footed job before,' Manton said, 'one which did not take a man into action. But like all those Australians who were supposed to be in fairly safe jobs, the drivers took a pride in showing what they could do when they came into the thick of it.'* He added: *"even the animals came to know when a shell was coming close; and if, when halted, the horses heard the whine of an approaching salvo, they would tremble and sidle closer to their drivers, burying their muzzles in the men's chests."*

On the 30th of December 1916, Driver Tompkins crossed over to France, departing from Southampton. After disembarking in Havre, France, the Artillery moved by train to Bailleul in Northern France and then marched the 10 kilometres to Strazeele.

On the 6th of January 1917, a reorganization of Divisional Artillery was conducted. As part of that reorganization, a section from the 109th Battery was transferred to the 7th Field Artillery Brigade (FAB). Driver William Tompkins was one of the soldiers transferred, and was allocated to the 107th Howitzer Battery, equipped with 4.5-inch howitzers.

The 107th first saw action near Armentieres in early 1917, as part of 3rd AIF Division commanded by General Monash. Subsequently, the Battery saw action at Ypres, Ploegsteert, Zonnebeke, Passchendaele, Hell Fire Corner, and Messines. It was at Messines on the 23rd of July 1917, that Driver William Tompkins was Killed in Action a few days before his 27th birthday. He had been a member of the AIF for one year and 10 months and was laid to rest at Kandahar Farm Cemetery (Plot II, Row C, Grave No. 10), Neuve-Eglise, Belgium. His death was reported in the 331st List of Australian Casualties, published on the 18th of August 1917.

William Edward Tompkins' name is inscribed on The Australian War Memorial Roll of Honour (Panel 15), The City of Townsville WW1 Honour Roll, and The Townsville Cenotaph. Tomkins Street, Cluden, Townsville, is named in his memory (note misspelling of the name Tompkins).

21430 Driver William Edward Tompkins served on the Western Front in WW1. He was awarded the British War Medal (no:40303) and the Victory Medal (no:39899). A memorial scroll and a memorial plaque were passed to his mother in 1922 and 1923, respectively.

The following notices were published in the *Townsville Daily Bulletin* (Qld) on the 23rd of July 1918:

TOMPKINS. — In loving memory of my dear brother, Driver W. E. Tompkins, who was Killed in Action in France on the 23rd July 1917.

> In a hero's grave he lies.
> You are always in our thoughts dear Willie,
> 'Tis sweet to breathe your name.
> In life we loved you dearly,
> In death, we do the same.

(Inserted by his sorrowing sister, Mrs. G. Morris.)

TOMPKINS. — In loving remembrance of my dear son and our brother, Driver William Edward Tompkins, who was Killed in Action in France, on the 23rd July 1917. Aged 27 years on the 25th July.

For Britain's home and freedom.
And to right a dreadful wrong.
He chose the path of duty,
And joined the khaki throng.
A better son never lived,
Nor one more true or kind,
His equal in this weary world,
We very rarely find.
His heart was always in his home,
He toiled and did his best.
Until there came that sudden call,
And God called him home to rest.

(Inserted by his loving parents, brothers, and sisters.)

320 PRIVATE JAMES ERNEST YATES (1893 – 1917)

41st Infantry Battalion

James Ernest Yates was born on the 15th of April 1893 in Townsville, to James and Ellen Yates (née Robinson). He was the third of five children and had one older brother and three sisters. James Yates Jr. hardly knew his father as James Yates Sr. died on the 5th of November 1899 at the age of 46. Ellen (Nellie) was left as a widow with five young children, all under the age of nine.

In 1903, Nellie Yates and the family lived in a property on Elizabeth Street, Aitkenvale, Townsville. James Yates grew up in Townsville and attended the Mundingburra State School. In 1905 Nellie Yates married Thomas Tobin, a horse team driver who worked the road between Townsville and Georgetown. After his school years were completed, James Yates worked as a labourer until the 3rd of January 1916 when he applied, at Townsville, to enlist in the Australian Imperial Force (AIF).

His older brother, William Yates, had joined in October 1915 and was serving in Brisbane with the 41st Battalion. James Yates declared that he had no previous military experience and that he was 22 years old, single and working as a labourer. His medical examination noted that he was 5'9" (175 cm) tall, weighed 10st 11lbs (68kg), had a dark complexion, brown eyes, and dark brown hair. He identified as Church of England and nominated his mother, Ellen Tobin, as his next of kin and simply gave her address as Aitkenvale Townsville.

Accepted into the AIF, James Yates was allocated service number 320, assigned to the Infantry and posted to the 35th Infantry Battalion. Two weeks later he was posted as an original member to "A" Company of the 41st Battalion. His brother was in "C" Company of the same unit.

The 41st Battalion was raised in February 1916 with recruits from Brisbane, northern Queensland, and the northern rivers district of NSW. The battalion completed the basic training requirements at Enoggera in Brisbane. After initial training, Private James Yates embarked from Sydney on the 18th of May 1916, bound for France via England.

The unit sailed as part of Convoy No. 21 aboard HMAT A64 *Demosthenes,* which carried 1,612 troops on the voyage. The entire convoy of 18 ships transported 25,333 troops. They travelled via Cape Town and the Cape Verde Islands before arriving at Plymouth in July 1916. Once disembarked, the battalion moved with the 3rd Division to training grounds on Salisbury Plain at Larkhill, the site of the current Royal School of Artillery.

Three months later, in November 1916, the 41st Battalion sailed for the Western Front from Portsmouth, England. They disembarked at Le Havre, France, and moved towards the border area with Belgium near Armentières.

On Christmas Eve of 1916, the 41st Battalion including Private Yates entered the frontline at the Mushroom Salient, situated just east of La Chapelle-d'Armentières. Throughout the harsh winter of 1916-17, the soldiers of the 41st alternated between duties in the front line and training and labouring in the rear areas. The battalion's rotations took them through the frontline areas around Armentieres in France and across the Lys River in Belgium for much of the first half of 1917. Tragically, on the 15th of April 1917 Private William Yates, older brother of Private James Yates, was Killed in Action due to shellfire.

During the Battle of Messines in June 1917, the 41st Battalion played a key role in securing the entire 11th Brigade's frontage until the mines were detonated, starting the assault.

After this success, the battalion engaged in smaller "Bite and Hold" operations to advance further. One notable action was the assault on the 'Windmill' position, a German stronghold near Warneton, Belgium. The mission expanded beyond its original scope, requiring control of a larger area and dominance over new positions.

The waterlogged terrain made trench digging nearly impossible, forcing the battalion to use flooded shell craters as a defensive line. Despite the difficult conditions, they held their ground and repelled multiple German counterattacks over the next 18 days.

Amidst the harsh terrain and persistent enemy efforts to reclaim ground, the 41st Battalion maintained their position. Private James Yates took part in these gruelling battles.

At the beginning of August 1917, the 41st was engaged in reoccupying a post. Five soldiers were killed, and 18 wounded during this action. One of those casualties was Private James Ernest Yates, who was Killed in Action on the 1st of August 1917. He was 24 years old and had served with the AIF for 18 months.

His body was laid to rest in Kandahar Farm Cemetery (Plot II, Row F, Grave No. 19), Neuve-Eglise, Belgium. His death was notified in the 333rd list of Australian Casualties released on the 27th of August 1917.

James Ernest Yates' name is inscribed on The Australian War Memorial Roll of Honour (Panel 134), The City of Townsville WW1 Honour Roll, and The Townsville Cenotaph. Yates St Nelly Bay, Magnetic Island, and Yates St Railway Estate, Townsville, are named in his memory and in memory of his brother William James Yates.

320 Private James Ernest Yates served on the Western Front in The Great War. He was awarded the British War Medal (no:39974) and the Victory Medal (no:39588). A memorial scroll and a memorial plaque were also passed to his mother in 1922 and an additional one pound (£1/-/-) per fortnight was paid to her for the loss of her second son. This combined amount of two pounds (£2/-/-) per fortnight was the equivalent of one quarter of the wage paid to two soldiers of the AIF in 1917.

The following tribute was published in the *Townsville Daily Bulletin* on the 24th of August 1917:

YATES. — Private James Ernest Yates, son of Mrs. T. Tobin, Aitkenvale, Killed in Action on 1st August 1917. Aged 24 years.

He died for his King and Country.

7314 PRIVATE WILLIAM F.F.A.D. SHEEHAN (1896 – 1917)

15th Infantry Battalion

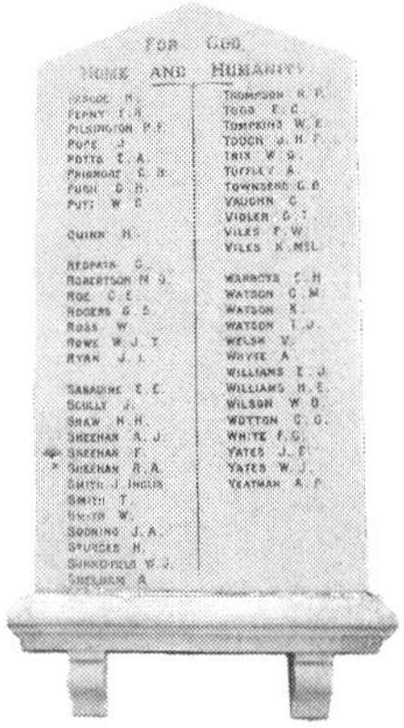

William Frederick Francis Alexander David Sheehan (Fred) was born on the 30th of April 1896 in Townsville, Queensland. He was the only son of six children born to David and Winifred Sheehan (née Clair). David Sheehan worked as a butcher in Townsville, and in 1903, the family lived in Alexandra Street, Townsville. Fred Sheehan grew up in Townsville and attended the Christian Brothers School. After completing his schooling, he started work as a clerk with the Railways Department.

On the 7th of October 1916, Fred Sheehan applied at Townsville to enlist in the Australian Imperial Force (AIF). He declared that he was single, 20 years old and employed as a clerk. He also declared that he had four years' service in the Reserves. His enlistment medical examination noted that he was 5'10" (178 cm) tall and weighed 9st. 7lbs. (61 kg). He had a dark complexion, yellow-brown eyes, and black hair. He identified as Roman Catholic and nominated his mother as his next of kin, giving her address as Alexander Street, Townsville.

Accepted into the AIF, Fred Sheehan was allocated service number 7314, assigned to the Infantry and posted to the 15th Battalion, 24th Reinforcements. The 15th Battalion was fighting on the Western Front when Fred Sheehan enlisted. The 24th Reinforcements spent the four months after enlistment in Australia undergoing basic training.

Once trained and equipped, they embarked for England from Sydney as part of Convoy No. 29 on the 24th of January 1917. They sailed aboard HMAT A33 *Ayrshire*, which carried 1,746 troops. The entire convoy of eight ships carried a total of 10,679 troops on the voyage.

Private Fred Sheehan and the other reinforcements disembarked at Devenport, England, on the 12th of April 1917. From there, they moved to the 4th Training Battalion at Codford, south of Salisbury Plain Valley in Wiltshire, England. Three months of training followed at Codford before Private Sheehan crossed to France on the 9th of July 1917. Another two weeks of training at Étaples was completed before the Reinforcements, including Private Sheehan, were taken on the strength of the 15th Battalion on the 27th of July 1917.

The Battalion was resting and training in billets in the vicinity of Vieux-Berquin in northern France when Private Fred Sheehan marched in. That day, the 15th Battalion was inspected by the 2nd Army Commander, General (later Field Marshal) Herbert Plumer, who had won an overwhelming victory over the German Army at the Battle of Messines, one month earlier.

On the 4th of August, the 15th Battalion relocated to new positions on Messines Ridge. During the following week, the Battalion was subjected to heavy shelling, and it was in this period that Private William F.F.A.D. Sheehan was Killed in Action on the 8th of August 1917. He was 21 years and three months old and had been a member of the AIF for 10 months at the time of his death. His passing was reported in the 334th list of Australian casualties, published on the 1st of August 1917. His service record records that he was initially *"buried on the side of a hedge about 100 yds S.S.W of Midar Farm, 100 yds W. of Road just N.W. of MESSINES FRANCE."* His body was later reinterred and laid to rest at Messines Ridge British Cemetery (Plot V, Row A, Grave No. 9), Belgium.

Private William F.F.A.D. Sheehan's name is inscribed on The Australian War Memorial Roll of Honour (panel 77), The City of Townsville WW1 Honour Roll, The Northern Division Railway Honor Board in Townsville, and The Townsville Cenotaph. Sheehan Street, Hermit Park, Townsville is named in his memory as well as the memory of Ray Allen Sheehan and Albert John Sheehan.

7314 Private William F.F.A.D. Sheehan was awarded the British War Medal (no:55475) and the Victory Medal (no:54635). A memorial plaque and a memorial scroll were passed to his father in 1922.

The following notice was published in the *Townsville Daily Bulletin* on 24th August 1917:

"SHEEHAN. — Private William Frederick, beloved only son of Mr. and Mrs. D. Sheehan, Alexandra Street, North Ward, Townsville, killed in action, in France, 8th August 1917. Aged 20 years."

3411 CORPORAL JOHN POPE (1897 – 1917)

47th Infantry Battalion

John Pope was born on the 15th of March 1897 in Bristol, Gloucestershire, England. He was the first-born son of Arthur and Florence Pope (née Dodsworth). His parents had one other son born two years after John. Arthur Pope was a bootmaker.

John Pope attended Henleage College, Westbury, Bristol, England. In 1906, the Pope family emigrated to Australia. They arrived in Sydney and later settled in Queensland, first residing in Toowoomba and then moving to Townsville. John Pope completed his education at the Rangerite Technical School in Toowoomba.

After completing school, John Pope joined the Bank of New South Wales in Charters Towers on the 4th of September 1914. In March 1915, he was transferred to Bowen as a ledger-keeper. He left the bank on the 2nd of August 1915 and applied to enlist in the Australian Imperial Force (AIF) at Brisbane.

At 18 years and five months, John received written consent from both parents to enlist. He declared that he was single and had previous military service with the Kennedy Regiment in Bowen. His medical examination noted that he was 5'4½" (164 cm), 132 lbs (60 kg), with a fresh complexion, brown eyes, and dark brown hair. He identified as Church of England and listed his father as next of kin, with an address at Queen's Hotel, Townsville.

Accepted into the AIF, John Pope was allocated service number 3411, assigned to the Infantry, and posted to the 15th Battalion, 11th Reinforcements. After 80 days of training in Australia, Private John Pope and his fellow Draft of Reinforcements embarked on the 21st of October 1915 from Brisbane bound for Egypt. They sailed aboard HMAT A48 *Seang Bee*, which carried 1,013 troops, as part of Convoy No.13. The 11-ship convoy transported a total of 11,731 troops and 400 horses.

Once in Egypt, Private Pope was transferred to the newly raised 47th Battalion on the 6th of March 1916 at Tel-el-Kebir. The 47th Battalion was established on the 24th of February 1916 through the "doubling" of the AIF. It consisted of a core of seasoned Gallipoli veterans from the 15th Battalion and fresh reinforcements from Australia. The composition of the 15th Battalion was mirrored in the new 47th Battalion, with a significant number of men recruited from Queensland and Tasmania.

Shortly after being transferred to the 47th, Private John Pope was hospitalized with mumps. He spent three weeks there and rejoined the battalion in late March 1916.

On the 2nd of June 1916, the 47th Battalion sailed from Alexandria, Egypt, to France. They arrived on the 9th of June and entered the Western Front trenches on the 3rd of July. The battalion's first major battle was at Pozières. During the 2nd Division's attack on the 4th of August, the 47th provided working parties and later defended the captured territory. Private Pope endured two tough tours in the Pozières trenches.

It was during the fighting around Pozières that Private John Pope was Wounded in Action on the 12th of August 1916 and admitted to the hospital suffering from shell shock. He returned to the battalion three days later but was sent back to the hospital two days afterward, this time suffering from mumps. Private Pope once again returned to the battalion in early September 1916 and was promoted to Corporal on the 16th of December 1916.

After Pozières, Corporal Pope and the 47th alternated between frontline duties and rest periods. On the 11th of April 1917, the battalion assaulted the fortified village of Bullecourt, part of the Hindenburg Line. The attack, lacking surprise and reliant on faulty tank support, ended in failure.

As the year progressed, the focus of the AIF's operations shifted to the Ypres sector in Belgium. During this time, the 47th Battalion played significant roles in the battles of Messines and Passchendaele.

On the 12th of August 1917, the 47th Battalion was located east of Wytschaete in the vicinity of Messines in Belgium. An unfortunate shell fell on Company Headquarters that day, resulting in the death of seven soldiers and wounding

10. Corporal John Pope was one of those Killed in Action. The 47th Battalion War Diary states that it was "*Rather unfortunate as we have had no casualties for 3 days, having completely silenced snipers.*"

Corporal John Pope was 20 years and five months old at the time of his death and had been a member of the AIF for two years and one week. His death was notified in the 338th list of Australian casualties released on the 15th of September 1917. He was laid to rest at Derry House Cemetery No. 2 (Plot II, Row C, Grave No. 2), Wytschaete, Belgium.

Corporal John Pope's name is inscribed on various memorials, including the Australian War Memorial Roll of Honour (Panel 144), The Bowen War Memorial, Toowoomba St Luke's Church WW1 Honour Roll, The Bank of New South Wales Honour Roll, The Townsville Honour Roll, and the Townsville Cenotaph. Pope St, Aitkenvale, Townsville, is named in his memory.

3411 Corporal John Pope was awarded the 1914/15 Star (no:18128), the British War Medal (no:22342), and the Victory Medal (no:22175). A memorial scroll and a memorial plaque were passed to his father in 1921 and 1922, respectively.

The following notice was published in the *Townsville Daily Bulletin* on the 4th of September 1917:

"*The Great War. Mr and Mrs Arthur E. Pope, of Townsville, have received news from the Base Records Office, that their eldest son, Corporal John Pope, was Killed in Action in France, on August 12.*"

16486 SAPPER RICHARD JOHN HAYES (1884 – 1917)

2nd Field Company Engineers

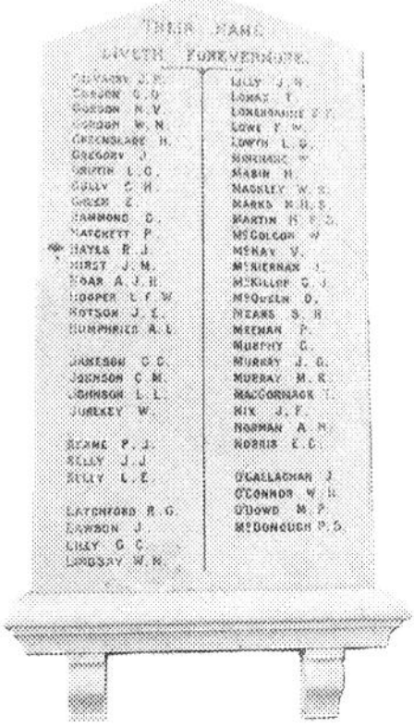

Richard John Hayes was born on 19th May 1884 in Townsville, Queensland, to Timothy and Mary Hayes (née Merrick). He was the third son of a large family. Richard Hayes qualified as a blacksmith and lived in a property on Main Street, Townsville, in 1913. Later, he moved to Mackay in Queensland, where he lived with his widowed sister, Mary Quillinan, whose husband died in November 1913.

On the 2nd of September 1916, Richard Hayes applied in Mackay to enlist in the Australian Imperial Force (AIF). He declared that he was 32 years and two months old, single, and qualified as a blacksmith. Richard's medical examination recorded that he was 5'9" (175 cm) tall, weighed 150 lbs (68 kg), had a fresh complexion, blue eyes, and light brown hair. He identified as Roman Catholic and nominated his sister Mary as his next of kin.

Accepted into the AIF, Richard Hayes was allocated service number 16486, assigned to the Infantry, and initially posted to the 15th Battalion, 22nd Reinforcements. He was sent to Enoggera, Brisbane, for training for overseas service. In October 1916, his blacksmithing skills were recognized, and he was reassigned as a Sapper to the Field Engineers. After completing basic training, Sapper Hayes embarked on the 2nd of December 1916 from Sydney, New South Wales, as part of Convoy No. 28, bound for England. He sailed aboard HMAT A67 *Orsova*, which carried 738 troops on the voyage. The entire convoy of 13 ships transported a total of 2,132 troops.

The *Orsova* docked in England in February 1917, and Sapper Hayes continued his training at the Engineer Training Depots at Perham Downs on the edge of the Salisbury Plain in Wiltshire and at Brightlingsea in Essex. Training complete, Sapper Hayes crossed the channel to France and joined his unit, the 2nd Field Company, on the 23rd of May 1917. His unit was in a period of rest and training stationed behind the lines on the Somme when he marched in. The months of May and June 1917 were spent in training.

In August, the 2nd Field Company moved north to take up positions at Doulieu near Ypres in Belgium and began work preparing the ground for the opening battles of the Ypres campaign of 1917. Apart from the normal tasks assigned to the Engineers, the unit also supplied 2,684 man-hours to farmers in assisting to bring in the harvest.

In September, the unit was located at Hooge, east of Ypres, and was engaged in improving trenches, constructing a light rail from the Hooge supply dump to the front lines through Chateau Wood, and establishing forward dressing stations. It was during these operations that Sapper Richard Hayes was tragically Killed in Action by shell fire on the 12th of September 1917. He was returning to camp with other unit members when a shell exploded amongst the group, killing him instantly and wounding another man. Sapper Richard Hayes was 33 years and three months old at the time of his death. He had served in the AIF for one year. He has no known grave. His death was notified in the 341st list of Australian Casualties issued on 19th October 1917.

Richard Hayes' name is inscribed on The Australian War Memorial Roll of Honour (Panel 24), The Ypres (Menin Gate) Memorial (Panel 7), Belgium, The City of Townsville WW1 Honour Roll, and The Townsville Cenotaph.

16486 Sapper Richard Hayes was awarded the British War Medal (no:57065) and the Victory Medal (no:57065). There was some difficulty in locating his next of kin and it is not recorded to whom these medals, the memorial scroll, and the plaque were passed. Hayes Street, North Ward, Townsville, was named in his memory in 1926.

The following was published in local newspapers in October 1922:

"NEXT OF KIN WANTED. The next of kin or persons knowing the whereabouts of the next of kin of the undermentioned deceased soldiers are requested to communicate with the Officer in Charge Base Records, Melbourne, as early as possible. 16486, Spr. R. J. Hayes. 2nd F.C.R."

5821 PRIVATE JAMES DORAN (1898 - 1917)

25th Infantry Battalion

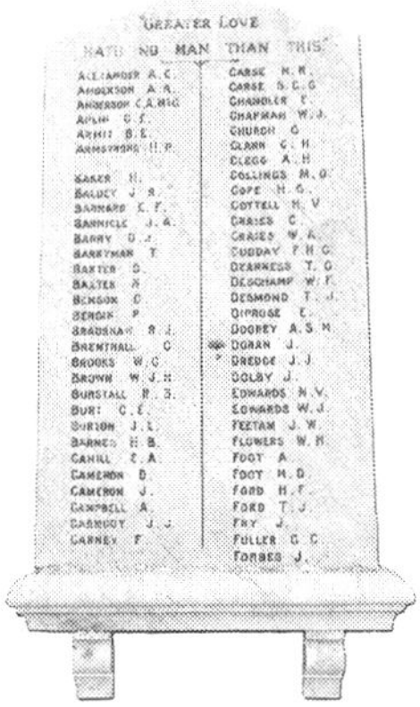

James Doran was one of eight children born to John and Eliza Doran (née McLean) on the 5th of August 1898 in Hughenden, Queensland. He had four brothers and three sisters. John Doran worked as a shop assistant and later as a bookkeeper. It is not clear where James Doran spent most of his childhood or where he attended school. Eliza Doran lived in a property on Tully Street, Townsville in 1913 while John Doran worked as a bookkeeper in Muttaburra.

James Doran took a job as a printer at the *Townsville Daily Bulletin* after finishing school and was working there when he applied to enlist in the Australian Imperial Force (AIF) in Townsville on the 7th of August 1916, two days after his eighteenth birthday. He declared that he was single, employed as a printer, and had previously served for four years with the cadet forces. His enlistment medical noted that he was 5'9½" (177 cm) tall, weighed 9st 3lbs (59 kg), had a ruddy complexion, black eyes, and brown hair. He identified as Roman Catholic and nominated his mother as his next of kin, giving her address as Castling St, West End, Townsville.

Accepted into the AIF, James Doran was allocated service number 5821, assigned to the Infantry, and posted to the 25th Battalion, 16th Reinforcements. The 25th Battalion was formed at Enoggera in March 1915 from Queensland volunteers. After fighting at Gallipoli, the 25th was redeployed to the Western Front, landing on the 19th of March 1916 as the first AIF battalion to arrive in France. The 25th was heavily engaged in its first major battle at Pozières when James Doran enlisted.

After completing 10 weeks of basic training in Australia, Private James Doran and his fellow reinforcements joined Convoy No. 25 on the 21st of October 1916 and embarked from Brisbane, bound for England. They sailed aboard HMAT A36 *Boonah*, which carried 1,235 troops. The entire convoy of 12 ships carried a total of 12,397 troops and 5,200 horses. Twelve weeks later, they disembarked at Plymouth and were transferred to the 7th Training Battalion located at Rollestone (near Amesbury on Salisbury Plain in Wiltshire). After another four months of training, Private James Doran crossed to France and was taken on the strength of the 25th Battalion on the 2nd of May 1917.

The battalion was stationed in reserve positions at Bullecourt, and the day after Private Doran joined the unit, the 5th and 6th Brigades of the 2nd Division launched an attack on the formidable Hindenburg defences. During this assault, the 25th Battalion deployed a small force, but they sustained heavy casualties. While the battalion played a supporting role during the second battle of Bullecourt, it did not engage in a major offensive again until the 20th of September 1917, when it led the first wave in the battle of Menin Road in Belgium. This battle was part of the larger Third Battle of Ypres on the Western Front. The 1st and 2nd Divisions suffered 5,013 casualties during this intense operation. Among the fallen was Private James Doran, who was Killed in Action on the 20th of September 1917. There are no recorded details of his death, and he has no known grave. At the time of his death, he was 19 years and one month old and had been serving in the AIF for 13 months. He had been with the 25th Battalion for 20 weeks, and his death was reported in the 345th list of Australian Casualties released on the 24th of October 1917.

James Doran served on the Western Front in WW1. His name is inscribed on The Australian War Memorial Roll of Honour (panel 104), The Ypres (Menin Gate) Memorial (Panel 23), Belgium, The City of Townsville WW1 Honour Roll, and The Townsville Cenotaph. Doran Street, Mysterton, Townsville, is named in his memory.

5821 Private James Doran was awarded the British War Medal (no:50021) and the Victory Medal (no:49421). A memorial plaque and a memorial scroll were passed to his father in 1923 and 1922, respectively.

Eliza Doran was granted a war pension of one pound and 15 shillings (£1/15/-) per fortnight for the loss of her son. This was about 40% of the wage, four pounds and four shillings, paid to a soldier in the AIF in 1917.

The following tribute was published in the *Townsville Daily Bulletin* on the 20th of September 1919:

"DORAN— In sad but loving Memory of our dear son and brother, Pte James Doran, who was Killed in Action on the 20th of September 1917.

> Though in lonely grief I sigh
> for one I loved no longed nigh,
> Submissively, I will reply.
> "Thy will be done."

(Inserted by his loving Parents, Brothers, and Sisters)"

374 SERGEANT GERALD DESMOND GORDON (1897 – 1917)

25th Infantry Battalion

Gerald Desmond Gordon (Gerry) was born on the 18th of January 1897 in Charters Towers, Queensland. He was the second child and oldest son of five children born to Gerald and Rose Gordon (née Lawson). He had one brother and three sisters. His father was a Mine Director in Charters Towers and, 10 years later, became the manager of the Townsville offices of Burns Philp. The family lived in Paxton Street, Townsville. Gerry Gordon completed his schooling at the prestigious Knox College in Sydney.

After finishing school, Gerry Gordon worked as a bank clerk. On the 8th of February 1915, with his parents' consent, he applied to enlist in the Australian Imperial Force (AIF). He declared that he was single, eighteen years old, employed as a bank clerk, and had served in the Senior Cadets for over three years. His enlistment medical examination noted that he was 5'8½" (170 cm) tall, weighed 9st 2lbs (75kg), had a fresh complexion, blue eyes, and light brown hair. He identified as Presbyterian and nominated his father as his next of kin giving his address as Paxton St, Townsville.

Accepted into the AIF, Gerry Gordon was allocated service number 374, assigned to the Infantry and posted to the soon-to-be-raised 25th Battalion as an original member of 'B' Company. Formed in March 1915 at Enoggera in Queensland, the 25th Battalion was allocated to the 7th Brigade. While most of its personnel were recruited in Queensland, the battalion also comprised a small group of men from Darwin.

Private Gerry Gordon was a capable soldier and was soon promoted to Sergeant. After basic training in Australia, the unit embarked from Brisbane on board HMAT A60 *Aeneas* on the 29th of June 1915, bound for Egypt. The *Aeneas* carried 1,720 passengers on the voyage. The 25th Battalion trained in Egypt during August 1915, and by early September, they were manning trenches at Gallipoli. At Gallipoli, the 7th Brigade reinforced the depleted New Zealand and Australian Division. The 25th Battalion had a quiet time on Gallipoli as the last major Allied offensive had been launched and turned back in the previous month.

While serving on Gallipoli, Gerry Gordon contracted hepatitis (recorded as jaundice in his documents) in November 1915 and was evacuated to No. 2 AGH in Heliopolis, Egypt. After his release from convalescence in December, he was attached to the overseas base and employed as a storeman, a role he continued until September 1916 when he sailed to England with the Australian Stores Depot.

In December 1916, Sergeant Gerry Gordon re-joined the 25th Battalion on the Somme in France. He arrived shortly after the early snows had brought the Battle of the Somme to an end on the 18th of November 1916. With the onset of winter, large-scale offensive operations ceased, and both the Allies and Central Powers dug in to await the arrival of spring. After serving in a reserve role during the second battle of Bullecourt, the 25th Battalion was not involved in a major offensive until September 1917, when it participated in the battle of Menin Road in Belgium as part of the 2nd Division's first wave. On the 20th of September 1917, the battalion advanced under the cover of a protective barrage, successfully capturing the first objective, the Red Line, near Hanebeek to the left of Polygon Wood. They then dug in to consolidate their position. During this attack, Sergeant Gerry Gordon, acting as Sergeant Major at the time, was Killed in Action.

In 1918, his mother (Gerry's father died in 1915) requested the Australian Red Cross Society to carry out an investigation of the circumstances of Gerry's death. In his statement taken in April 1918, Private E.G. Sugrue said that: *"Sergeant Gordon was killed outright by shellfire, and he had helped bury him where he fell the same evening that he was killed, and that a small cross was erected."* A Memorial Cross was later erected in the Perth China Wall Cemetery but was removed after the Menin Gate Memorial was completed. Sergeant Gerry Gordon was 28 years and eight months old at the time of his death and had been a member of the AIF for two years and eight months. His death was notified in the 345th list of Australian Casualties released on the 24th of October 1917.

Gerald Gordon served at Gallipoli and on the Western Front in WW1. His name is inscribed on The Australian War Memorial Roll of Honour (Panel 104), The City of Townsville Honour Roll, The Townsville Cenotaph, Townsville West State School Memorial Gate, and The Ypres (Menin Gate) Memorial (Panel 23), Belgium. Gorden Street, Garbutt, Townsville, is named in his and N.V. and W.N. Gordon's memory (NB The spelling of the street name i.e., Gorden instead of Gordon, has been attributed to an error. Townsville City Council minutes clearly record that the street was to be renamed Gordon St from McAllister Street).

374 Sergeant Gerald Desmond Gordon was awarded the 1914/15 Star (no:21218), the British War Medal (no:16628), and the Victory Medal (no:16561). These medals, as well as a memorial scroll and a memorial plaque, were passed to his mother. Rose Gordon was granted a war pension of £2/13/- per fortnight for the loss of her son. This was a little more than half of the fortnightly wage paid to a private soldier in 1917.

6789 PRIVATE CECIL HENRY GULLEY (1897 – 1917)

9th Infantry Battalion

Cecil Henry Gulley was born on the 26th of March 1897 in Townsville, Queensland, to William and Laura Gulley (née Rimes). He was the middle child of a large family. The Gulley family lived at Sellheim, west of Townsville, where William Gulley worked as a Horse Driver.

Cecil Gulley attended the Sellheim State School, and when his school years were completed, he worked as an Engine Driver and a Storeman. The family moved from Sellheim and relocated to Townsville, taking up residence at Honeymoon Road.

Cecil Gulley's older brother, Charles, enlisted in the Australian Imperial Force (AIF) in June 1915 and was serving overseas with the 9th Battalion when Cecil Gulley applied to enlist in the AIF at Townsville on the 27th of September 1915. He declared he was single, working as a storeman, and 18 years and six months old. His medical examination noted that he was 5'9½" (176 cm) tall, weighed 10st 6lbs (66 kg), had a dark complexion, grey eyes, and black hair. He identified as Church of England and nominated his mother as his next of kin, giving her address as Ralston Street, West End, Townsville.

Accepted into the AIF, Cecil Gulley was allocated service number 374, assigned to the Infantry, and, like his brother, was posted to the 9th Battalion, 22nd Reinforcements. Following enlistment, Cecil was transferred to Enoggera in Brisbane to commence his basic training with the 7th Depot Battalion. Training was interrupted from June to October 1916 while Private Cecil Gulley was hospitalized. While he was undergoing treatment, his brother was Killed in Action in France during the Battle of Mouquet Farm in August 1916.

Private Cecil Gulley recovered and was released from the isolation camp at Lytton. After completing basic training in Australia, Cecil joined Convoy No. 26 on the 27th of October 1916 and embarked from Brisbane, bound for England. He sailed aboard HMAT A74 *Marathon*, which carried 1,078 troops. The entire convoy of 13 ships carried a total of 12,381 troops and 2,416 horses on the voyage.

Disembarking in England in February 1917, Private Gulley was transported to the 3rd Training Battalion located at Durrington on the Salisbury Plain. He contracted mumps and was admitted to the hospital about a month after his arrival in the Training Battalion and was kept in the hospital for two weeks before rejoining his unit. That was not to be Private Gulley's only hospital admission. Three weeks later, he was readmitted to the hospital, suffering from scabies. This time, he did not return to the unit for three weeks.

Private Cecil Gulley committed three military offences during his time in England. As a result, he spent a total of 21 days in military prison. The first offence was Absenting Himself Without Leave (AWOL) for two days, the second was AWOL for three days, and the third was Absent from a Defaulters Parade. Private Gulley received 168 hours (one week) of detention for each offence and lost a total of 31 days' pay.

Private Gulley crossed to France after his release from detention and finally joined the 9th Battalion in the field on the 5th of August 1917, along with 54 other reinforcements. The 9th Battalion was located at Affringues in the north of France when Private Gulley joined them and was carrying out trench-to-trench attack training. Training continued until mid-September when the 9th Battalion started the route marches to return to Belgium and the front lines.

During the period 19-21st of September 1917, the 9th Battalion was deployed in the Battle of Menin Road with the objective of gaining control over sections of the meandering ridge to the east of Ypres, intersected by the Menin Road. Despite facing robust German defensive positions, including interconnected concrete pillbox strongholds designed to mutually support one another, the assault achieved success along its entire front. The Allied forces also endured determined German counter-offensives during this challenging operation.

It was during this battle that Private Cecil Gulley was Killed in Action. He was initially reported as wounded and then missing; however, his body was located and buried near Chateau Wood. A Court of Enquiry confirmed his death. He was 20 years old and had served in the AIF for two years at the time of his death.

Private Cecil Gulley was notified as 'missing and wounded' in the 381st list of Australian Casualties, which was subsequently amended to 'Killed in Action' in the 382nd list of Australian Casualties released on the 15th of February 1918.

Cecil Henry Gulley served on the Western Front in WW1. His name is inscribed on The Australian War Memorial Roll of Honour (Panel 56), The Ypres (Menin Gate) Memorial (Panel 17), Belgium, The City of Townsville WW1 Honour Roll, The Townsville Cenotaph, and the Sellheim State School WW1 Honour Roll. Gully Street, Stuart, Townsville is named in his memory.

6789 Private Cecil Henry Gulley was awarded the British War Medal (no:51462) and the Victory Medal (no:60782). A memorial scroll and a memorial plaque were passed to his father in 1921 and 1922, respectively. Laura Gulley received a war pension of two pounds (£2/-/-) per fortnight for each of her sons Killed in Action. This was the equivalent of half of the wages paid to her sons before their death.

The following notice was published in the *Townsville Daily Bulletin* (Qld) on the 20th of September 1921:

"GULLEY. — In loving memory of our dear son and brother, Private Cecil Henry Gulley, No. 6789, 9th Battalion, Killed in Action in Belgium, 20th September 1917.

The Supreme Sacrifice." (Inserted by his loving father and mother, brothers and sisters. W. H. and L. C. Gulley.)

(Inserted by his loving parents, W. H. and L. C. Gulley and Family.)

5672 PRIVATE JAMES SCULLY (1887 – 1917)

26th Battalion

James Scully was born on the 20th of May 1887, in Townsville, Queensland. He was the second child of seven born to William and Marjory Scully (née Soutar). He had two brothers and three sisters. His older sister, Annie, died as a child when James was four years old. This left James as the oldest child and first-born son in the family.

After attending State School, James Scully started working with the Railways in 1909. In 1913, he worked as a sapphire miner in Emerald, Queensland.

On the 3rd of April 1916, James Scully applied to enlist in the Australian Imperial Force (AIF) at Townsville, Queensland. He was one month short of his 29th birthday. He declared that he was single, employed as a labourer, and had no previous military experience. His medical examination noted that he was 5'9½" (176 cm) tall, weighed 10st 9lbs (68 kg), had a fresh complexion, grey eyes, and dark brown hair. He identified as Church of England and nominated his mother as his next of kin giving her address as Bayswater Terrace, Hyde Park in Townsville.

Accepted into the AIF, James Scully was allocated service number 5672, assigned to the Infantry and posted to the 26th Battalion, 15th Reinforcements. The 26th was in France on the Western Front, after serving in Gallipoli, when James Scully enlisted. He commenced his military training with the 11th Depot Battalion at Enoggera near Brisbane. The next five months were spent preparing for overseas service.

On the 7th of September 1916, Private James Scully and his fellow reinforcements joined Convoy No. 24 and embarked from Brisbane, bound for England. They sailed aboard HMAT A46 *Clan Macgillivray,* which carried 1,083 troops. The entire convoy of 13 ships carried a total of 12,785 troops on the voyage.

Disembarking at Plymouth in November 1916, Private Scully and the rest of the 15th Reinforcements then moved to Rollestone on the Salisbury Plain, where training continued for another three months. About two weeks after arrival at Rollestone, Private Scully was hospitalized with mumps and spent two weeks in the hospital.

Once training was completed in England, Private Scully and the other reinforcements crossed to the 2nd Australian Division's Base Depot in France in early February 1916. A few days later, on the 10th of February 1917, they were taken on the strength of the 26th Battalion located in the Somme Valley.

Two weeks after joining the 26th, Private Scully was admitted to the hospital suffering from Trench Feet, which is a serious condition resulting from feet being wet for too long due to fighting in cold, wet conditions in the front line. After a month in the 5th Dressing Station, he was released back to the 26th Battalion on the 26th of March 1917, in time to participate in the Battalion attack at Lagincourt.

On the 3rd of May 1917, the 2nd Division launched a successful assault on the Hindenburg defences at Bullecourt. During this battle, Private James Scully was wounded in the back by shrapnel on the 7th of May. He was evacuated for treatment; however, the wound did not prove to be serious, and he was back on duty 11 days later.

In September, the 26th Battalion moved to Ypres in Belgium and, on the 20th of September 1917, acted in reserve at the battle of the Menin Road. Enemy shelling of the reserve positions was at times heavy, and the 26th Battalion provided carrying parties to bring ammunition and supplies to the 28th Battalion, one of the attacking units.

It was during this phase of the operations that Private James Scully was Killed in Action on the 20th of September 1917. He was 30 years old at the time of his death and had served with the AIF for 17 months. He has no known grave.

His death was notified in the 309th list of Australian Casualties released on the 11th of June 1917. Private James Scully's name is inscribed on The Australian War Memorial Roll of Honour (Panel 109), The Ypres (Menin Gate)

Memorial, Belgium, The City of Townsville WW1 Honour Roll, and The Townsville Cenotaph.

5672 Private James Scully served in WW1 on the Western Front. He was awarded the British War Medal (no:472299) and the Victory Medal (no:46678). A memorial scroll and a memorial plaque were passed to his father in 1921 and 1922, respectively.

4292 PRIVATE HENRY STURGES (1893 – 1917)

9th Infantry Battalion

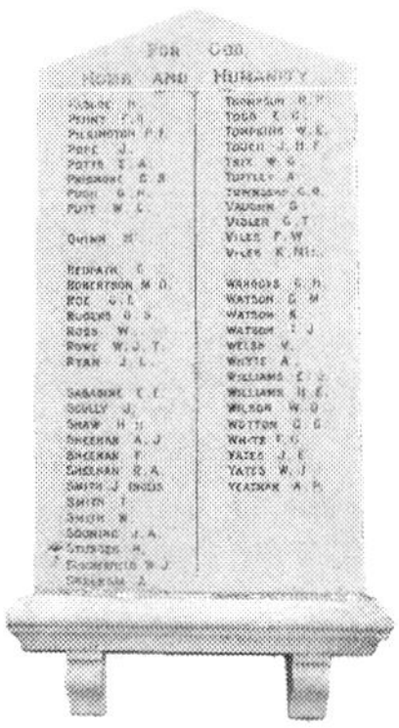

Henry (Harry) Sturges was born on the 26th of December 1893 in Charters Towers, Queensland. He was the youngest of six children born to James and Mary Sturges (née Quigley). He had four brothers and one sister. The family lived in Charters Towers, and Harry Sturges' brothers worked as miners. In 1905, James Sturges was paralysed and did not work. Harry Sturges' mother died in 1904, and his father died in 1911.

Harry Sturges attended Queenton State School as a boy, and his siblings, especially George Sturges, raised him after his parents died. When his school years were finished, Harry worked as a blacksmith's striker, first in Charters Towers and then later in Townsville, where he worked at the Cleveland foundry. At some time before 1915, he became engaged to Miss Linda Peterson of Hughenden and included her as an equal beneficiary with his brother George in his Will.

Harry Sturges was living with his brother George in Townsville when he applied to enlist in the Australian Imperial Force (AIF) on the 30th of August 1915. He was 19 years old and declared that he was single, had no previous military experience, and was working as a blacksmith's striker. His enlistment medical examination noted that he was 5'11¼ " (181 cm) tall, weighed 9st 12lbs (63 kg), had a dark complexion, grey eyes, and dark brown hair. He identified as Church of England and nominated his brother George as his next of kin, giving his address as Hanran St, Hermit Park, Townsville.

Accepted into the AIF, Harry Sturges was allocated service number 4292, assigned to the Infantry and posted to the 9th Battalion, 13th Reinforcements. Soon after enlistment, he was transferred to Enoggera in Brisbane to commence his basic training, which lasted until the end of 1916.

On the 3rd of January 1916, Private Harry Sturges and his fellow Reinforcements joined Convoy No. 16 and embarked from Brisbane, bound for Egypt. They sailed aboard HMAT A55 *Kyarra,* which carried 904 troops. The 11-ship convoy carried a total of 10,859 troops on the voyage. They disembarked at Alexandria on the 19th of February 1916.

Training continued in Egypt until the 9th of May 1916, when the 9th Battalion embarked from Alexandria bound for France to join the British Expeditionary Force on the Western Front. The 9th Battalion and the rest of the 3rd Brigade marched into the 1st Division's Base Depot at Étaples in France. It was here that Private Sturges fell ill with severe gastritis and was invalided to England for treatment and admitted to the 3rd Southern General Hospital. He did not return to the 9th Battalion, deployed in Flanders, until the 7th of September 1916.

About a month after his return from hospital, Harry Sturges was struck down with influenza, which was sweeping through the AIF. The influenza developed into phthisis (tuberculosis), and he needed to be invalided to England for a second time for treatment and convalescence. After his recovery, he was transferred back to France at the end of May 1917. It had taken seven months for Private Sturges to be returned to duty after contracting tuberculosis.

The 9th Battalion was resting in Northern France when Private Harry Sturges re-joined them. They remained there training and getting ready for the battles to come. The 9th returned to Belgium in September 1917 and fought the Battle of Menin Road, which commenced on the 20th of September 1917.

It was during this engagement that Private Harry Sturges was Killed in Action. The Australian Red Cross Investigation into his death records the following details of his death in a statement made by 5991 Private S.J. Beldan, 9th Battalion:

"I saw him hit by a shell and killed instantly near the Racecourse on the edge of Polygon Wood. His pay book and other things were found on him, which we handed over to Lieut. Humphreys of B.C.F. Cpl. Williams buried him in the open just behind the trenches."

Private Harry Sturges was 23 years and nine months old at the time of his death. He had served in the AIF for two years. Despite the report of his burial in the field, his body was not recovered, and he has no known grave.

Private Henry (Harry) Sturges served on the Western Front in WW1. His name is inscribed on The Australian War Memorial Roll of Honour (Panel 57), The Ypres (Menin Gate) Memorial (Panel 17), Belgium, The City of Townsville WW1 Honour Roll, and The Townsville Cenotaph. Sturges Street in Hermit Park, Townsville, is named in his memory.

4292 Henry (Harry) Sturges was awarded the British War Medal (no: 28534) and the Victory Medal (no: 28354). A memorial scroll and a memorial plaque were passed to his brother George in 1922.

The following notice was published in the *Townsville Daily Bulletin* on the 22nd of October 1917:

"George Sturges, Hermit Park, received the sad news on Saturday that his youngest brother, Private Harry Sturges, was Killed in Action in France on the 20th of September. He enlisted in Townsville in August 1915 and was previously employed at the Cleveland foundry.

5756 PRIVATE GEORGE OLUF TOWNSEND (1885 – 1917)

9th Infantry Battalion

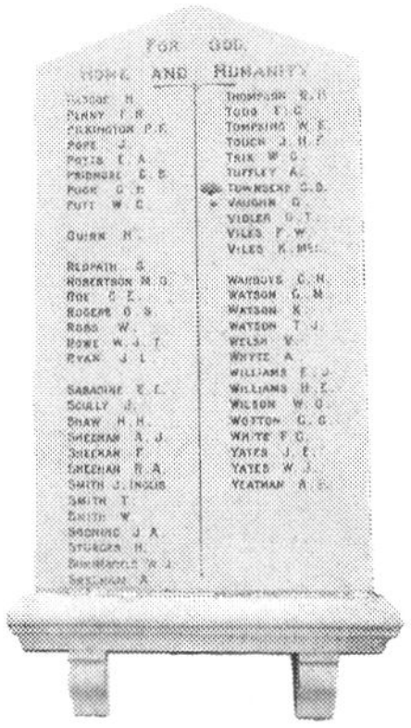

George Oluf Townsend was born on the 31st of July 1885 in Charters Towers, Queensland, to George and Elevine (Minnie) Townsend (née Ingelbretsen). He was the fourth of ten children and the third-born son, with five brothers and four sisters.

In 1908, the Townsend family lived in Charters Towers, where George Townsend Snr worked as a miner until 1911, when he passed away. George Townsend Jnr worked as a labourer. In 1913, Minnie Townsend moved to Townsville, taking up residence on Seventh Street. George Townsend lived with his mother and continued to work as a labourer until the 7th of January 1916, when he applied at Townsville to enlist in the Australian Imperial Force (AIF).

George Townsend declared that he was single, 28 years old, and without previous military experience. His medical examination noted that he was 5'4¼" (163 cm) tall, weighed 8st 4lbs (53 kg), had a fair complexion, blue eyes, and brown hair. He identified as Church of England and listed his widowed mother as his next of kin, providing her address as Tenth Avenue, South Townsville.

Accepted into the AIF, George Townsend was allocated service number 5756, allotted to the Infantry and posted to the 9th Battalion, 18th Reinforcements. The 9th Battalion was among the first infantry units raised for the AIF during the First World War. It was the first battalion recruited in Queensland, and along with the 10th, 11th, and 12th Battalions, it formed the 3rd Brigade. On the date that George Townsend enlisted, the 9th was in Egypt, rebuilding after the Gallipoli campaign. George Townsend was transferred to Enoggera in Brisbane soon after enlistment to commence his basic training with the 11th Depot Battalion.

On the 4th of May 1916, Private George Townsend and his fellow Reinforcements joined Convoy No. 22 and embarked from Brisbane bound for Egypt.

They sailed aboard HMAT A49 *Seang Choon*, which carried 875 troops. The entire convoy of 13 ships boarded a total of 8,462 troops on the voyage.

After several months of further training in Egypt, Private Townsend embarked from Alexandria on the 29th of July 1916, bound for England to join the British Expeditionary Force. Once in England, more training was undertaken at the 3rd Training Battalion, located at Durrington on the Salisbury Plain. This lasted until the 14th of October 1916, when the Reinforcements crossed to France and marched into the 1st Australian Division Base Depot (1st ADBD) at Étaples.

Private George Townsend contracted mumps and was admitted to the hospital about a month after his arrival at the Base Depot. He remained in the hospital for 21 days before being sent to a convalescence unit in Boulogne, where he stayed until the 8th of December 1916. Another six weeks at the 1st ADBD followed before Private Townsend joined the 9th Battalion in the field on the 17th of January 1917. The 9th was deployed near Bresle on the Somme. It was bitterly cold, and the unit diary records that snow was six inches deep. Private Townsend was allocated to 9 Platoon, 'C' Company.

Throughout early 1917, the 9th Battalion remained stationed on the Somme, following the German retreat to the Hindenburg Line. They saw action at Lagnicourt in April and Bullecourt in May.

By September, the battalion moved to Belgium, participating in major engagements such as the Battle of Menin Road, which began on the 20th of September 1917. This operation sought to gain control of parts of the ridge east of Ypres, crossed by the Menin Road. It marked the entry of Australian forces (1st and 2nd Divisions AIF) into the developing Third Battle of Ypres.

Despite facing heavily fortified German defences, including interlinked concrete pillbox positions that provided mutual support, the assault succeeded across the entire front. However, the advancing AIF troops also encountered determined German counterattacks.

It was during this engagement that Private George Townsend was initially reported as missing. A Court of Enquiry held in April 1918 confirmed that he was Killed in Action on the 20th of September 1917. In April 1918, 5660 J

Cassidy, 9th Battalion, stated the following: *"The last I saw of Private Townsend was about 600 yards in front of our objective; he was then dead, and he appeared to be very badly knocked about in the head and body."*

George Townsend was 32 years old and had served in the AIF for one year and eight months on the day that he was killed. He has no known grave, and his death was notified in the combined 397th and 398th Casualty Lists released on the 8th of May 1918.

George Oluf Townsend served in Egypt and on the Western Front in WW1. His name is inscribed on The Australian War Memorial Roll of Honour (Panel 57), The Ypres (Menin Gate) Memorial (Panel 17), Belgium, The City of Townsville WW1 Honour Roll, and The Townsville Cenotaph. Townsend St, Mysterton, Townsville, is named in his memory.

5756 Private George Oluf Townsend served on the Western Front in The Great War. He was awarded the British War Medal (no: 39151) and the Victory Medal (no: 38830). A memorial scroll and a memorial plaque were passed to his mother in 1922. Minnie Townsend was granted a war pension of two pounds (£2/-/-) per fortnight for her son's sacrifice. This was about half of the wage paid to an AIF soldier in 1917.

10052 PRIVATE WILLIAM JOSEPH CHAPMAN (1893 – 1917)

14th Field Ambulance

William (Bill) Joseph Chapman was born on the 22nd of November 1893 at Charters Towers, Queensland. He was the first of four children born to Joseph and Jane Chapman (née Endersby). Joseph Chapman worked as a miner, and Bill Chapman grew up in Charters Towers, attending the local State School.

By 1913, the family had moved to Townsville and, after his school years were completed, Bill Chapman worked as a salesman with Plumridges Ltd. He married Frances Alice Bretherton on the 1st of May 1915, and on the 30th of July 1915 their only child, Irene, was born. On the 20th of September 1915, Bill Chapman and his brother Joseph Chapman applied at Townsville to enlist in the Australian Imperial Force (AIF).

Bill Chapman declared that he was married, 21 years and 10 months old, employed as a salesman, and had served for 10 months with the voluntary cadets. Frances Chapman provided a letter of consent for Bill's application to enlist. Bill Chapman's enlistment medical examination noted that he was 5'8" (172 cm) tall, weighed 9st 4lbs (59kg), had a dark complexion, brown eyes, and dark hair. He identified as a Methodist and nominated his wife as his next of kin, giving her address as 6th Avenue, South Townsville.

Accepted into the AIF, Bill Chapman allocated was service number 10052 and, along with his brother, allotted to the Australian Army Medical Corps. Bill Chapman was posted, initially, to the 7th Field Ambulance. Later, he was re-posted to the 14th Field Ambulance. A Field Ambulance was not a vehicle; it was a mobile front line medical unit manned by troops of the Australian Army Medical Corps.

After six months of basic training and preparing for deployment, Private Bill Chapman and the other members of his unit joined Convoy No. 19 on the 31st of March 1916. They sailed aboard HMAT A16 *Star of Victoria* (later renamed *Port Melbourne*), from Sydney bound for Egypt. Their ship carried 1,550 troops, and the entire convoy of 10 ships carried a total of 8,075 troops and 1,168 horses on the voyage.

Private Chapman marched into Tel-el-Kebir on the 5th of May 1916 and was taken on strength of the 14th Field Ambulance one week later. After three months of training in Egypt, Private Bill Chapman embarked with the 14th Field Ambulance from Alexandria for service in France. He contracted influenza on the voyage and in November 1916 ended up being evacuated to England for treatment. After treatment, and a period of furlough, he eventually re-joined the 14th Field Ambulance in April 1917. The 14th was located near Becordel, France, near Albert and was responsible for facilitating casualty evacuation and delivering essential surgical care to the 14th Brigade.

The field ambulance troops were assigned the duty of transporting wounded soldiers from the Regimental Aid Post, situated immediately behind the front lines, to a more advanced dressing station. This journey covered about one to three miles (1.6 to 4.8 km) and usually demanded approximately six hours to complete. The 14th Field Ambulance was heavily involved in the Battle of Passchendaele also known as the Third Battle of Ypres. On the 20th of September 1917, the 5th Division took over from the 1st Division following the Battle of Menin Road. It was during this advance that Private Bill Chapman was Killed in Action on the 22nd of September 1917. He was 23years and 10 months old and had served with the AIF for two years.

An Australian Red Cross report into his death found that he was killed by a piece of shell that landed near the stretcher he was carrying at Glencourse Wood. Wounded, he was taken back to a dressing station but later Died of Wounds He was laid to rest at the Hooge Crater Cemetery, Flanders, Belgium. His death was notified in the 343rd. list of Australian Casualties released on the 26th of October 1917.

William Chapman's name is inscribed on The Australian War Memorial Roll of Honour (Panel 182), The City of Townsville WW1 Honour Roll, and The Townsville Cenotaph. Chapman Street, Mysterton, Townsville is named in his memory.

10052 Private William Joseph Chapman was awarded the British War Medal (no:35062) and the Victory Medal (no:34805). A memorial plaque was passed to his widow, Alice, in 1922. A pension of one pound (£1/-/-) per fortnight each was paid to Alice and Irene Chapman their loss. The total amount of two pounds was approximately half of the amount paid to an AIF soldier in 1917.

William Chapman's brother, 10058 Private Joseph Chapman, survived the war and returned to Australia on the 6th of June 1918.

The following tribute was published in the *Townsville Daily Bulletin* (Qld) 22nd September 1919:

CHAPMAN. —Sacred to the memory of our dear husband and father, William Joseph Chapman, of the 14th Field Ambulance, Killed in Action in France, September 21st, 1917.

At Rest.

Fighting in France the evil hordes of hell.
Dim visions of gold wattle and blue skies
Shut out the northern bleakness of his bier.
Voices of love and thoughts of friends and home.
We may believe God granted at the last,
Not lonely did he die, God saw to that
And filled him with the peace we do not know.

(Inserted by his loving wife, Alice, and little daughter, Irene.)

30105 GUNNER HAROLD GEORGE COPE (1896 – 1917)

2nd Divisional Ammunition Column

Harold George Cope was born in Townsville on the 18th of September 1896. He was the second of six children born to Harold and Elizabeth Cope (née Schatzler). Harold Cope Sr. worked as a plumber and passed away in 1907 when Harold Jr. was eleven years old. Harold Cope Jr. grew up in Townsville, and the Cope family lived in premises on Flinders Street. Elizabeth Cope remarried in 1911 to Cornelius Schatzler when Harold Cope was 15 years old.

After completing his school years, Harold Cope worked as a labourer. He also joined the militia and at the age of 17 and served in the Kennedy Regiment before enlisting on the 14th of August 1914 in the Australian Naval and Military Expeditionary Force (AN&MEF). This force was raised to participate in the capture of German New Guinea. Private Harold Cope was stationed on Thursday Island for garrison duty. He departed the island on the 16th of August 1914 aboard the troopship *Kanowna* and officially joined the AN&MEF on the 6th of September 1914. The ship, however, returned to Townsville on the 18th of September due to a dispute with its firemen, without reaching New Guinea. The soldiers aboard the *Kanowna,* including Private Harold Cope were discharged upon their return. Harold Cope resumed working as a labourer.

Eleven months later, on the 30th of August 1915, just before his 19th birthday, Harold Cope applied to enlist in the Australian Imperial Force (AIF). His medical examination noted that he was 5'6½" (169 cm) tall, weighed 9st 12lbs (63kg), had a fair complexion, grey eyes, and dark brown hair. He identified as Church of England. Accepted into the AIF, Harold Cope was assigned to the infantry and posted to the 5th Light Horse Regiment/15th Reinforcements. His training took place at a camp near Brisbane. However, Harold struggled with training and faced serious charges, including multiple instances of illegal absence, disobedience, and creating a disturbance in a house in Brisbane. As a result, he was labelled as *"undesirable"* and recommended for discharge. Following this, Harold's mother, Elizabeth, contacted the Army to state that she had not given her under-age son permission to enlist. Consequently, Harold Cope was discharged from the AIF on the 21st of January 1916.

Six months later, on the 5th of June 1916, Harold Cope applied to re-enlist in the AIF, this time with the consent of both his mother and stepfather, Cornelius Schatzler. He nominated his mother as his next of kin, providing her address as Mrs. E. Schatzler, Macintosh Street, Hermit Park, Townsville. Despite his previous service record and an adverse medical finding in September 1915, which recommended discharge due to varicocele, Harold Cope was accepted into the AIF. He was allocated service number 30105, assigned to the Artillery and posted to the 9th Field Artillery Brigade, 6th Reinforcements. Gunner Cope was deployed from Townsville to Sydney for further training, during which time he married Olive Victoria Barker.

Eleven months after his reenlistment, Gunner Cope and his fellow reinforcements embarked from Sydney on the 9th of November 1916, bound for England. They sailed as part of Convoy No. 27 aboard HMAT A24 *Benalla*, which carried 1,591 troops. The entire convoy of 13 ships transported 15,589 troops. The *Benalla* arrived at Devenport in England on the 9th of January 1917. The next day, the contingent of fresh reinforcements was transported to Wiltshire and marched into the Australian Artillery Training Depots at Larkhill and then at Bulford.

During his time at the training Depot, Gunner Harold Cope was admitted twice to the hospital. On the first occasion, he was suffering from a cough and was hospitalized for 19 days. On the second occasion, he was suffering from myalgia (muscular rheumatism) and was hospitalized for nine days. Once recovered from illness and with his training completed, Gunner Cope crossed to France from Southampton on the 19th of June 1917 and marched into the Australian General Base Depot (AGBD) at Rouelles, France. After five days of training, Gunner Cope joined the 2nd Division Artillery and was allocated to the 2nd Divisional Ammunition Column (2nd DAC), joining his unit by at the end of June 1917.

The DAC's key task was transporting ammunition from storage areas in the "*Third Line*" to the Front (*"First Line"*), where artillery was stationed. This dangerous job used motorized and horse-drawn vehicles, along with rail systems and tramways. A steady ammunition supply was crucial, and depots and routes near the Front Line were prime targets for enemy artillery and later air attacks. Horse-drawn transport posed dual risks: horses were vulnerable to fire, and damage to their ammunition could cause severe destruction.

The battle for Menin Road commenced on the 20th of September. The 2nd DAC was in the vicinity of Busseboom, near Poperinghe, Belgium. On the 21st of September 1917, Gunner Cope was tasked with transporting ammunition by wagon. While completing this task, he was Killed in Action by enemy shellfire. He was the only casualty sustained by the 2nd DAC on that day. Gunner Harold Cope was 21 years old and had served with the AIF for 15 months. His remains were not recovered, and he has no known grave.

Harold George Cope's name is inscribed on The Australian War Memorial Roll of Honour (Panel 21), The Ypres (Menin Gate) Memorial, The City of Townsville WW1 Honour Roll, and The Townsville Cenotaph. Cope Street, Stuart, Townsville is named in his memory. Olive Cope was granted a fortnightly pension of two pounds (£2/-/-) for the loss of her husband. This was nearly half of the fortnightly wage paid to a soldier in the AIF in 1917.

30105 Gunner Harold George Cope was awarded the British War Medal (no:53371) and the Victory Medal (no:52616). A memorial scroll and a memorial plaque were passed to his widow in 1921 and 1922, respectively.

The following notice was published in the *Townsville Daily Bulletin* (Qld.) on the 17th of December 1917:

COPE— Gunner Harold George. 2nd D.A.C., beloved son of Mrs. E. A. Schatzler, Townsville. Killed in Action in France on 21st September, Age 21 years.

He answered his country's call

(Inserted by his loving Mother, Brother, and Sisters.)

4691 LIEUTENANT JOSEPH MARK HIRST (1892 – 1917)

3rd Field Artillery Brigade

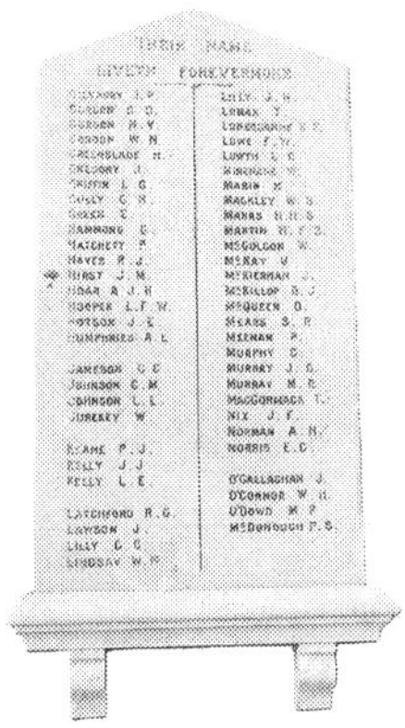

Joseph Mark Hirst was born about September 1892 in Melbourne, Victoria, to Joseph and Ellen Hirst (née Bealin). He was the eldest son of seven children. He had three sisters and two surviving younger brothers. Joseph Hirst senior was a schoolteacher, and the family lived in several towns across Queensland, New South Wales, and Victoria. In 1903, the Hirst family lived in Umbiram, Queensland, where Joseph Hirst senior was teaching. After finishing his early school years, Joseph Hirst junior trained as a clerk and worked for the Australian Mutual Provident Society (A.M.P.). In 1915, he was living at Waldby's Hotel (Victoria Bridge Hotel) in Townsville.

On the 23rd of February 1915, Joseph Hirst applied to enlist in the Australian Imperial Force (AIF) in Townsville. He declared that he was 22 years old, single, employed as a clerk, and was still serving with the Volunteer Artillery Garrison. Joseph Hirst's medical examination noted that he was 5'5½" (166 cm) tall, weighed 9st (57kg), had a fair complexion, blue eyes, and dark brown hair. He identified as Church of England and nominated his father as his next of kin, giving his address as the State School Sarina. Accepted into the AIF, Joseph Hirst was allocated service number 46791, assigned to the Artillery, and posted to the 3rd Artillery Brigade, 7th Reinforcements.

Following enlistment, Joseph Hirst was deployed to Enoggera, Brisbane, for training for overseas service. His previous service must have been apparent early in training as he was promoted to Acting Sergeant in April 1915. Five months of basic training followed enlistment until Acting Sergeant Joseph Hirst and his fellow Reinforcements embarked from Melbourne for Egypt on the 10th of August 1915. They sailed as part of Convoy No.9 aboard RMS *Persia.* Their ship boarded 499 troops, and the entire convoy of eight ships transported 5,723 troops and 328 horses on the voyage.

After disembarking in Egypt in late September 1915, the Reinforcements underwent further training at Tel-el-Kebir. In February 1916, Acting Sergeant Hirst was posted to the 11th Field Artillery Brigade (FAB), 41st Battery, and in March, he was promoted from Acting Sergeant to substantive Sergeant. The 11th FAB embarked at Alexandria on the 1st of June 1916, bound for France and the Western Front to join the British Expeditionary Force (BEF). They disembarked at Marseilles on the 10th of June 1916, and 10 days later Sergeant Hirst was promoted to 2nd Lieutenant and seconded for duty with the 4th Medium Trench Mortar Battery.

The Australians used the 2-inch Medium Trench Mortar. Mortar crews earned the moniker *'the shoot and scoot mob'* among the front-line troops. Following the mortar crew's departure shortly after firing, the infantry was left to endure the retaliatory fire. For this reason, mortar crews were unpopular with the infantry. October 1916 brought promotion to Lieutenant (Lt) for Joseph Hirst. The next few months saw Lt Hirst attached for temporary duty to several Artillery Units, including the 4th Divisional Ammunition Column and the 11th Field Artillery Brigade.

On the 18th of September 1917, Lieutenant Joseph Hirst was transferred to the 10th Field Artillery Brigade, which was actively involved in the Battle of Menin Road. This engagement marked the debut of Australian units (1st and 2nd Divisions AIF) in the Third Battle of Ypres. The objective was to secure segments of the curved ridge east of Ypres, which the Menin Road traversed. Despite the presence of fortified German pillboxes, the assault achieved success along its entire front. The battle was characterized by intense artillery barrages from both British and German forces.

It was during this offensive that Lieutenant Joseph Hirst was Killed in Action on the 21st of September 1917. He was 26 years old and had been a member of the AIF for two years and six months. Lieutenant Hirst's body was recovered, and he was laid to rest in the Birr Cross Roads Cemetery (Plot IV, Row C, Grave No. 8), Zillebeke, Belgium. There was some miscommunication in respect to Joseph Hirst's Date of Death. He was originally reported as killed on the 4th of September 1917. It was his father who questioned this with the Army and, by producing letters from Joseph sent after the 4th of September, finally had the records amended to indicate the correct date of death.

Joseph Hirst's death was notified in the combined 371st and 372nd list of Australian Casualties issued on the 16th of December 1917. Joseph Hirst's name is inscribed on The Australian War Memorial Roll of Honour (Panel 16), The City of Townsville WW1 Honour Roll, and The Townsville Cenotaph. Hirst Street, Hermit Park, Townsville, was named in his memory in 1926.

Lieutenant Joseph Mark Hirst was awarded the 1914/15 Star (no:3023), the British War Medal (no:17457), and the Victory Medal (no:17389). A memorial scroll and a memorial plaque were passed to his father in 1921 and 1922, respectively.

Joseph Hirst's younger brother, 3762 Corporal William Hirst, also enlisted in the AIF, and he returned to Australia on the 20th of November 1918, having won the Military Medal.

The following Family Notices were published in The Brisbane Courier on the 4th of September 1917, 8th October 1917, 12th October 1917, and the 20th of September 1918.

HIRST. —Sacred to the memory of Lieutenant Joseph Mark Hirst, who was Killed in Action on September 4, 1917, aged 23 years 11 months. "Thy will be done." Inserted by his grieved brother-in-law and sister, John and Lily Leitch.

HIRST. —Lieutenant Joseph Mark Hirst, son J. W. Hirst, Sarina, and late of A. M. P. Office, Townsville, killed in action, 4th September 1917.

407 PRIVATE WILLIAM FRALEY DESCHAMPS (1890 – 1917)

25th Infantry Battalion

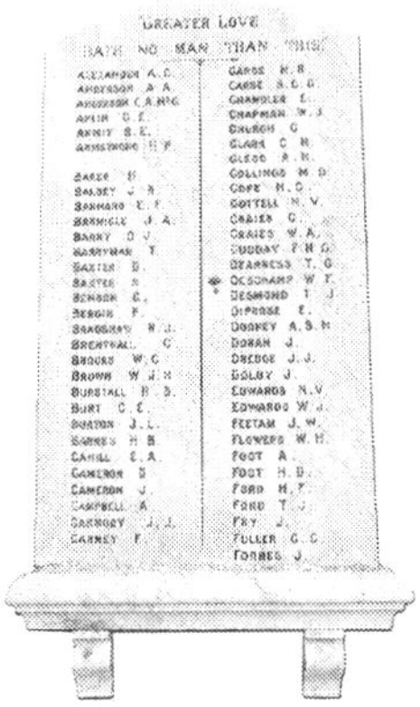

William Fraley Deschamps was born on the 8th of April 1890, in Townsville, Queensland. He was the youngest son of Auguste and Rose Deschamps (née Hind) and had four brothers and five sisters. Auguste Deschamps worked as a wharf labourer, and the family lived on Sturt Street in Townsville.

William Deschamps grew up in Townsville and attended Ross Island State School. Rose Deschamps passed away in 1908 when William was 18 years old. After completing his schooling, he took up work as a carter, driving a horse-drawn vehicle for transporting goods.

On the 23rd of October 1916, William Deschamps applied at Townsville to enlist in the Australian Imperial Force. He stated his age as 24 years and six months; however, Queensland birth records indicated he was 26 years old. He declared himself single, with no prior military experience, and employed as a carter. His enlistment medical examination noted that he was 5'3½" (161 cm) tall, weighing 8st 6 lbs (54 kg), with a ruddy complexion, grey eyes, and auburn hair. He identified as Church of England and nominated his oldest sister, Mary Rose Hitchcock, as his next of kin.

Accepted into the AIF, William Deschamps was allocated service number 6407, assigned to the Infantry, and posted to the 25th Battalion, 18th Reinforcements. The 25th Battalion was formed in March 1915 from Queensland volunteers and fought at Gallipoli before being redeployed in March 1916 to France to join the British Expeditionary Force.

Private Deschamps underwent two months of basic training in Australia before embarking from Sydney aboard HMAT A64 *Demosthenes* on December 23, 1916, bound for England as part of Convoy No. 27. After a voyage of 14 weeks, the reinforcements disembarked at Plymouth, England, and were transferred to the 7th Training Battalion at Rollestone, Wiltshire. While undergoing advanced training at Rollestone, Private Deschamps contracted mumps and spent most of March 1917 hospitalized.

On the 15th of June 1917, he crossed to France, where, after further training at Le Havre, he joined the 25th Battalion, located in a tented camp at Bapaume, on the 7th of July 1917. The 25th had been in France since the 19th of March 1916 and had suffered 785 casualties at the Battle of Pozières in 1916. The battalion's next major engagement was on the 20th of September 1917, during the Battle of Menin Road. Despite encountering heavy resistance, the assault achieved complete success, although casualties were significant.

On the 21st of September 1917, near Westhoek, the battalion was ordered to assault the "Red Line" objective. The attack succeeded, but casualties were high, including Private William Deschamps, who was Killed in Action at the age of 27 years and five months, after serving in the AIF for 11 months. He was laid to rest in the Tyne Cot British Cemetery, Passchendaele, Plot 25, Row H, Grave 20.

Private William Deschamps' death was notified in the 345th list of Australian Casualties released on the 24th of October 1917. His name is inscribed on The Australian War Memorial Roll of Honour (Panel 104), The City of Townsville WW1 Honour Roll, and The Townsville Cenotaph (n.b. incorrectly inscribed as Deschamp). Deschamps Street, Gulliver, Townsville, is named in his memory.

6407 Private William Fraley Deschamps was awarded the British War Medal (No: 54339) and the Victory Medal (No: 53499). A memorial plaque was presented to his next of kin in 1922.

The following notice was published in the *Townsville Daily Bulletin* (Qld.) on the 21st of September 1920:

DESCHAMPS. —In loving memory of William Fairley (sic), who was Killed in Action on 21st September 1917.

(Inserted by his loving father, sisters, and brother.)

6385 PRIVATE ALBERT JOHN SHEEHAN (1895 – 1917)

25th Infantry Battalion

Albert John Sheehan was born on the 28th of November 1895 in Cunnamulla, Queensland. He was the second son and second-born child of William and Mary Sheehan (née Ambrose). He had two brothers and one sister. William Sheehan worked as a miner at Monteagle in Queensland in 1903, and by 1913, he was a contractor living in Charters Towers Road, Townsville.

Albert Sheehan attended State School in Cunnamulla and started work as a farmer later in life. Albert Sheehan's older brother, Roy Sheehan, had enlisted in January 1916 and was serving with the artillery in France. On the 23rd of October 1916, Albert Sheehan decided to enlist as well and applied at Townsville to join the Australian Imperial Force (AIF) when he was 20 years and 10 months old. He declared that he had no military experience, was single, and working as a farmer. His medical examination noted that he was 5'8¾" (175 cm) tall, weighed 9st 1lbs (58kg), had a muddy complexion, grey eyes, and black hair. He identified as Roman Catholic and nominated his mother as his next of kin, giving her address as Short St, Townsville.

Accepted into the AIF, Albert Sheehan was allocated service number 6385, assigned to the Infantry, and posted to the 25th Battalion 18th Reinforcements. The 25th Battalion was deployed near Ypres in France when Albert Sheehan enlisted. Two months of basic training in Australia followed enlistment.

Basic training completed, Private Albert Sheehan and his fellow Draft of Reinforcements joined Convoy No. 27 and embarked from Sydney bound for England on the 23rd of December 1916. They sailed aboard HMAT A64 *Demosthenes*, which boarded 1,240 troops. The entire convoy of 13 ships carried 15,278 troops and 311 other passengers on the voyage. The Reinforcements disembarked at Plymouth 14 weeks later and were soon moved to the

7th Training Battalion located at Rollestone (near Amesbury on Salisbury Plain in Wiltshire).

While he was undergoing training at Rollestone, Private Albert Sheehan suffered a bout of mumps and was hospitalized from early March until the end of April 1916. Three weeks after his discharge from the hospital, Albert received the news that his brother, Gunner Roy Sheehan, had been Killed in Action on the 12th of May 1917.

Now it was Private Albert Sheehan's turn to join the battle, and on the 14th of June 1917, he crossed to France. After a couple of weeks of further training at Havre, Private Sheehan was taken on the strength of the 25th Battalion on the 3rd of July 1917, located in a tented camp at Bapaume. The 25th had been in France since the 19th of March 1916 and had suffered 785 casualties at the Battle of Pozières in 1916. While it acted in a support role during the Second Battle of Bullecourt, the 25th Battalion's next significant offensive engagement occurred on the 20th of September 1917, when it formed a part of the initial wave during the Battle of Menin Road.

The objective of the battle was to secure sections of the curved ridge east of Ypres, encompassed by the Menin Road. The assault achieved full success along its entire front, despite the presence of fortified German pill-box strongpoints, but at a heavy cost. On the 21st of September 1917, the 25th Battalion was deployed in the vicinity of Westhoek, which had been captured by the BEF on the 10th of August 1917. From here, the 25th was ordered to attack the first objective (the "Red Line"). The attack was successful, and the objective was taken; however, the casualty rate was high, with 43 killed or died of wounds, 138 wounded, and four reported missing. It was during this action that Private Albert Sheehan was Wounded in Action on the 21st of September 1917 and evacuated back to the 1st Australian Field Ambulance. He succumbed the next day, and on the 22nd of September 1917, he Died of Wounds received in action.

Albert Sheehan was 21 years and 10 months old at the time of his death and had been a member of the AIF for 11 months. He was laid to rest in The Huts Cemetery (Plot V, Row D, Grave No. II), Dickebusch, Belgium. His death was notified in the 343rd list of Australian casualties released on the 19th of October 1917.

Private Albert John Sheehan served on the Western Front in WW1. His name is inscribed on the Australian War Memorial Roll of Honour (Panel 106), the City of Townsville WW1 Honour Roll, and the Townsville Cenotaph. Sheehan Street, Hermit Park, Townsville, is named in his memory as well as the memory of Ray Allen Sheehan and William Frederick Francis Alexander David Sheehan.

6385 Private Albert John Sheehan was awarded the British War Medal (no:54356) and the Victory Medal (no:53514). A memorial scroll and a memorial plaque were passed to his mother in 1922. Mary Sheehan was granted a pension of one pound and 10 shillings (£1/10/-) per fortnight for the loss of her son. This amount was in addition to the 14 shillings per fortnight already granted in respect of Roy Allan Sheehan. The total amount of two pounds and four shillings per fortnight (£2/4/-) was about one quarter of the wage paid to two AIF soldiers in 1917.

The following notice was published in the *Townsville Daily Bulletin* on 9th October 1917:

"The Great War. Mr. and Mrs. W. J. Sheehan, Townsville, have received word that their son, Private Albert John Sheehan, has died of wounds. Private Sheehan was just over 21 years of age and is the second son to pay the supreme sacrifice within the last six months. Private Roy Sheehan being Killed in Action about five months ago."

471 PRIVATE WILLIAM PATRICK MEENIN (1895 – 1917)

47th Infantry Battalion

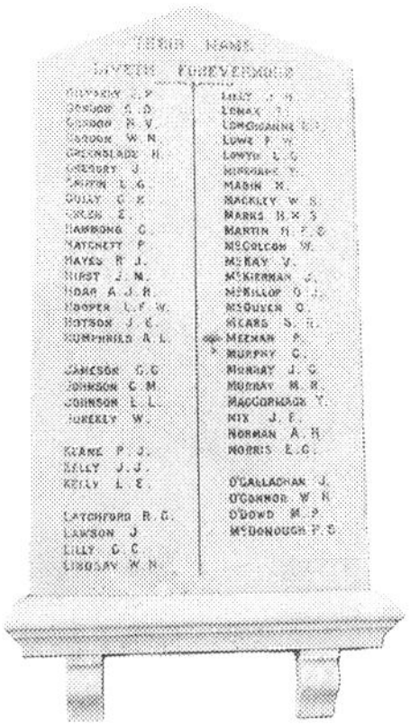

William Patrick Meenin (Patrick) was born on the 23rd of November 1895, in Mackay, Queensland. He had two sisters and was the youngest child and the only son of James and Mary Meenin (née McGuire). After leaving Mackay, the Meenin family moved to Townsville.

After his school years, Patrick Meenin worked in blacksmithing. On the 18th of October 1915, he applied in Townsville to join the AIF. He declared that he was one month short of his twentieth birthday, employed as a blacksmith's striker, and single. He also declared that he had no previous military experience. His father provided written consent for him to join the AIF.

Patrick Meenin's medical examination noted that he was 5'6½" (169 cm) tall, weighed 117 lbs (53 kg), had a dark complexion, brown eyes, and black hair. He identified as Roman Catholic and nominated his father as his next of kin, giving his address as Flinders St, Townsville.

Accepted into the AIF, Patrick Meenin was allocated service number 471, assigned to the Infantry, and posted as an original member to the 41st Battalion, B Company, which was established at Bell's Paddock Camp in Brisbane in February 1916. The battalion was manned by recruits from Brisbane, northern Queensland, and the northern rivers district of New South Wales. The 41st, along with the 42nd, 43rd, and 44th Battalions, was allocated to the 11th Brigade, which was part of the 3rd Australian Division.

Private Patrick Meenin spent the next seven months in Enoggera, Brisbane, training for overseas service. On the 18th of May 1916, his company embarked from Sydney, bound for France via England. They sailed as part of Convoy No. 21 aboard HMAT A64 *Demosthenes*, which carried 1,612 troops on the voyage. The 18-ship convoy transported 25,333 troops. The 41st Battalion disembarked in England two months later, on the 20th of July 1916.

Training continued for another two months at Rollestone on the Salisbury Plain. It was during this training period that Private Patrick Meenin was transferred to the 47th Battalion, which had suffered heavy casualties in fighting at Pozières and Mouquet. The 47th Battalion was part of the 12th Brigade of the 4th Division.

Private Meenin crossed to France at the end of September 1916 and joined the 4th Australian Division Base Depot (4th ADBD) at Étaples for further training. He was taken on strength with the 47th Battalion on 17 October 1916, which was resting and training in Flanders at that time.

In November 1916, the 47th moved to the Somme Valley. Throughout the freezing winter of 1916/17, they laboured behind the lines and rotated in and out of the front line a few miles east of Pozières. Conditions were extreme, and in late November 1916, Private Meenin was hospitalized and evacuated to England due to debility, exhaustion, and trench foot. He was admitted to St Thomas' Hospital in London, where 200 beds were designated for military use.

Four months after he was first admitted, Private Patrick Meenin was discharged from hospital and underwent a period of convalescence and furlough in London. It was during this time that he met 21-year-old Daisy Miles. The couple married at All Saints' Church, Highgate, in April 1917. They only spent a few weeks together before Private Patrick Meenin returned to France at the end of May 1917. After a couple of weeks refresher training at the 4th ADBD he rejoined the 47th on the 13th of June 1917.

The 47th was located at Messines at this time, and on the day that Private Meenin arrived, it was subjected to heavy shelling. The battalion had already taken significant casualties in the Battle of Messines (7-14 June 1917). The 47th endured two stints in the heavily contested trenches of Pozières (23 July - 3 September 1917), as well as a period in reserve.

On the 26th of September 1917, the 47th Battalion served as divisional reserve at the Battle of Polygon Wood. A few days after the successful advance, the battalion entered the consolidation phase of the operation. It was during this phase that Private Patrick Meenin was Killed in Action on the 29th of September 1917. He was buried on Anzac Ridge at map reference Sheet

28NE.J.3.c.0.8. His body was not reinterred and today he is recorded as having no known grave.

At the time of his death, Patrick Meenin was 21 years and 10 months old and had been a member of the AIF for nearly two years. His death was notified in the 353rd list of Australian casualties released on 19 November 1917.

William Patrick Meenin's name is inscribed on The Australian War Memorial Roll of Honour (Panel 144), The Ypres (Menin Gate) Memorial, Belgium, The City of Townsville WW1 Honour Roll, and The Townsville Cenotaph (n.b. incorrectly inscribed as Meenan). Meenan Street, Garbutt, Townsville, is named in his memory (also incorrectly spelt).

471 Private William Patrick Meenin was awarded the British War Medal (no: 39877) and the Victory Medal (no: 39493). A memorial scroll and a memorial plaque were passed to his widow in 1922. A fortnightly war pension of two pounds (£2/-/-) was paid to Daisy Meenin for the loss of her husband. This amount was nearly half of the fortnightly wage paid to an AIF soldier in 1917.

The following tribute was published in the *Townsville Daily Bulletin* on the 30th of September 1918:

MEENIN. —In loving remembrance of Private Patrick Meenin, who was

Killed in Action somewhere in France, 29th September 1917.

> Lord, take his soul into Thy keeping,
> And stay mourner's bitter weeping.
> He freely offered his life and all
> In answer to his country's call.
> He fought for honor and for right,
> He helped to show Australia's might.

(Inserted by his loving father, mother and sisters, Mr. and Mrs. J. Meenin, Mrs. T. Camp, and Olive Meenin.)

4519 PRIVATE ALFRED JAMES RICHARD HOAR (1895 – 1917)

47th Infantry Battalion

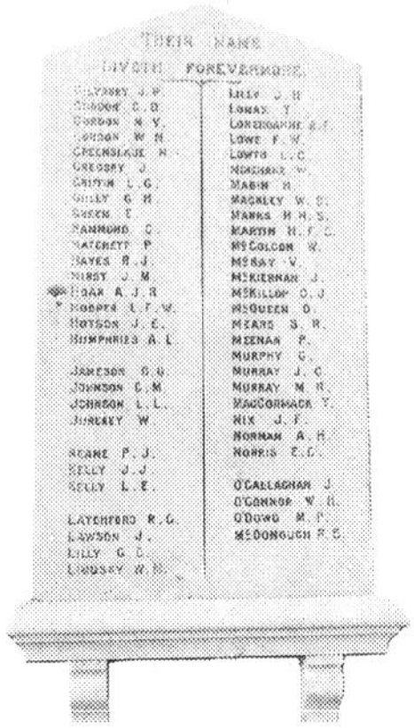

Alfred James Richard Hoar was born on the 25th of March 1895, in Cooktown, Queensland, to William and Henrietta Hoar (née Murgatroyd). He was the third of four children, having one older brother and two sisters. William Hoar was a seaman, and the family lived on the Esplanade in Cooktown. After completing his schooling, Alfred Hoar moved to Townsville, took up residence in Stuart Street, and worked as a cook's assistant. His sister, Ada, also lived in Townsville.

On the 13th of September 1915, Alfred Hoar applied to enlist in the Australian Imperial Force (AIF). He declared that he was 20 years and five months old, single, working as a cook's assistant, and had no previous military experience. His medical examination noted that he was 5'4½" (164 cm) tall, weighed 120 lbs (54.5 kg), had a dark complexion, brown eyes, and brown hair. He identified as Church of England and nominated his father as his next of kin, giving his address as Grassy Hill, Cooktown.

Accepted into the AIF, Alfred Hoar was allocated service number 4519, assigned to the Infantry, and posted to the 15th Battalion, 14th Reinforcements. The 15th Battalion was originally raised as part of the all-volunteer AIF in September 1914. Drawing personnel from volunteers from Queensland and Tasmania, it formed part of the 4th Brigade, along with the 13th, 14th, and 16th Battalions. The battalion was fighting on the Gallipoli Peninsula when Alfred Hoar enlisted.

After five months in Enoggera, Brisbane, completing basic training, Private Alfred Hoar and his fellow reinforcements joined Convoy No.18 and embarked from Brisbane on the 31st of January 1916, bound for Egypt. They sailed aboard HMAT A62 *Wandilla*, which carried 997 troops. The entire convoy of eight ships carried a total of 8,443 passengers and 670 horses on the voyage.

Soon after arriving in Zeitoun in March 1916, Private Hoar contracted mumps and was admitted to the hospital. It was not until early April that he had recovered and returned to duty. He was then transferred to the 47th Battalion. The 47th was raised in Egypt on the 24th of February 1916, as part of the "doubling" of the AIF. About half of the battalion's new recruits were Gallipoli veterans from the 15th Battalion, and the other half were fresh reinforcements from Australia. The battalion was composed mostly of men recruited in Queensland and Tasmania.

In June 1916, the 47th Battalion departed from Egypt and proceeded to France to join the fighting on the Western Front. Upon their arrival in France on the 9th of June, the 47th entered the trenches of the Western Front for the first time on the 3rd of July. They participated in their first major battle at Pozières. Initially, the battalion provided working parties during the 2nd Division's attack on the 4th of August and subsequently, with their own division, defended the captured ground. The 47th endured two stints in the heavily contested trenches of Pozières, as well as a period in reserve.

It was during this action that Private Alfred Hoar received a gunshot wound to the foot on the 19th of August 1916. He was evacuated back to England and, following treatment, rejoined the 47th in December 1916. After Pozières, the battalion spent the period up until March 1917 alternating between duty in the trenches and training and rest behind the lines. On the 11th of April 1917, it took part in the attack mounted against the heavily defended village of Bullecourt, part of the formidable Hindenburg Line to which the Germans had retreated during February and March. Devoid of surprise and dependent upon the support of unreliable tanks, the attack failed.

In June 1917, the 47th Battalion took part in the Battle of Messines. In this action, Private Alfred Hoar was wounded for the second time, this time a gunshot wound to the knee. Once again, he was evacuated back to England and did not return to the 47th Battalion until the 11th of September 1917, at which time the 47th Battalion was heavily committed to the Battle of Passchendaele in Belgium.

Private Alfred Hoar was only back in the battalion for three weeks before he was Killed in Action on the 30th of September 1917, near Westhoek. A subsequent Red Cross investigation took statements from three witnesses to his death. All stated that he, and two others, were killed instantly by a shell burst on 'ANZAC Ridge'. They also reported that there were few remains. Private Alfred Hoar was laid to rest at Aeroplane Cemetery (Plot II, Row C, Grave No. 36), Belgium. He was 22 years and six months old and had been a member of the AIF for two years and two weeks. His death was notified in the 353rd list of Australian casualties released on the 19th of November 1917.

Alfred Hoar's name is inscribed on the Australian War Memorial Roll of Honour (Panel 143), the City of Townsville WW1 Honour Roll, and the Townsville Cenotaph.

4519 Private Alfred James Richard Hoar was awarded the British War Medal (no: 31316) and the Victory Medal (no: 31092). A memorial scroll and a memorial plaque were passed to his father in 1921 and 1922, respectively. A fortnightly war pension of two pounds (£2/-/-) was paid to Alfred's mother for the loss of her son. This was nearly half of the fortnightly salary paid to an AIF soldier in 1917.

The following tribute was published in the *Townsville Daily Bulletin* on the 31st of October 1917:

HOAR.—Killed in Action in France on September 30th, 1917, Private Alfred James Richard Hoar, of the 47th Battalion, the younger beloved son of Mrs. W. Hoar, of Cooktown. Age 21 years.

He died a hero's death fighting for the flag.

1861 PRIVATE JOHN HENRY LILLY (1897 – 1917)

41st Infantry Battalion

John Henry Lilly was born on the 24th of April 1897, in Townsville, Queensland, to John and Louisa Lilly (née Donnithorne). He was the third surviving son of a large family. John and Louisa Lilly emigrated from Cornwall, England, and arrived in Queensland in 1889. John Lilly Sr. worked as a labourer, and the family lived on Macrossan St, South Townsville. John Lilly Jr. attended South Townsville State School and, after finishing his school years, worked as a painter.

On the 21st of February 1916, John Lilly applied at Townsville to join the Australian Imperial Force (AIF). He declared that he was single, 18 years and 10 months old and had two years' previous military experience with the TAIA. His medical assessment noted that he was 5'4½" (164 cm) tall, weighed 8st 11lbs (56 kg), had a fresh complexion, hazel eyes, and dark brown hair. He identified as Church of England and nominated his mother as his next of kin, giving her address as Main St, off Macrossan St, South Townsville.

Accepted into the AIF, John Lilly was allocated service number 1861, assigned to the Infantry, and posted to the 41st Battalion/2nd Reinforcement. The 41st Battalion was raised at Bell's Paddock Camp in Brisbane in February 1916 with recruits from Brisbane, northern Queensland, and the northern rivers district of New South Wales. It formed part of the 11th Brigade of the 3rd Australian Division. Private John Lilly spent the next seven months in Australia preparing for overseas service.

On the 16th of August 1916, Private Lilly and his fellow reinforcements joined Convoy No. 23 and embarked from Brisbane, bound for England. They sailed aboard HMAT A42 *Boorara*, which boarded 1,707 troops. The entire convoy of 12 ships carried a total of 11,246 troops on the voyage. Eight weeks later, they disembarked at Plymouth, England. Further training followed in England until the reinforcements crossed to France in December 1916. After spending a week at the 3rd Australian Division Base Depot at Étaples, Private John Lilly marched into the 41st Battalion on the 26th of January 1917.

The 41st had arrived in France on the 25th of November 1916. It spent the bleak winter of 1916-17 alternating between service in the front line and training and labouring in the rear areas. It was during this period that Private Lilly was wounded in the forehead by gunshot on the 15th of April 1917. He was evacuated and, after two weeks of treatment and recovery, Private Lilly rejoined the 41st Battalion in May. In June 1917, he was attached to the 11th Machine Gun Company.

At the end of June 1917, the 11th Brigade, including the 41st Battalion, was tasked with establishing a new front-line west of Warneton, fully exposed to the Germans. Despite heavy shellfire, work continued day and night, and this period became known to the battalion as 'the 18 days.' By early August 1917, the 41st Battalion was holding ground captured by two of its sister battalions during a feint attack on the 31st of 1917. Enduring relentless rain, flooded trenches, and heavy shelling, many of the battalion's platoons dwindled from 35 men to fewer than 10.

It was during this fighting that Private John Lilly was wounded on the 29th of September 1917, near Poperinghe during an aerial attack. He was evacuated to the 2nd Canadian Casualty Clearing Station but succumbed to his wounds and Died of Wounds on the 1st of October 1917. He was 20 years and seven months old at the time of his death and had been a member of the AIF for 19 months. He was laid to rest at Lijssenthoek Military Cemetery (Plot XXV, Row E, Grave No. 14), Belgium. Private John Lilly's death was reported in the 346th Casualty List released on the 27th of October 1917.

John Lilly's name is inscribed on The Australian War Memorial Roll of Honour (Panel 134), The 41st Battalion Roll of Honour, The City of Townsville WW1 Honour Roll, and The Townsville Cenotaph. Lily Street, Hermit Park, Townsville, is named in his and his brothers' memory (n.b., the street name has been spelled Lily, Liley, Lilley, and occasionally correctly; Lily is now the accepted spelling). Louisa Lilly was granted a pension of two pounds (£2/-/-) per fortnight for the loss of her son. This was nearly half of the amount paid to an AIF soldier in 1917.

1861 Private John Henry Lilly was awarded the British War Medal (no:45381) and the Victory Medal (no:44860). A memorial scroll and a memorial plaque were passed to his father in 1921 and 1922, respectively.

3206 Private George Charles Lilly, the older brother of John Lilly, was Killed in Action on the 12th of October 1917, while fighting with the 49th Battalion.

The following tribute was published in the *Townsville Daily Bulletin* on the 10th of November 1917:

LILLY — On the 1st of October, died of wounds, Private John Henry Lilly, third eldest son of Mr and Mrs J. H. Lilly, South Townsville, aged 20 years and seven months.

> His country called, and honor bade him go,
> To battle 'gainst a grim and deadly foe,
> He helped to bring Australia into fame,
> To build for her a never dying name,
> Foremost was he in thickest strife,
> For King and Country laid down his life.
> At the dawn of day,
> The shadows flee away.

Inserted by his sorrowing mother, father, brothers and sisters.

5551 PRIVATE NORMAN REGINALD CARSE (1893-1917)

25th Infantry Battalion

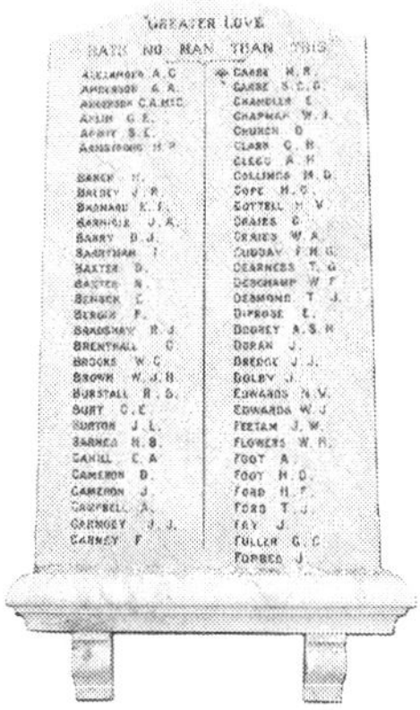

Norman Reginald Carse was born on the 21st of December 1893 at Charters Towers, Queensland. He was the son of John and Hannah Carse (née Greig). John Carse was a storekeeper and established Carse's General Drapery in Charters Towers and Townsville. Hannah Carse died in July 1894 when Norman was 18 months old. Raised by his father, Norman attended State School and, when his schooling years were completed, became a labourer and a sugar farmer.

Norman Carse was 21 years old, single, employed as a labourer, and without previous military experience when he applied on the 13th of December 1915, to enlist in the Australian Imperial Force (AIF). His medical examination noted that he was 5'7½" (171 cm) tall, weighed 9st 12lbs (52kg), had a fair complexion, blue eyes, and brown hair. He identified as Presbyterian and nominated his father as his next of kin, giving his address as Flinders St, Townsville. Accepted into the AIF, Norman Carse was allocated service number 555, assigned to the Infantry and posted to the 25th Battalion, 15th Reinforcements. The 25th Battalion was raised at Enoggera in Queensland in March 1915 as part of the 7th Brigade, 2nd Division. Although predominantly composed of men recruited in Queensland, the battalion also included a small contingent of men from Darwin. When Norman Carse enlisted, the 25th was in the process of being evacuated from Gallipoli. The battalion left the peninsula on the 18th of December 1915.

After completing basic training in Australia, Private Norman Carse and the other reinforcements embarked from Brisbane bound for England on the 7th of September 1916. They sailed with Convoy No. 24 aboard HMAT A46 *Clan MacGillivray* , which carried 1,084 troops. The entire convoy of 13 ships transported 12,600 troops on the voyage. Eight weeks later, in November 1916, they disembarked at Plymouth in England. From there, they were transported to the 7th Training Battalion at Rollestone on the Salisbury Plain for further training. During his time at Rollestone, Private Carse absented himself without leave (AWOL) for five days. When he returned, he was charged, found guilty, and sentenced to five days Field Punishment No. 2 (F.P. No. 2). There were two categories of field punishment. Field Punishment No. 1 consisted of heavy labouring duties, possibly being restrained in handcuffs or fetters, and being tied to a post or wheel. Field Punishment No. 2 differed in that the offender was not liable to be attached to a fixed object. In both forms of field punishment, the soldier was also liable for loss of pay. In November, Norman received the tragic news that his brother, Sydney, had been Killed in Action at Flers on the 5th of November 1916, while serving with the 26th Battalion.

Once the training in Rollestone was completed, Private Norman Carse and his fellow reinforcements embarked from Folkestone and crossed to France on the 13th of December 1916. On their arrival in France, the reinforcements were marched into the 2nd Australian Division Base Depot at Étaples on the 14th of December. After four more days of training, Private Norman Carse joined the 25th Battalion in France one week before Christmas Day, on the 18th of December 1916. Having transitioned from service in Gallipoli and Egypt, the 25th Battalion had arrived in France on the 19th of March 1916, making it the initial AIF battalion to reach the continent. Now integrated into the 2nd Division, it had engaged in its inaugural significant battle at Pozières from the 25th of July to the 7th of August 1916, resulting in a total of 785 casualties.

While playing a supporting role during the second battle of Bullecourt, the 25th Battalion was not engaged in a major offensive until the 20th of September 1917, when it was part of the first wave at the Battle of Menin Road in Belgium. On this date, it led the assault aimed at securing control over parts of the winding ridge east of Ypres, traversed by the Menin Road. This engagement marked the entry of Australian forces (1st and 2nd Divisions AIF) into the Third Battle of Ypres. The assault was successful across its entire frontline, despite facing well-fortified German defensive positions with interlocking concrete pillbox strongholds. These strongholds created formidable barriers for the advancing diggers of the AIF. Additionally, the diggers had to withstand determined German counter-offensives. During this intense conflict, the 1st and 2nd Divisions suffered a total of 5,013 casualties. This victory was followed by the capture of Broodseinde Ridge.

It was during this battle that Private Norman Carse was Killed in Action on the 4th of October 1917, at Zonnebeke. He was 23 years old at the time of his death and had served with the AIF for one year and nine months. Private Norman Carse had been a member of the 25th Battalion for nine months. There are no precise details of his death, and he has no known grave. His death was notified in the 352nd list of Australian casualties released on the 15th of November 1917.

Norman Carse's two brothers also joined the AIF, one of whom survived and one who was Killed in Action. Their details are: 21128 Driver David Andrew Carse, 12th Field Artillery Brigade, returned to Australia on the 1st of August 1919; 4801 Private Sydney Charters Carse, 26th Battalion, Killed in Action on the 5th of November 1916.

Norman Carse's name is inscribed on The Australian War Memorial Roll of Honour (Panel 104), The Ypres (Menin Gate) Memorial, Belgium, The City of Townsville WW1 Honour Roll, and The Townsville Cenotaph. Carse Street, Hyde Park, Townsville, is named in his and his brother Sydney's memory.

5551 Private Norman Reginald Carse was awarded the British War Medal (no: 46726) and the Victory Medal (no: 46146). A memorial scroll was passed to his father in 1921 and a memorial plaque in 1922.

The following article was published in the *Townsville Daily Bulletin* (Qld.) on 27th October 1917:

"The Great War.

The sad news has been received that Pte Norman Reginald Carse, son of Mr John Carse of Carse and Co., has been killed in action. Another son, Pte S.C.G. Carse, was Killed in Action last November and there is still another son fighting."

The following was published in Family Notices in the *Townsville Daily Bulletin* (Qld.) on the 4th of October 1921:

"CARSE. -In loving memory of our son and brother, Pte Norman Reginald Carse, 25th Battalion who was Killed in Action at Zonnebeke, 4th October 1917.

"Until the day breaks and the shadows flee away"

5874 PRIVATE PERCY SAMUEL McDONOUGH (1891 – 1917)

25th Infantry Battalion

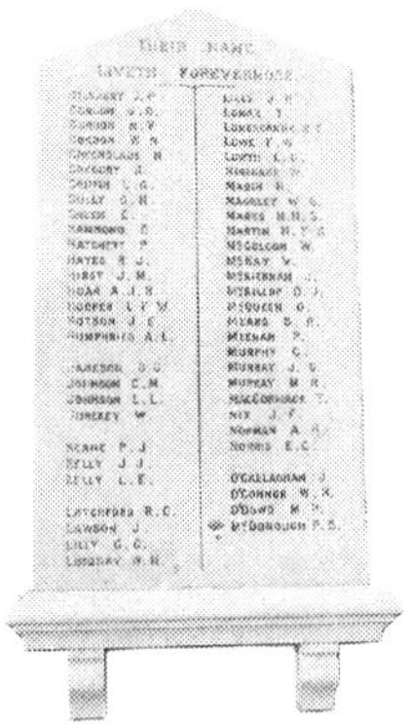

Percy Samuel McDonough was born on the 13th of July 1891 in Ravenswood, Queensland, to John and Katherine McDonough (née Miller). He was the fifth of nine children, and he had four brothers and four sisters. The McDonough family worked at mining and labouring in Ravenswood.

Percy McDonough moved to Townsville prior to WW1 and worked as a baker's carter. On the 25th of March 1916, he applied in Townsville to join the AIF. His older brother, James McDonough, had enlisted in Cairns on the previous day. Percy declared that he was 24 years and eight months old, single, employed as a carter and had no previous military service. His medical examination noted that he was 5'6" (168 cm) tall and weighed 9st 3lbs (58kg). He had a dark complexion, brown eyes and dark brown hair. He identified as Roman Catholic and listed his mother as his next of kin giving her address as 7th Street, Chillagoe, Queensland.

Accepted into the AIF, Percy McDonough was allocated service number 5874, assigned to the Infantry and posted to the 25th Battalion, 16th Reinforcements. For the next six months, Percy and the rest of the Reinforcements completed basic training in Australia before they embarked from Brisbane, on the 21st of October 1916 bound for England. They travelled as part of Convoy No. 25 aboard HMAT A36 *Boonah* which carried 1,237 passengers. The entire convoy of 12 ships carried a total of 12,513 passengers and 5,600 horses on the voyage.

The Boonah docked in Plymouth in early January 1917 and Private Percy McDonough and the other Reinforcements were transported to Rollestone on Salisbury Plain in Wiltshire, for further training with the 7th Training Battalion. Percy suffered from ear infections during his time at Rollestone and received treatment at Fargo Hospital. It was not until the 14th of June 1917 that Private McDonough boarded a transport ship at Southampton and made the one-day trip to Le Havre to join the 2nd Australian Division Base Depot (2nd ADBD) for further training. While he was at the 2nd ADBD, Private Percy McDonough was charged for *"being improperly dressed on parade"*. He was found guilty by the Commanding Officer and was awarded the punishment of being confined to camp for seven days.

After completing his training, Private McDonough joined the 25th Battalion in France on the 2nd of June 1917. The battalion was stationed at Senlis, about 40 kilometres north of Paris, focusing on training and rebuilding for upcoming battles.

The 25th's next major action was on 20 September 1917 during the Battle of Menin Road in Belgium, as part of the 2nd Division's first wave. This victory was followed by the successful capture of Broodseinde Ridge on the 4th of October 1917. Despite mud and rain, the 25th Battalion was in position by 0400 hr (4:30 am). The rain and cloud cover concealed their movements from German observers. The attack began before dawn, with heavy shelling causing casualties even before the assault.

Although facing German troops, who had coincidentally chosen the same morning to launch their own attack, the Australians advanced and secured all objectives along the ridge. German pillboxes inflicted heavy losses, and despite their victory, the Australians suffered 6,500 casualties.

One of the casualties was Private Percy McDonough, Killed in Action on the 4th of October 1917. His body was recovered, and he is now laid to rest at Tyne Cot Cemetery (Plot XXII, Row G, Grave No. 7), Passchendaele, Belgium. Percy McDonough was 26 years and two months old at the time of his death and had served with the AIF for 18 months. His death was notified in the 423rd Australian casualty list, released on 17 August 1917.

Percy Samuel McDonough's name is inscribed on The Australian War Memorial Roll of Honour (panel 105), The City of Townsville WW1 Honour Roll and The Townsville Cenotaph.

5874 Private Percy Samuel McDonough was awarded the British War Medal (no:52232) and the Victory Medal (no:51550). A memorial scroll and a memorial plaque were passed to his mother in 1921 and 1922, respectively.

Percy McDonough's brother, 5380 Sapper James William, 1st Tunnelling Company, Died of Wounds on 29th September 1918 in France.

The following notice was published in the *Townsville Daily Bulletin* on 13th November 1917:

"Mrs J. Sherrington, of Plume Street, South Townsville has been advised that her brother Percy McDonough is reported missing. He enlisted early last year and joining the Signalling Corps in Brisbane. Before enlisting he was in the employ of Mr C. Boyle, Baker of South Townsville."

The following notice was published in the *Cairns Post* on 29th September 1919:

"McDONOUGH — In loving memory of our dear brother Percy, Killed in Action 4th October 1917, the beloved son of Mr and Mrs John McDonough of Mareeba."

> "What though in lonely grief we sigh,
> For friends beloved no longer nigh,
> Submissive still we wish reply,
> Thy Will be Done."

3152 PRIVATE GEORGE REDPATH (1896 – 1917)

42nd Infantry Battalion

George Redpath was born on the 1st of February 1896 in Ravenswood, Queensland, the sixth child and fourth son of Robert and Elizabeth Redpath (née Surtees). He had six brothers and three surviving sisters, with his parents raising 11 children in 15 years.

The family lived in Ravenswood, where Robert Redpath worked as a miner until his death in 1914, when George was 18. George Redpath attended Ravenswood State School and later worked as a clerk. After her husband's death, Elizabeth Redpath moved to Townsville, remarried in 1916 to Joseph Close, and relocated to Yungaburra on the Atherton Tablelands.

On the 6th of December 1916 George Redpath applied, in Townsville, to enlist in the Australian Imperial Force (AIF). He declared that he was 19 years and nine months old, single, and a clerk with no prior military experience. His medical examination noted that the was 5'8¾" (175 cm) tall, weighed 9st 2lbs (58kg), with a medium complexion, grey eyes, and brown hair. He identified as Church of England and named his mother, now Elizabeth Jane Close, as his next of kin.

Accepted into the AIF, George Redpath was allocated service number 3152, assigned to the Infantry, and posted to the 42nd Battalion, 7th Reinforcements. He spent December 1916 to February 1917 completing basic training at the 11th Depot Battalion located at Enoggera, Brisbane.

Private George Redpath and his fellow reinforcements embarked from Sydney, bound for England, as part of Convoy No. 23 on the 7th of February 1917. They sailed aboard HMAT A18 *Wiltshire*, which boarded 1,746 troops. The entire convoy of eight ships carried a total of 10,679 troops on the voyage. The *Wiltshire* arrived in Devenport England and disembarked passengers on the 11th of April 1917.

Following three months of further training at Durrington on the Salisbury Plain, Private George Redpath crossed to France for final training. On the 11th of August 1917, he joined the 42nd Battalion, then stationed near Ypres, Belgium.

At the end of August, the 42nd Battalion moved to positions near Poperinge, eight miles west of Ypres, preparing for upcoming battles. On the 4th of October 1917, they took part in the Battle of Broodseinde Ridge. At 4:30 a.m., the Australian troops launched their attack but faced heavy shelling at the starting point, causing significant casualties even before the assault began.

The Australians then encountered advancing German troops, who had chosen the same morning for their own attack. Despite this, the Australians pushed through the German assault and reached their ridge objectives. Fighting in the second wave of the 11th Brigade's advance from the ruins of Zonnebeke railway station, the 42nd Battalion secured part of the red line facing Tyne Cot and dug in.

The battle was not without cost. The Australian divisions suffered 6,500 casualties. The 42nd Battalion's casualties were 331 killed or wounded. One of these casualties was Private George Redpath, Killed in Action on the 4th of October 1917. There are no recorded details of his death, and he has no known grave. He was 20 years and eight months old and had been a member of the AIF for 10 months. His death was notified in the 352nd list of Australian casualties released on the 15th of November 1917.

Private George Redpath served on the Western Front in WWI. His name is inscribed on the Australian War Memorial Roll of Honour (panel 136), the Ypres (Menin Gate) Memorial (panel 27), Belgium, the 42nd Infantry Battalion AIF Roll of Honour, the Atherton War Memorial, the Malanda Eacham Memorial Gates, the Townsville Honour Roll, and the Townsville Cenotaph. Redpath Street, North Ward, Townsville, is named in his memory.

3152 Private George Redpath was awarded the British War Medal (no: 56352) and the Victory Medal (no: 551130). These medals were passed to his mother, as well as a memorial scroll and a memorial plaque, which were also passed to her in 1921 and 1922, respectively.

The following tribute was published in the *Townsville Daily Bulletin* on 4th October 1918:

REDPATH. —In loving memory of Private George Redpath, who was killed in action, October 4th, 1917. Aged 20 years and 8 months. (Fourth and dearly beloved son of Mrs. J. Close.)

> Oh, little I thought when I bid him good-bye
> It would be the last parting between George and I.
> I loved him in life, he is dear to me still.
> But although a hard blow, it was God's holy will.

457 PRIVATE CHARLES FREDERICK BRENTNALL (1876 – 1917)

23rd Australian Machine Gun Company

Charles Frederick Brentnall was born the first son of four children to William and Emilia Brentnall (née Carthew) at Hokitika, New Zealand, in October 1876. The Brentnall family moved to Australia prior to 1913 and lived in Narrabeen in New South Wales. From there, Charles Brentnall moved to Townsville where he worked as a labourer, a cane cutter, and finally as a forwarding clerk (wharf foreman) with Samuel Allen and Sons. He lived at the Victoria Bridge Hotel in South Townsville.

On the 6th of July 1916, Charles Brentnall applied to enlist in the Australian Imperial Force (AIF). He declared that he had no military experience, was single, 39 years and nine months old, and employed as a forwarding clerk. His enlistment medical examination noted that he was 5'8" (173 cm) tall, weighed 10st 8lbs (67 kg), had a fair complexion, blue eyes, and brown hair. He identified as Church of England and nominated his widowed mother, Miss Emily Brentnall, as his next of kin giving her address as Wetherill St, Narrabeen New South Wales.

Accepted into the AIF, Charles Brentnall was allocated service number 457 and assigned to the Australian Machine Gun Corps. He was initially posted to the 7th Machine Gun Company, 7th Reinforcements, and later served with the 23rd Machine Gun Company. The 21st to 25th Machine Gun Companies were established as the fourth machine gun company for each of the soon-to-be-raised Divisional Machine Gun Battalions. These units were equipped with the Vickers medium machine gun, which required a three-person crew and tripod mounting. Due to its bulkiness, the Vickers was typically sited in predetermined fixed positions.

After three months of initial training, Private Charles Brentnall and the rest of the Reinforcements embarked from Melbourne, bound for England, as part of Convoy No. 26 on the 25th of October 1916. They sailed aboard HMAT

38 *Ulysses* which boarded 2,040 troops. The entire convoy of 13 ships carried a total of 12,381 troops and 2,416 horses on the voyage. The ship arrived in Plymouth at the end of December 1916.

From Plymouth, the machine gunners were transported to Perham Downs, located north of Salisbury, for further training. In February 1917, Private Brentnall was transferred to the 3rd Australian Division Machine Gun Company and in May 1917, he was temporarily attached to the 9th Training Battalion. Following the completion of training and his temporary detachment, Private Brentnall crossed to France with the 23rd Machine Gun Company from Southampton in September 1917.

Once disembarked in France, the unit moved to Le Havre and spent a week in the Rest Camp before heading to St. Omer. By September 27, 1917, they were stationed along the Belgian border at Godewaersvelde in northern France. Here, they consolidated and prepared for the upcoming Battle of Broodseinde. At 4:30 a.m. on October 4, 1917, the Australian troops initiated their attack but were heavily shelled at their starting point, resulting in significant casualties even before the assault began. They encountered a line of advancing German troops, as the Germans had chosen the same morning to launch their own attack. Despite this, the Australians persevered through the German assault and successfully achieved their ridge objectives.

It was during this battle that Private Charles Brentnall was wounded in the abdomen and buttock by gunshot at Zonnebeke Ridge. He was evacuated by the 11th Field Ambulance to the 3rd Australian Casualty Clearing Station on the 4th of October and Died of Wounds on the 5th of October 1917. He was 41 years old at the time of his death and had served in the AIF for one year and eleven months. His death was notified in the 350th list of Australian Casualties released on the 6th of November 1917.

Charles Brentnall's younger brother, 257 Private George Vivian Brentnall, 54th Battalion, also served and was Killed in Action in France on the 21st of May 1918.

457 Private Charles Frederick Brentnall is buried at Nine Elms British Cemetery Poperinghe, Belgium. His name is inscribed on The Australian War

Memorial Roll of Honour (panel 177), The City of Townsville WW1 Honour Roll, and The Townsville Cenotaph. He was awarded the British War Medal (no:51566) and the Victory Medal (no:50886). These medals, as well as a memorial scroll and a memorial plaque, were passed to his mother. Brentnall Street, Mysterton, Townsville is named in his and his brother's memory.

Emilia Brentnall was granted a war pension of two pounds (£2/-/-) per fortnight for the loss of her son Charles and two pounds, thirteen shillings and threepence (£2/12/3) per fortnight for the loss of George. The combined sum of four pounds, 12 shillings and threepence (£4/12/3) were approximately half of the wages paid to two soldiers in the AIF in 1917.

The following tribute was published in The *Sydney Morning Herald* on the 5th of October 1918:

BRENTNALL. - Died of Wounds in France, October 5, 1917, Charles Frederick, eldest son of late W. S. Brentnall (New Zealand), and of Mrs. E. A, Brentnall (Narrabeen), and brother of late Addie, Shard, and Viv.

2925 PRIVATE WALTER NICOL LINDSAY (1885 – 1917)

41st Infantry Battalion

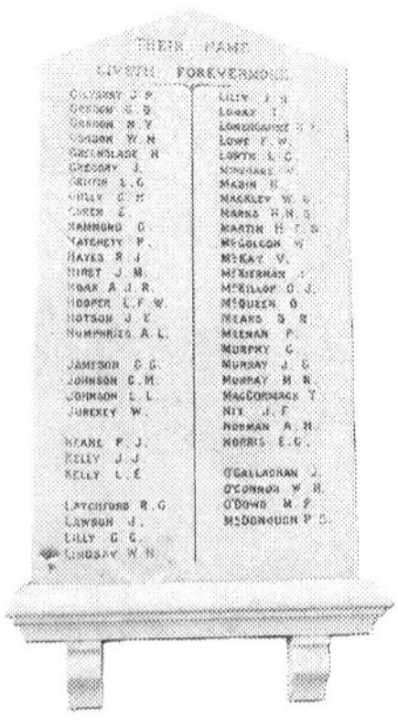

Walter Nicol Lindsay was born on the 9th of November 1885 in Kirriemuir, Angus, Scotland to Walter and Eliza Lindsay (née Winter). He was the third child, and he had two brothers and one sister. Walter attended the Public School in Kirriemuire. When he was 26 years old, Walter migrated to Australia from Scotland in 1911. By 1913, Walter had moved to Townsville and lived in a property on Stanley St. He was employed as a draper.

On the 21st of October 1916, Walter Lindsay applied in Townsville to enlist in the Australian Imperial Force (AIF). He was 30 years and 11 months old at the time and declared that he was single, employed as a draper and had no previous military experience. His medical assessment noted that he was a small man at 5' 2½" (159 cm) tall, weighed 8st 5lbs (53kg), had a fair complexion, grey eyes, and fair hair. Like most men of Scottish descent, he identified as Presbyterian and he nominated his father as his next of kin giving his address as 55 Roods St, Kirriemuir, Scotland.

Accepted into the AIF, Walter Lindsay was assigned service number 2925, allocated to the Infantry and posted to the 41st Battalion, 6th Reinforcements. The 41st Battalion was raised in Brisbane in February 1916 and formed part of the 11th Brigade of the 3rd Australian Division. The 41st was fighting on the Western Front when Walter Lindsay enlisted.

After completing basic training with the 11th Depot Battalion in Australia, Private Walter Lindsay and the rest of the Draft of Reinforcements embarked from Sydney, on board HMAT A64 *Demosthenes* on the 23rd of December 1916, bound for England. The *Demosthenes* was part of Convoy No. 27, and she boarded 1,240 troops. The entire convoy of 13 ships carried 15,278 troops and 311 other passengers on the voyage.

Nine weeks later in March 1917, the Reinforcements disembarked at Plymouth in England and were transported to Lark Hill on the Salisbury Plain for another three months of training. Private Walter Lindsay and the other reinforcements crossed to France in June 1917 and, after completing further training in France, marched into the 41st Battalion in Belgium on the 13th of July 1917.

The 41st Battalion had been in action in France and Belgium for over a year when Private Lindsay joined them at Messines. At the end of June 1917, the 11th Brigade, including the 41st Battalion, was ordered to establish a new front-line west of Warneton, in full view of the Germans. Work carried on night and day under heavy shellfire, and the period became known to the battalion as *"the 18 days"*.

The start of August found the 41st holding ground captured by two of its sister battalions in a feint attack on 31st July 1917. Enduring continual rain, flooded trenches, and heavy shelling, many of the battalion's platoons dwindled from 35 men to less than 10.

Private Lindsay was admitted to the hospital in early August 1917 suffering from trench foot and then again with influenza as well as scabies. He re-joined the 41st at the end of September. On the 4th of October 1917, the 41st Battalion was committed to the attack on Broodseinde ridge and captured its objectives with little difficulty.

It was on the day after the attack, on the 5th of October 1917, that Private Walter Lindsay was declared as Wounded in Action. It was not until January 1918 that Walter was declared Killed in Action. An Australian Red Cross Investigation found that he had been killed instantly by shellfire. He was 31 years and 10 months old at the time of his death. He had been a member of the AIF for 11 months. He has no known grave.

Walter Lindsay's name is inscribed on The Australian War Memorial Roll of Honour (panel 134), The Menin Gate Memorial Belgium (panel 27), The 41st Battalion Roll of Honour, The Townsville No.37 Magnetic Lodge P.A.F.S.O.A. Honor Board, The City of Townsville WW1 Honour Roll, and The Townsville Cenotaph. Lindsay Street, Rosslea, Townsville is named in his memory.

2925 Private Walter Nicol Lindsay was awarded the British War Medal (no:54908) and the Victory Medal (no:54068). A memorial scroll and a memorial plaque were passed to his father in 1922.

309 PRIVATE ERNEST JOHN WILLIAMS (1891 – 1917)

41st Infantry Battalion

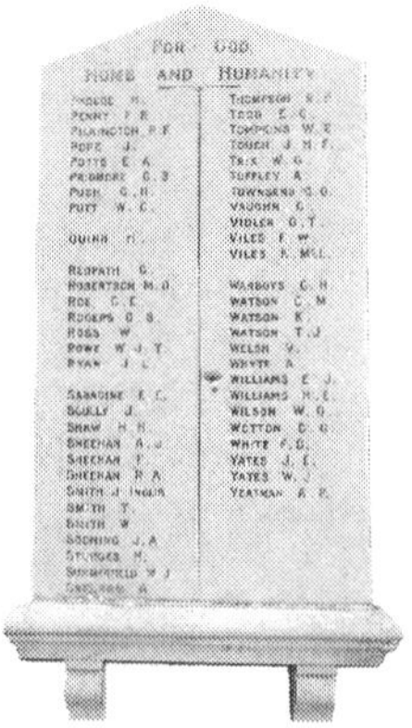

Ernest John Williams was born on the 16th of October 1891 in Townsville, Queensland. He was the first of five children born to Abraham (Abe) Williams and his wife Ada (née Hawkins). He had two brothers and two sisters. Abe Williams worked as a labourer, and the family lived in Sixth Street, South Townsville.

Ernest Williams grew up in Townsville and attended the Ross Island State School. After his school years were over, Ernest worked as a labourer, and by 1913 he lived on Magnetic Island. He was also a member of the Masons and belonged to the Loyal North Kennedy Lodge No. 79.

On the 29th of November 1915, Ernest Williams applied at Townsville to join the AIF. He declared that he was 24 years old, single, and a labourer without any previous military experience. His medical assessment noted that he was 5'4¾" (164 cm) tall, weighed 8st 13lbs (57 kg), had a dark complexion, brown eyes, and brown hair. He identified as Church of England and nominated his mother as his next of kin, giving her address as Sixth Street in Townsville.

Accepted into the AIF, Ernest Williams was allocated service number 309, assigned to the Infantry and posted as an original member to 'A' Company of the 41st Battalion. The 41st Battalion was raised in February 1916 with recruits from Brisbane, northern Queensland, and the northern rivers district of NSW. Ernest William's younger brother, George Williams, enlisted in March 1916 and joined his older brother in 'A' Company of the 41st Battalion.

Recruit training started soon after enlistment in Brisbane and continued through to May 1916. The 41st Battalion embarked from Sydney on board HMAT A64 *Demosthenes*, as part of Convoy 21, on the 18th of May 1916, bound for England. The *Demosthenes* boarded 1,240 troops. The entire convoy of 18 ships carried 25,287 troops on the voyage, which lasted nine weeks, docking at Plymouth, England, on the 20th of July 1916.

After completing another four months of training at Lark Hill in Wiltshire, the 41st Battalion arrived in France on the 25th of November 1916. Private Ernest Williams contracted influenza in late December 1916 and was ill enough to be hospitalized, firstly at the 10th Australian Field Ambulance and then the 10th Australian Casualty Clearing Station. He re-joined the 41st Battalion in January 1917 and spent the bleak winter of 1916-17 alternating between service in the front line and training and labouring in the rear areas.

On the 7th of June 1917, the 3rd Division was committed to the Battle of Messines. The 41st Battalion was held in Divisional Reserve for this battle but was still under enemy fire. It was during this operation that Private Ernest Williams was wounded in the arm by shrapnel. He was evacuated to the 9th Australian Field Ambulance and then on to the 56th General Hospital located at Étaples in France. He was discharged five weeks later, after treatment and rest, and re-joined the 41st Battalion on the 13th of July 1917.

On the 25th of June 1917, Ernest William's brother, George, was wounded by gunshot in the right shoulder and arm. He was evacuated back to England for treatment and eventually returned to Australia as unfit for further military service. The start of August found the 41st holding the ground captured by two of its sister battalions in a feint attack on the 31st of July 1917. Enduring continual rain, flooded trenches, and heavy shelling, many of the battalion's platoons dwindled from 35 men to less than 10.

On the 4th of October 1917, Private Williams took part in the 41st's attack on Broodseinde Ridge. The Battalion captured its objectives with little difficulty. Six days after the attack, the 41st Battalion was deployed in the vicinity of Friezenberg Ridge and was resting and remaining in old trenches to escape observation. During daylight, the Battalion relieved the support line and approached in a single file along the Ypres-Zonnebeke Road with 50 yards (46 metres) between platoons. The weather was snowy with bad visibility. The casualties were recorded as *"slight"* however, one of those casualties was Private Ernest Williams, who was Killed in Action.

Private Ernest Williams was *"killed outright by shell fire"* on the 10th of October 1917 and laid to rest in the field where he fell, *"about 1500 yards northwest of Zonnebeke."* He was one month short of his 26th birthday at the time of his death and had been a member of the AIF for 22 months. Private Ernest Williams has no known grave. His death was notified in the combined 318th and 319th list of Australian casualties released on the 6th of July 1917.

Ernest John Williams' name is inscribed on the Australian War Memorial Roll of Honour (Panel 78), the Ypres (Menin Gate) Memorial (Panel 25), Belgium, the 41st Battalion Roll of Honour, the City of Townsville WW1 Honour Roll, and the Townsville Cenotaph.

309 Private Ernest John Williams was awarded the British War Medal (no. 39962) and the Victory Medal (no. 39576). A memorial scroll and a memorial plaque were passed to his father in 1922.

The following notice was one of several published in the *Townsville Daily Bulletin* (Qld.) on the 10th of October 1918:

WILLIAMS. — In sad and loving Memory of my dear son, No. 309, Private Ernest John Williams, 41st Battalion, who was Killed in Action on the 10th October 1917. Aged 26 years.

Somewhere in France.
You sleep not in your native land,
But under foreign skies,
Far from those who loved you best,
But in a hero's grave you lie.
I cannot clasp your hand, dear son,
Your face I cannot see,
But let this little token tell
I'll always remember thee.

(Inserted by his sorrowing Father, Abraham Williams.)

3206 PRIVATE GEORGE CHARLES LILLY (1893 – 1917)

49th Infantry Battalion

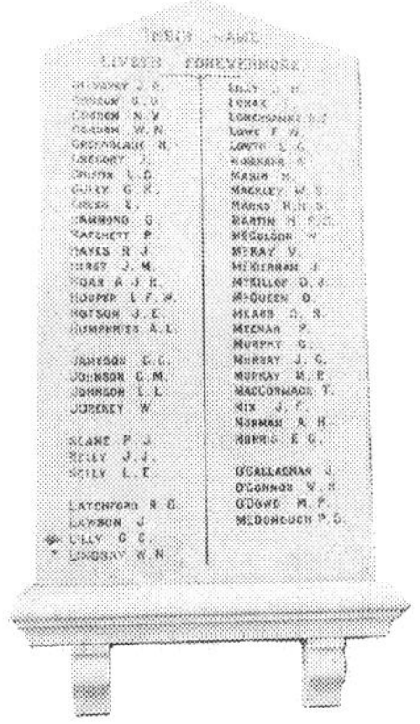

George Charles Lilly was born on the 2nd of May 1893 in Townsville, Queensland, to John and Louisa Lilly (née Donnithorne). He was the second son of a large family. George's parents emigrated from Cornwall in England and arrived in Queensland in 1889. John Lilly worked as a labourer, and the family lived in premises on Macrossan St, South Townsville.

George Lilly was employed as a labourer when, on the 30th of October 1916, he applied at Charters Towers to enlist in the Australian Imperial Force (AIF). His younger brother John had already enlisted in January 1916 and was serving with the 41st Battalion. George Lilly gave his postal address as c/- Post Office Winton. He was 23 years and five months old at the time, but he stated that he was 22 years and four months old on his attestation paper. He also declared that he was single and had no previous military experience.

George Lilly's medical assessment noted that he was 5'6" (168 cm) tall, weighed 138 lbs (62.5 kg), had a fair complexion, blue eyes, and dark brown hair. He identified as Church of England and nominated his father as his next of kin, giving his address as Cornwall, England. This was later amended to read Macrossan St, South Townsville. Later in his service, George nominated his mother as the only beneficiary in his will and gave her address as Main St, Townsville. He also allotted three shillings per day (half of his pay) to his mother.

Accepted into the AIF, George was allocated service number 3206, assigned to the Infantry and posted to the 49th Battalion, 8th Reinforcements. Two months of basic training in Australia followed enlistment. Once completed, Private George Lilly and his fellow reinforcements embarked from Sydney on the 23rd of December 1916, bound for England as part of Convoy No. 27. They sailed aboard HMAT A64 *Demosthenes*, which boarded 1,240 troops. The entire convoy of 13 ships carried 15,278 troops and 311 other passengers on the voyage.

Ten weeks later, the *Demosthenes* disembarked its passengers at Plymouth, England. The reinforcements were then transported to Codford in Wiltshire, where they completed an additional three months of training. In June 1917, Private George Lilly crossed to France and joined the 4th Division's Base Depot in Harfleur, near Le Havre. Three weeks later, on the 4th of July 1917, he joined the 49th Battalion. The battalion was deployed in the Ypres sector in Belgium, having participated in the Battle of Messines in June. Throughout July and August, the 49th alternated between the front lines and reserve positions in that sector. By early September 1917, the 49th Battalion relocated to a camp behind Ypres in preparation for upcoming operations.

The next action that the 49th was engaged in was the Battle of Polygon Wood on the 26th of September 1917. The fighting was bloody. German concrete pillboxes often blocked the Australians' progress, and many men fell under shell and machine-gun fire. However, with heavy artillery support, the objectives were taken, and enemy counterattacks held off. Private George Lilly's younger brother, Private John Henry Lilly, fighting with the 41st Battalion, Died of Wounds received in battle on the 1st of October 1917.

On the 12th of October 1917, the 49th Battalion was committed to an attack against the village of Passchendaele. In the face of heavy fire, the men fought in the mire while struggling to keep up with their artillery barrages. Ground was taken, but it could not be held. In wretched conditions, with casualties mounting at an appalling rate, the Australians had to fall back. It was here on that date that Private George Lilly was Killed in Action. Initially, he was reported as wounded; however, a later inquiry determined that he was killed. He was 24 years and five months old and had been a member of the AIF for one year. He has no known grave.

George Lilly's name is inscribed on The Australian War Memorial Roll of Honour (panel 148), The Ypres (Menin Gate) Memorial (panel 29), Belgium, The City of Townsville WW1 Honour Roll, and The Townsville Cenotaph. Lily Street, Hermit Park, Townsville, is named in his and his brothers' memory (n.b. the street name has been spelled Lily, Liley, Lilley, and occasionally correctly, Lily is now the accepted spelling).

3206 Private George Charles Lilly was awarded the British War Medal (no: 54859) and the Victory Medal (no: 53977). A memorial scroll and a memorial plaque were passed to his father in 1923.

The following tribute was published in the *Townsville Daily Bulletin* on the 13th of October 1919:

LILLY.—In loving memory of our dear sons and brothers, Private John Henry Lilly, Died of Wounds on October 1st, 1917, and Private George Charles Lilly, Killed in Action on October 12th, 1917.

(Inserted by their loving mother and father, sisters and brothers.)"

3525 PRIVATE FRANCIS WILLIAM LOWE (1894 - 1917)

47th Infantry Battalion

Francis (Frank) William Lowe was born on the 10th of March 1894 in Townsville, Queensland, to John and Mary Lowe (née Snell). He was the oldest child of a large family. John Lowe was a cab proprietor, and the family lived in Hodel St and later in Francis St, Townsville. Frank grew up in Townsville and attended Townsville West State School. After his school years were completed, he became a draper.

In August 1914, Frank Lowe, who was a member of the Citizen Forces at the time, was mobilised for service and embarked for Cairns and Thursday Island for Garrison Duty on the 8th of August 1914. Six days later, Private Frank Lowe enlisted for service outside Australia. He re-embarked at Thursday Island on HMAT *Kanowna* on the 16th of August 1914 to take part in the capture of German New Guinea as part of the Australian Naval and Military Expeditionary Force. The AN&MEF was tasked with the Australian takeover and occupation of the Pacific colony of German New Guinea in September–November 1914.

As events turned out, Private Lowe did not participate in the capture of German New Guinea on account of "*trouble caused by the firemen on troopship.*" Private Lowe returned to Townsville on the 18th of September 1914 and was discharged from the AN&MEF on the same day.

Frank Lowe returned to his civilian employment for a little over two years until he applied at Townsville to enlist in the Australian Imperial Force (AIF) on the 9th of October 1916. He declared that he was single and 22 years and seven months old. Frank's medical examination noted that he was 5'10" (179 cm) tall, weighed 9st 1lb (57.6 kg), had a fair complexion, blue eyes, and brown hair. He identified as Church of England and nominated his father as his next of kin, giving his address as Francis St, West End, Townsville.

Accepted into the AIF, Frank Lowe was allocated service number 3525, assigned to the Infantry and posted to the 47th Battalion, 9th Reinforcements. The 47th Battalion was raised in Egypt in 1916 as part of the expansion of the AIF that took place following the Gallipoli campaign and was fighting on the Western Front when Frank Lowe enlisted. Three months of training at the 11th Depot Battalion at Enoggera in Brisbane followed enlistment.

On the 10th of January 1917, Frank Lowe married Annie Georgena Donley in Queensland. It was to be a short honeymoon as 14 days after his wedding day, Private Frank Lowe and the rest of the Reinforcements embarked from Sydney on the 24th of January 1917, bound for England. They sailed as part of Convoy No. 29 aboard HMAT A33 *Ayrshire*, which boarded 948 troops. The entire convoy of eight ships carried a total of 10,679 troops on the voyage. Private Lowe was promoted to Corporal for the voyage.

The reinforcements disembarked in Devonport, England, in April 1917 and were transported to the 12th Training Battalion at Codford, near the Salisbury Plains. Codford has a long history with ANZAC soldiers, and during World War I, large training and transfer camps were established there for the tens of thousands of troops waiting to move to France. Upon arrival at Kandahar Barracks, Frank Lowe reverted to the rank of Private. Training continued at Codford before the contingent crossed to France, finally joining the 47th Battalion in Belgium on the 16th of September 1917. At that time, the 47th was located near Ypres, resting and preparing for future operations.

Ten days after Private Frank Lowe joined the 47th, the Battle of Polygon Wood began. The battalion initially served as a reserve until the 12th of October 1917, when they advanced to fight for Passchendaele ridge. Monash's Division started with 5,800 soldiers, but by day's end, only 2,600 remained due to heavy casualties. The battle gained little ground, which was soon lost. Torrential rain turned the battlefield into mud, contributing to the failure. This battle has since symbolized the futility of such sacrifices.

After the Passchendaele ridge devastation, Private Frank Lowe was declared Wounded in Action on the 12th of October 1917. It was not until the 11th of March 1918 that a Court of Enquiry found that he had been Killed in Action. An Australian Red Cross Investigation found that Frank was killed when he *"left the protection of a shell hole, which he was sharing with his mate Charlie*

Klezel, to go to the aid of a man wounded by a sniper. As he was leaning over the man, a sniper shot Frank, killing him instantly." Private Frank Lowe's body was later recovered and "buried 1000 yards SW of Passchendaele and 1000 yards NE of Zonnebeke". His body was not reinterred and he is recorded as having no known grave. Private Frank Lowe was 23 years and seven months old and had been a member of the AIF for one year at the time of his death..

Francis William Lowe's name is inscribed on The Australian War Memorial Roll of Honour (Panel 144), The Ypres (Menin Gate) Memorial (Panel 27), Belgium, The State School Townsville West Honour Board, The City of Townsville WW1 Honour Roll, and The Townsville Cenotaph. Lowe Street, Stuart, Townsville, is named in his memory.

3525 Private Francis William Lowe was awarded the British War Medal (no:55309) and the Victory Medal (no:54469). A memorial scroll and a memorial plaque were passed to his wife in 1922.

The following tribute was published in the *Townsville Daily Bulletin* on the 12th of October 1918:

LOWE. —In fond and loving Remembrance of our dear son and brother, Francis William (Frank), who was reported Killed in Action at Passchendaele Ridge, the 12th of October 1917.

We pictured you dear Frank, returning.
We longed for a clasp of your hand;
But God has postponed our meeting,
We shall meet in a better land.
You are always in our thoughts, dear Frank,
It is sweet to speak your name;
In life we loved you dearly,
In death we do the same.
While at the post of duty.
What more could he do?
Than die for King and country,
A soldier brave and true.

(Inserted by his ever loving Father and Mother, Brothers and Sisters, J. H. and M. I. Lowe and Family.)

4098 PRIVATE HERBERT MABIN (1897-1917)

31st Infantry Battalion

Herbert Mabin was born on the 23rd of January 1897 in Charters Towers, Queensland, to William and Helen Mabin (née Hunter). He was the eighth of 10 children and had five brothers and four sisters. The family lived in Charters Towers where William Mabin worked as a tailor.

In 1904, William died at the age of 45 years when Herbert was only seven years old. After William's death, Helen Mabin stayed in Charters Towers until about 1909 when she moved with her family to Townsville and lived on the Strand.

Herbert Mabin attended state school, and after his school years were finished, he took up a five-year watchmaker apprenticeship with Horn & Peterson Jewellers in Flinders Street. Soon after completing his apprenticeship, Herbert Mabin applied in Townsville to enlist in the Australian Imperial Force (AIF) on the 13th of March 1916. He declared that he was single, 19 years and two months old, and that he had four years of previous service in the Senior/Citizen Cadets. His medical assessment noted that he was 5'10" (178 cm) tall, weighed 9st 5lbs (59 kg), had a fresh complexion, brown eyes, and dark brown hair. He identified as Presbyterian and nominated his sister, Isabella, as his next of kin, giving her address as the corner of Mitchell and Gregory Streets, Townsville.

Accepted into the AIF, Herbert Mabin was allocated service number 4098, assigned to the Infantry, and posted to the 31st Battalion, 10th Reinforcements. The 31st Battalion was raised as part of the 8th Brigade at Enoggera, Brisbane, in August 1915. When Herbert Mabin enlisted, the 31st was in Egypt preparing to proceed to France. Six months of training in Australia followed enlistment for Herbert Mabin and the contingent.

After completing basic training, Private Herbert Mabin and his fellow reinforcements joined Convoy No. 25 on the 21st of October 1916 and embarked from Brisbane, bound for England. They sailed aboard HMAT A36 *Boonah*, which boarded 1,235 troops. The entire convoy of 12 ships carried a total of 12,397 troops and 5,200 horses on the voyage.

The Boonah disembarked the contingent at Plymouth, England, on the 10th of January 1917. The voyage had lasted about 12 weeks, and the troops disembarked during the freezing winter. Further training occurred at the 8th Training Battalion located at Hurdcott Camp, near Fovant, Wiltshire, England. In April 1917, Private Herbert Mabin crossed to Étaples in France and was taken on strength of the 31st Battalion, located in the vicinity of Bapaume, on the 12th of April 1917. In early 1917, the German Army withdrew to the Hindenburg Line, allowing the British front to advance, and the 31st Battalion participated in the follow-up operations.

Private Herbert Mabin fought in the major battle of Polygon Wood between the 26th of September and the 3rd of October 1917. The Australians advanced behind creeping artillery barrages. The aim was to cover 1,000–1,500 yards (900m to 1400 meters) and secure reverse slopes for observation. Swift preparations countered German counterattacks by consolidating captured ground with deep defences. The attack inflicted significant losses on the German 4th Army, endangering their hold on Broodseinde Ridge. Favourable weather aided the Allied advance, and subsequent German counterattacks failed.

It was during these engagements that Private Herbert Mabin was Wounded in Action on the 9th of October 1917. He was evacuated to the 3rd Canadian Casualty Clearing Station, suffering from gunshot wounds to his back, which penetrated his abdomen. He survived his wounds for three days after admission but Died of Wounds on the 12th of October 1917.

Private Herbert Mabin was 20 years and eight months old at the time of his death and had served with the AIF for 19 months. He was buried at the Lijssenthoek Military Cemetery (Plot XXI, Row A, Grave No. 10A), Belgium. His death was notified in the 352nd list of Australian Casualties released on the 14th of November 1917.

Herbert Mabin's name is inscribed on The Australian War Memorial Roll of Honour (panel 119), The City of Townsville WW1 Honour Roll, and The

Townsville Cenotaph. Mabin Street, Mundingburra, Townsville, is named in his memory.

4098 Pte Herbert Mabin was awarded the British War Medal (no: 50269) and the Victory Medal (no. 49629). A memorial scroll and a memorial plaque were passed to his sister in 1922. Isabella Mabin was granted a war pension of 15 shillings per fortnight for the loss of her brother. This was the equivalent of three days' pay for an AIF soldier in 1917.

Herbert's older brother, George Morrison Mabin, enlisted on the 1st of February 1917. He served in the 4th Pioneer Battalion and returned to Australia on the 3rd of May 1919.

The following tributes were published in the *Townsville Daily Bulletin* on 12th October 1918:

"MABIN. — In loving Memory of my dear brother, Private Herbert Mabin, who Died of Wounds in France, on the 12th of October 1917.

> Sleep on, dear brother, in a soldier's grave,
> Your life for your country you nobly gave.
> No one stood near to say good-bye.
> But safe in God's keeping now you lie.

(Inserted by his loving Brother, George, O.A.S. France)

MABIN. —In loving Memory of our dear brother, Private Herbert Mabin, who Died of Wounds in France on the 12th of October 1917.

> His heart was good, his spirit brave,
> His resting place a hero's grave:
> And so, tho' sore at heart, we say
> O Lord, Thy will be done.

(Inserted by his loving Sisters and Brothers)

21493 GUNNER WILLIAM McCOLGAN (1890 – 1917)

9th Field Artillery Brigade

William McColgan was born on the 15th of October 1890 in Townsville, Queensland, to William McColgan and Ellen Herbert. He was the third-born child, and he had four brothers and one sister. William snr. was a storekeeper, and the family lived in Railway Estate in Townsville. William jnr. grew up in Townsville and attended Catholic school. After his school years were completed, he qualified as a wood machinist after serving a five-year apprenticeship.

On the 6th of September 1915, William McColgan applied in Townsville to enlist in the Australian Imperial Force (AIF). He declared that he was 24 years and 10 months old, single, employed as a wood machinist, and had four years of militia service with the Australian Garrison Artillery. His medical examination noted that he was 5'8¾" (175 cm) tall, weighed 10st 7lbs (67 kg), had a ruddy complexion, grey eyes, and brown hair. He identified as Roman Catholic and nominated his father as his next of kin, giving his address as Railway Estate, but this was later changed to the Royal Oak Hotel, Townsville.

Accepted into the AIF, William McColgan was allocated service number 21493 and assigned to the Artillery. In April 1916, he was posted as a Driver to the 9th Field Artillery Brigade (9th FAB), which was raised in Egypt in early 1916 to support the newly formed 3rd Division. A Driver (Dvr) was a military rank in the AIF equivalent to the rank of Private. This rank was used in the Artillery for the men who drove the teams of horses that pulled the guns and ammunition resupply wagons. It was a hazardous job, and Drivers and their teams were high-value targets for enemy artillery.

After enlistment, William McColgan was moved to Sydney for basic training. Eight months after enlistment, Gunner William McColgan and the 9th Field Artillery Brigade embarked from Sydney on the 11th of May 1916, as part of Convoy No. 21. They sailed aboard HMAT A8 *Argyllshire*, which boarded

1,763 troops for the 10-week journey to Plymouth, England. The entire convoy of 18 ships carried a total of 25,333 troops on the voyage.

Once disembarked, more training followed at Larkhill, Salisbury Plains, England. Gunner William McColgan was hospitalized in October 1916, suffering from an infection, and consequently, he did not leave for France until late December.

Further hospitalization followed in France due to the same infection, and Gunner McColgan was transferred to the 7th Field Artillery Brigade. He was taken on the strength of the 107th Howitzer Battery in February 1917. The 107th spent the next four months in the Armentières sector and, on the 7th of June 1917, fired in support of the 3rd Division in the Battle of Messines.

Soon after this action, Gunner McColgan was gassed on the 25th of June 1917. He was evacuated to the 13th Australian Field Ambulance and eventually moved back to No. 3 Rest Camp to recuperate. He returned to his battery at the end of July 1917.

At this time, the 107th was located close to Messines, and it remained there until the end of the month. The next move was to a camp in the vicinity of Ypres to prepare for the planned operations to come.

On the 17th of September 1917, Gunner William McColgan, at his own request, re-mustered to Gunner from Driver. The 107th Battery fired in support of the Australian attacks at Ypres in September–October 1917.

It was during this fight that Gunner William McColgan was Killed in Action on his 27th birthday, on the 15th of October 1917. He had been a member of the AIF for two years and one month. His remains were buried near where he fell, and the burial site map reference was recorded as Sheet 28NE.D.25.0. His body was not reinterred, and today he has no known grave. William McColgan's death was notified in the combined 367th and 368th list of Australian casualties, issued on the 10th of December 1917.

William McColgan's name is inscribed on The Australian War Memorial Roll of Honour (Panel 15), The Ypres (Menin Gate) Memorial (Panel 7), Belgium, The City of Townsville WW1 Honour Roll, and The Townsville Cenotaph.

McColgan Street, Hyde Park, Townsville, was named in William's memory. The street was closed in 1969.

21493 Gunner William McColgan was awarded the British War Medal (No: 40284) and the Victory Medal (No: 39881). A memorial scroll and a memorial plaque were passed to his father in 1922.

William's younger brother, 12463 Gunner Joseph Sigmas McColgan, enlisted in the AIF, and he returned to Australia on the 14th of January 1919, having also been gassed.

The following Family Notice was published in *Townsville Daily Bulletin* on the 14th of October 1944.

McCOLGAN. — In loving memory of our beloved brother, William McColgan, Killed in Action in France, October 15, 1917.

(Inserted by his brother and sister, Joe and Julia.)

585A/5852 CORPORAL FREDERIC MICHAEL CARNEY (1885 - 1917)

11th Light Horse Regiment

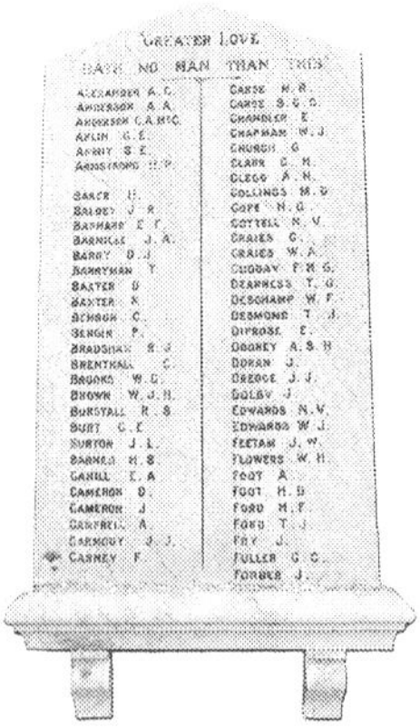

Frederic* (Teddy) M. Carney was born on the 18th of December 1885 in Bellingen, NSW. He was the youngest child of seven children born to Michael and Ann Carney (née Bushell). Michael Carney was a farmer, and the family lived at "Fermount," Bellinger Rd, Bellingen, NSW. Teddy Carney grew up in Bellingen and attended public school. After his school years were finished, he took up farming. Later in life, he attended the Salvation Army Training Garrison in Melbourne, Victoria, and graduated as a Salvation Army Officer. He was appointed as a Captain in the Salvation Army and transferred to Townsville, where he took up residence in the Peoples Palace, Sturt St, Townsville.

On the 25th of January 1915, Teddy Carney applied at Townsville to enlist in the Australian Imperial Force (AIF). He declared that he had no previous military experience, was 29 years old, single, and employed as a Salvation Army Officer. His medical examination noted that he was 5'9" (175 cm) tall, weighed 139 lbs (63 kg), had a fair complexion, blue eyes, and brown hair. He identified as Salvation Army and nominated his mother as his next of kin, giving her address as "Fermount," Bellinger Rd, Bellinger River, NSW.

Accepted into the AIF, Teddy Carney was assigned service number 585, allocated to the Australian Army Medical Corps, and posted to the Headquarters of the 11th Light Horse Regiment (11th LHR). The formation of the 4th LH Brigade, and the 11th LHR as part of it, was announced on the 11th of February 1915. Two squadrons of the 11th LHR were subsequently formed in Queensland, and a third in South Australia.

In April 1915, Private Teddy Carney was promoted to Corporal. On the 2nd of June 1915, Corporal Carney and his unit, after assembling, equipping, and training, departed from Brisbane bound for Egypt. They sailed aboard HMAT *Medic* as part of Convoy No. 6. Their ship boarded 519 troops, while the entire convoy of 15 ships transported 13,949 troops and 2,594 horses on the voyage.

Following a short stay in Egypt, Corporal Teddy Carney embarked from Alexandria bound for Gallipoli on the 25th of August 1915. Four days later, he was transferred to the 9th LHR and then to the 2nd LHR at Gallipoli. Corporal Teddy Carney remained on Gallipoli until the 26th of December 1915 when he returned to Egypt as part of the general evacuation. He came back from Gallipoli in poor health, suffering from mumps. He was admitted to a hospital in Heliopolis. Over the next two months, he was admitted to the hospital suffering from gastritis, inflamed stomach, indigestion, mammary abscess, and dysentery.

Corporal Carney continued to serve in Egypt and was promoted to Lance Sergeant in June 1916. Once recovered from his illnesses, Lance Sergeant Carney was transferred to the 11th LHR in October 1916. He served with the 11th until September 1917 when he was posted to the 4th Light Field Ambulance (LHFA). This unit was formed in February 1917 from some 60 men drawn out of all the other Light Horse formations in Egypt. In October 1917, Lance Sergeant Teddy Carney reverted to Corporal. A few weeks later, Corporal Carney was at Beersheba for the famed charge of the 4th LH Brigade on the 31st of October 1917. The 4th LHFA, Unit History Account of Beersheba, provides the following description of the unit's casualties on the 1st of November 1917:

"Having cleared the Dressing Station of patients, the Ambulance moved on to Beersheba and occupied a campsite in the Brigade area to the east of the town. The horses were watered, and tents pitched. Just as the animals returned from water, the attention of the men of the unit was attracted by the Machine Gun Squadron opening up fire. On looking skywards, an enemy plane, flying very low, was seen approaching the Ambulance. All dived to the ground for shelter, but it was then too late. The enemy had dropped his bomb in the midst of the bearer lines. The thunderous explosion was followed by an appalling spectacle. The horse lines were strewn with the intestines of eviscerated animals, and in the midst of this scene were lying many of our own personnel. Some had their limbs shattered, whilst others had their chest and abdomen torn away. Here and there were horses limping away on three legs. The scene was simply indescribable. The casualties were four killed and eight wounded, whilst 14 animals were also killed or had to be destroyed. These were the first casualties that were suffered since the inception of the Ambulance. The deceased soldiers were all buried in a gum tree plantation which was set apart from the Military Cemetery."

One of the four killed in this attack was Corporal Frederic M. Carney, recorded as being Killed in Action at Beersheba on the 1st of November 1917. He was 32 years and eleven months old and had served with the AIF for two years and nine months. He was one of the members of the 4th LHFA buried at the Beersheba War Cemetery. His death was notified in the 371st/372nd list of Australian casualties released on the 17th of December 1917. Frederic Michael Carney's name is inscribed on the Australian War Memorial Roll of Honour (panel 182), the City of Townsville WW1 Honour Roll, and the Townsville Cenotaph. Carney Street, Cluden, Townsville, is named in his memory.

585 Corporal Frederic Michael Carney served in WW1 in Egypt, Gallipoli and Palestine. He was awarded the 1914/15 Star (no:1228), the British War Medal (no:12974) and the Victory Medal (no:12916). A memorial scroll was

passed to his mother in 1921 and a memorial plaque was passed to his brother in 1922. Ann Carney was granted a war pension of two pound and 12 shillings (£2/12/-) per fortnight for the loss of her son. This was nearly half of the wage paid to an AIF soldier in 1917.

The following tribute was published in *The Daily Telegraph* (Sydney, NSW) on the 16th of November 1917:

ROLL OF HONOR.

CARNEY. — Reported Killed in Action at Beersheba, on the 1st of November, Corporal F. M. Carney, 4th Light Horse Field Ambulance (late 11th Light Horse). After 2 1/2 years' service. Loved youngest brother of Mrs. F. Sutton, of Fernmount, Bellinger River.

> He heard his country's call,
> And gave his life— his all.

*Sometimes recorded as "Frederick" however birth records and service documents show first name is "Frederic"

6265 PRIVATE NORMAN VICTOR GORDON (1894 – 1917)

15th Infantry Battalion

Norman Victor Gordon was born on the 2nd of October 1894 in Hillgrove, NSW. He was the youngest child of Charles and Caroline Gordon (née Evens). Charles left the family soon after Norman's birth. In a letter to the Army in 1923, Caroline Gordon states that, based on a newspaper report, Charles Gordon died in Western Australia in about 1897 or 1898. Caroline remarried Herbert Condor in 1900. The family lived in a property on Stagpole Street, West End, Townsville. Norman Gordon attended State School and worked as a labourer when his school years were finished.

On the 29th of November 1915, Norman Gordon applied at Townsville to join the AIF. He declared that he had no previous military service, was single, 21 years and two months old, and a labourer. His enlistment medical examination noted that he was 5'7¼" (171 cm) tall and weighed 9st. 6lbs. (60 kg). He had a dark complexion, hazel eyes, and dark brown hair. He identified as Church of England and nominated his mother as his next of kin, giving her address as Stagpole St, West End, Townsville.

Accepted into the AIF, Norman Gordon was allocated service number 6265, assigned to the Infantry, and posted to the 15th Battalion, 20th Reinforcements. The 15th Battalion was originally raised as part of the all-volunteer AIF in September 1914. Drawing personnel from volunteers from Queensland and Tasmania, it formed part of the 4th Brigade, along with the 13th, 14th, and 16th Battalions. Private Norman Gordon spent the next nine months in Australia undergoing basic training. On the 7th of September 1916, he and his fellow reinforcements embarked from Brisbane with Convoy No. 24 bound for England. They sailed aboard HMAT A46 *Clan MacGillivray* , which boarded 1,084 troops. The entire convoy of 13 ships transported 12,600 troops on the voyage.

Eight weeks later, in November 1916, the reinforcements disembarked at Plymouth in England. From there, they were transported to the 4th Training Battalion at Rollestone on the Salisbury Plain for further training. Private Norman Gordon crossed to France from Folkestone on the 28th of December 1916. Following a week spent at the 4th Australian Division Base Depot at Étaples, Private Gordon joined the 15th Battalion in the field on the 6th of January 1917. The 15th Battalion was occupying trenches in the Ypres salient, around St Eloi and Boorlartbeek where they spent the winter on the Western Front. During the worst winter in Europe in 40 years, the troops faced severe hardships and sickness as they rotated through defensive duties and labouring in the trenches.

The 15th Battalion, stationed in Gueudecourt, a village on the Somme battlefield in northern France, faced particularly harsh conditions. On the 1st of February 1917, the battalion launched an attack on *'Stormy Trench'*, located just 100 yards from their front line. Difficulties arose as uncut wire hindered progress on the right flank. However, the left-hand section reached the trench and bombed their way down to capture the rest. German counterattacks began at 0155hr (1:55 a.m.) on the 2nd of February, but during a subsequent attack at 4:00 a.m., the Australian SOS signals were not seen for fifty minutes. By the time the artillery responded, the 15th Battalion had been repulsed, suffering 144 casualties. After the battle for Stormy Trench, Private Norman Gordon was among those affected by the cold and was hospitalized with bronchitis for two weeks. He returned to the 15th Battalion on the 24th of February 1917.

In support of operations at Arras, the 4th Australian Division, including the 15th Battalion, and the 62nd British Division launched a hastily planned attack on Bullecourt on the 11th of April 1917. Despite the failure and destruction of supporting tanks, Australian infantry managed to breach the German defences. However, uncertainty about the progress of the assault led to the withholding of artillery support. Eventually, the Australians, now surrounded, were forced to retreat. The attack resulted in heavy losses for the two Australian Brigades, the 4th and 12th, with 3,300 casualties, including 1,170 taken prisoner—the largest number captured in a single engagement during the war.

It was after this action that Private Norman Gordon was reported as Missing in Action on the 21st of April 1917. His whereabouts remained unknown until May 1917 when he was permitted to send a postcard to Army authorities advising that he had been captured at Bullecourt on the 11th of April 1917 and was being held as a prisoner of war in Dulmer, Germany. In December 1917, a fellow prisoner of war, Sergeant T. Miller, wrote to Norman's mother advising her that Norman had been killed when struck by an electric train on the 17th of November 1917. Later details revealed that Norman had been walking back to camp along a narrow path alongside the railway lines. In stepping aside to avoid a train, he slipped and fell under the wheels. It was also stated that the occurrence was entirely accidental.

Private Norman Gordon was 23 years old at the time of his death and had been a member of the AIF for one month short of two years. He was initially buried at the Friedhof Golpa Cemetery but his body was later re-interred *"with every measure of care and reverence"* at Berlin South-Western Cemetery (Plot V, Row H, Grave No. 5), Stahnsdorf, Germany. His death was notified in the 378th and 379th list of Australian Casualties released on the 18th of January 1918.

Norman Victor Gordon's name is inscribed on The Australian War Memorial Roll of Honour (panel 76), The City of Townsville WW1 Honour Roll, The Gympie & Widgee War Memorial Gates, and The Townsville Cenotaph. Gorden Street, Garbutt Townsville is named in his and G.D. and W.N. Gordon's memory (n.b. The spelling of the street name, i.e., Gorden instead of Gordon, has been attributed to an error. Townsville City Council minutes clearly record that the street was to be renamed Gordon St from McAllister Street).

6265 Private Norman Victor Gordon was awarded the British War Medal (no:46623) and the Victory Medal (no:46085). A memorial plaque and a memorial scroll were passed to his mother in 1923.

3034 PRIVATE JAMES RAYMOND BALDEY (1892 – 1917)

23rd Australian Machine Gun Company

James Raymond Baldey was born in Tenterfield, NSW, on the 7th of July 1892. He was the oldest son of six children born to James and Louisa Baldey (née Coxall). James grew up in Tenterfield and attended the Tenterfield Public School as a boy. At some time before 1913, the Baldey family relocated to Ayr in Queensland where James Baldey snr. worked as an accountant and James jnr took up work as a hardware salesman (ironmonger). By 1916, the Baldey family had relocated to Townsville and lived in Morris St, West End.

On the 4th of November 1916, James Baldey jnr applied at Townsville to enlist in the Australian Imperial Force (AIF). He declared that he was 24 years and three months old, single, and an ironmonger. His medical examination noted that he was 5'6" (167 cm) tall, weighed 8st 3lbs (52 kg), had a fair complexion, brown eyes, and light brown hair. He identified as Methodist and nominated his father as his next of kin, giving his address as Morris St, (later changed to Garrick Street), West End Townsville.

Accepted into the AIF, James Baldey was assigned service number 3034, allocated to the Infantry and posted to the 15th Battalion, 20th Reinforcements. The 41st Battalion had been formed in February 1916 at Bell's Paddock near the current Gallipoli Barracks, Enoggera, in Brisbane, Queensland, with men drawn from Queensland and northern New South Wales. It was assigned to the 11th Brigade, 3rd Division.

When James Baldey enlisted, the 41st Battalion was training in England prior to crossing to France to fight on the Western Front. For the next three months, Private James Baldey and the rest of the Reinforcements completed basic training in Australia before they embarked from Sydney for England on the 7th of February 1917. They travelled as part of Convoy No. 29 aboard HMAT A18 *Wiltshire* which boarded 1,746 troops. The entire convoy of eight ships transported a total of 10,679 troops on the voyage. The Reinforcements disembarked at Devenport, England, on the 11th of April 1917 and were moved to Durrington Camp for further training with the 11th Training Battalion.

In June 1917, Private James Baldey was transferred to the 23rd Australian Machine Gun Company. He crossed to France and the Western Front in September 1917 and from there he was deployed to Flanders. In December 1917, James Baldey was in action at Ploegsteert Wood, located around the Belgian village of Ploegsteert. This was part of the Ypres Salient. The Ypres Salient in Belgium was the scene of several battles and an extremely important part of the Western front during the First World War. It was here that Private James Baldey was Killed in Action on the 13th of December 1917.

The details of Private James Baldey's death are recorded in the statement of LCpl J.S. Hanley taken from him in February 1918 by the Australian Red Cross. Hanley states, *"We were in the subsidiary line at Ploegsteert on December 13th. We were at our dugout. Baldey was just at the door making tea when a whizz bang killed him instantly. He did not move after. This was about 4.30 p.m."*

Private James Baldey was 25 and five months old at the time of his death. He had served in the AIF for 13 months and had been a member of the 23rd

Australian Machine Gun Company for six months. He was laid to rest at the Pont d'Achelles Military Cemetery, Nieppe, France and his death was notified in the 378th list of Australian Casualties released on the 12th of January 1918.

James Baldey's name is inscribed on The Australian War Memorial Roll of Honour (panel 177), The City of Townsville WW1 Honour Roll, and The Townsville Cenotaph. Baldey Street, Hyde Park, Townsville is named in his memory.

3034 Private James Raymond Baldey was the last Townsville soldier killed in 1917. He served in WW1 on the Western Front. He was awarded the British War Medal (no:56124) and the Victory Medal (no:55204). These medals, as well as a memorial scroll and a memorial plaque, were passed to his father.

James Baldey had one brother and one uncle who also enlisted in WW1, their details are:

Brother: 2574 Sergeant Clifford John Baldey, 9th Battalion, returned to Australia, 18th July 1919.

Uncle: 5 Private Thomas Henry Baldey, 3rd Divisional Cyclist Company, died of pneumonia, 13th December 1916.

PART 4 – 1918

In 1918, 44 men from Townsville and North Queensland lost their lives, each with a unique and tragic story. Among these men, seven were married, leaving behind spouses and, in some cases, children who had to carry on without them. The impact of these losses reverberated deeply across families and the wider community, as the war created widows and orphans whose lives would be forever marked by the absence of their loved ones.

The war claimed 36 of Townsville's men on the battlefields of France, a major theatre of conflict where Allied forces fought fiercely against the Germans. The conditions were gruelling, with trench warfare exposing soldiers to relentless artillery bombardments, gas attacks, and harsh environmental hazards.

Four other men died in England, from injuries sustained in battle or illnesses contracted during service. Medical facilities were often overwhelmed, and the medical science of the time could offer only limited treatment for the severe injuries common in World War I.

One soldier died of disease in New Guinea, and another lost his life in Palestine, part of the campaign against Ottoman forces. A serving merchant seaman also fell in the line of duty when his ship was sunk by German forces. Tragically, another of these men died in a Turkish prisoner-of-war camp, succumbing to the Spanish influenza sweeping the world.

The last Townsville soldier to lose his life was Private Patrick Keane who enlisted at the end of May in 1918. He was 18 years old and was allocated to the AIF Railway Unit. Tragically he succumbed to influenza and pneumonia in London two weeks before the war ended.

Each man's death was a profound loss to their families and a sobering reminder of the war's far-reaching impacts These 44 Townsville and Northern Queensland men came from all levels of society, but each shared a common bond in their service and sacrifice. Their stories that follow ensure that their memory endures.

Lest We Forget

2793 PRIVATE WILLIAM JEFFERSON SUMMERFIELD (1896 – 1918)

15th Infantry Battalion/47th Infantry Battalion

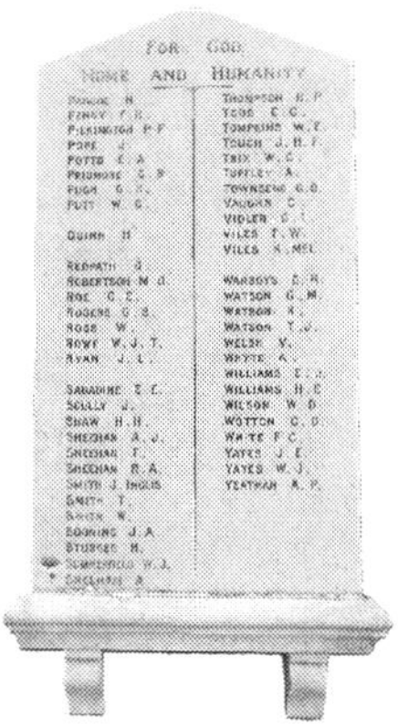

William Jefferson Summerfield was born on the 1st of September 1896 in Townsville, Queensland, the son of William and Mary Summerfield (née Jefferson). He was the youngest of three children, with one brother and one sister. His father, a grocer, and his mother, a schoolteacher, lived in a property on Park Lane, Hyde Park, Townsville.

William Summerfield grew up in Townsville and attended the West End State School (where his mother was a teacher) and later, Christian Brothers. After completing his school years, he worked as a clerk and served in the Citizen Military Forces. He spent four months on duty with the garrison on Thursday Island.

On the 10th of August 1915, William Summerfield applied to enlist in the Australian Imperial Force (AIF). He declared that he was single, one month short of his 19th birthday, and employed as a clerk. He also disclosed his previous military service in the Citizen Forces. His enlistment medical examination notes that he was 5'11" (180 cm) tall and weighed 141 lbs (64 kg). He had a dark complexion, grey eyes, and black hair. He identified as Roman Catholic and nominated his father as his next of kin, giving his address as simply Hyde Park, Townsville.

Accepted into the AIF, William Summerfield was allocated service number 2793, assigned to the Infantry, and initially posted to the 15th Battalion, 9th Reinforcements. The 15th Battalion was deployed on the Gallipoli Peninsula when William Summerfield enlisted. On his 19th birthday, the 1st of September 1915, Private William Summerfield embarked with his unit from Sydney, bound for Egypt. They joined Convoy No. 11 and sailed aboard HMAT A33 *Ayrshire*, which carried 326 troops. The entire convoy of nine ships carried a total of 7,336 troops on the voyage.

After a short stay in Egypt, Private Summerfield and the other reinforcements were deployed to the Gallipoli Peninsula and taken on strength of the 15th Battalion on the 13th of November 1915. In early December, Private Jefferson Summerfield (he had decided to use his second Christian name) was evacuated to the Hospital Ship H.S. *Oxfordshire*, suffering from mumps. In late December, he was transferred to Ghezireh Red Cross Hospital in Cairo, Egypt, suffering from haemorrhoids. He was moved to the 2nd Auxiliary Hospital in Heliopolis and then to a convalescent facility. It was not until the 27th of January 1916 that Private Summerfield was discharged to duty. By this time, the Gallipoli campaign was finished, and the ANZACs had been successfully evacuated back to Egypt.

Successful enlistment in Australia coincided with the return of soldiers from the Dardanelles, leading to a surplus of manpower that facilitated the expansion of the Australian Imperial Force (AIF). A strategic decision was made to divide existing battalions, incorporating experienced Gallipoli veterans, to establish new units. This process, known as the "doubling" of the AIF, resulted in the formation of two additional divisions while ensuring geographical diversity within battalions. The 47th Battalion can be considered the progeny of the 15th Battalion, predominantly comprising soldiers from Queensland and Tasmania, akin to its "parent" unit. Private Summerfield was among those transferred, joining the 47th Battalion on the 7th of March 1916.

The 47th Battalion left Egypt in early June 1916 to join the British Expeditionary Force (BEF) on the Western Front in France. Upon arrival in France on the 9th of June 1916, Private Jefferson Summerfield, with his battalion, entered the Western Front trenches on the 3rd of July 1916. The 47th Battalion's inaugural significant battle took place at Pozières. Initially, the battalion contributed labour during the 2nd Division's attack on the 4th of August, but later, it played a crucial role in defending the captured territory alongside its own division. It was during this action that Private Summerfield was Wounded in Action and evacuated, suffering from shell shock on the 8th of August 1916. He re-joined the 47th three days later.

On the 12th of October 1917, the 47th was tasked with attacking the enemy positions on the Passchendaele Ridge. They encountered heavy opposition and shellfire. It was during this operation that Private Jefferson Summerfield was Wounded in Action for a second time. On this occasion, he suffered a gunshot wound to his right hand. He was evacuated to No. 8 Stationary Hospital at Wimereux, near Boulogne. From there, he was transferred to No. 10 Convalescent Depot at Écault (8 km south of Boulogne). Following treatment, convalescence, and some time at the 4th Australian Division Base Depot, he re-joined the 47th Battalion in the field on the 26th of November 1917 at Béthencourt-sur-Somme. He was one of 40 men returned to the 47th after recovering from wounds.

In the winter of 1917-18, the 47th Battalion underwent rotations between periods on the front line and periods of rest. During the spring of 1918, the battalion contributed to halting the German Spring Offensive. It successfully defended against assaults near Dernancourt, France, from late March to early April 1918. It was during these engagements that Private Jefferson Summerfield was Killed in Action on the 28th of March 1918. He was one of the 18 soldiers who lost their lives in the 47th Battalion on that day.

In a statement made to the Australian Red Cross in August 1918, 4529 Private Arthur James said; *"I saw him killed instantly by a machine gun bullet whilst we were going up to the front line at Dernancourt near Albert. I could not give any particulars of burial, as we had to keep on going. I knew him well in the company. He came from Townsville, Queensland."*

Private Summerfield was 21 years and seven months old at the time of his death. He had served with the AIF for two years and seven months. He was laid to rest at Ribemont Communal Cemetery Extension (Henencourt Wood Cemetery Memorial), France. His death was notified in the 393rd list of Australian casualties released on the 25th of April 1918.

Private William Jefferson Summerfield's name is inscribed on The Australian War Memorial Roll of Honour (Panel 144), The City of Townsville WW1 Honour Roll, and The Townsville Cenotaph. Summerfield Street, Hermit Park, Townsville, is named in his honour.

2793 Private William Jefferson Summerfield was awarded the 1914/15 Star (no: 13316), the British War Medal (no: 18038), and the Victory Medal (no: 17968). A memorial scroll and a memorial plaque were passed to his father in 1921 and 1922, respectively.

3781 2ND LIEUTENANT WILLIAM CHARLES LOWRY BROOKS (1894 – 1918)

51st Infantry Battalion

2Lt William Brooks, 4th from the left in 2nd Row

William Brooks was the fifth of eight children born to George and Amelia Brooks (née Lowry) at Townsville, Queensland on the 13th of May 1894. George Brooks was a seaman, and the family lived at several addresses in Townsville. William Brooks attended the Ross Island State School, and upon completing his school years, he worked as a clerk. On the 5th of April 1915, William Brooks applied in Brisbane to join the AIF. He declared that he was 20 years old, single, qualified as a clerk, and possessed no military experience. His enlistment medical examination noted that he was 5'7" (170 cm) tall, weighed 123 lbs (56 kg), had a fair complexion, blue eyes, and brown hair. He identified as Church of England, and nominated his father as his next of kin, giving his address as Allen St, South Townsville. This address was subsequently changed several times in William Brooks' documents.

Accepted into the AIF, William Brooks was allocated service number 3781, assigned to the Medical Corps, and posted to the 13th Field Ambulance. Field ambulances were assigned to each of the AIF brigades. These were military units, not vehicles. They fulfilled a crucial function, prioritising casualties and rapidly transporting them back to treatment facilities. On the 24th of May 1915, seven weeks after enlistment, Private William Brooks embarked for Egypt from Brisbane. His contingent joined Convoy No. 7 and sailed aboard HMAT A11 *Ascanius*, which carried 1,746 troops. The entire convoy of four ships transported a total of 5,047 troops and twenty horses on the voyage.

By September 1915, Private William Brooks was stationed on Gallipoli as a member of the 7th Field Ambulance, the medical unit attached to the 7th Brigade of the 2nd Division. The 2nd Division was formed in Egypt and deployed to Gallipoli in the later stages of the campaign (August - September 1915). The 7th Brigade, along with the 7th Field Ambulance, served throughout the Gallipoli campaign until the general evacuation in December 1915. Once back in Egypt, Private William Brooks was transferred to the 13th Field Ambulance at Ismalia on the 24th of February 1916. The 13th operated in the vicinity of Serapeum near the Suez Canal until the 6th of June 1916 when the unit boarded ships at Alexandria bound for France and the Western Front.

William Brooks was promoted to Corporal in September 1916, then Acting Sergeant in March 1917. Soon after his promotion, Sergeant William Brooks was assigned duty at the hot baths near Albert. During this duty, he contracted mumps and was admitted to the 6th Field Ambulance and reverted to the rank of Corporal. He returned to the 13th Field Ambulance, located at Baupaume, on the 7th of April 1917. Corporal William Brooks continued to serve with the 13th Field Ambulance as the unit moved in support of the 13th Infantry Brigade. In early August 1917, while deployed in the vicinity of Steenwerck, Corporal William Brooks was selected to attend No. 6 Officers Cadet Battalion in Oxford, England. Having successfully completed the Officer's Course, Corporal William Brooks was commissioned as a Second Lieutenant (2Lt.) and posted to the general Infantry Reinforcements. He crossed to France on the 14th of December 1917 and joined the 4th Australian Division Base Dept at Havre five days later. On the 19th of December 1917, 2Lt. William Brooks was one of four 2nd Lieutenants assigned to the 51st Infantry Battalion. The

51st was stationed at Templeux-la-Fosse in the French region of Picardie. Another winter of trench routine followed for the Battalion.

After the Russian collapse in October 1917, a significant German offensive was anticipated on the Western Front for early 1918. This offensive was launched in France during late March. On the 30th of March 1918, the 51st Battalion received an operational order at 1600 hr (4:00 pm) from HQ 13th Brigade. The Battalion was tasked to relieve the 104th Brigade by taking over the front line from Buire to Dernancourt. The relief was successfully carried out by 0120 hr (1:20 am) on the 31st of March. Casualties for the operation were listed as one Officer and two other ranks killed, and six other ranks wounded. 2Lt. William Brooks was the Officer listed. He was Killed in Action by shell fire about 300 yards northwest of the village of Bresle on the 30th of March 1918. At the time of his death, he was 23 years, 10 months old and had served in the AIF for one month short of three years. His death was notified in the 392nd list of Australian Casualties released on the 22nd of April 1918.

William Charles Lowry Brooks' name is inscribed on The Australian War Memorial Roll of Honour (panel 152), The Australian National Memorial, Villers-Bretonneux, France, The City of Townsville WW1 Honour Roll, and The Townsville Cenotaph. Brooks Street in Railway Estate, Townsville, is named in his memory.

2nd Lieutenant William Charles Lowry Brooks served at Gallipoli and on the Western Front and was awarded the 1914/15 Star (no:3998), the British War Medal (no:12957), and the Victory Medal (no:12899). A memorial scroll and a memorial plaque were passed to his father in 1920.

The following tribute was published in the *Townsville Daily Bulletin* on the 17th of April 1918:

"ROLL OF HONOR.

BROOKS. —In loving memory of our dear son and brother, Lieut. William Charles Brooks, who was Killed in Action on March 30th, 1918.

> Under the flag of the five-starred cross.
> The pride of the southern sea.

He laid down his life in a deadly strife
To keep Australia free.
He sleeps not in his native land,
But beneath the foreign skies.
Far from those who loved him best.
In a hero's grave, he lies.

(Inserted by his loving father, sisters, brother and brother-in-law, Mr. G. Brooks, senr., Mr. and Mrs. G. E. Brooks, Mr. and Mrs. J. McRae, and Mr. and Mrs. A. Freis and Mrs. C. Gill.)"

3202 PRIVATE VICTOR WELSH (1898 - 1918)

41st Infantry Battalion / 52nd Infantry Battalion

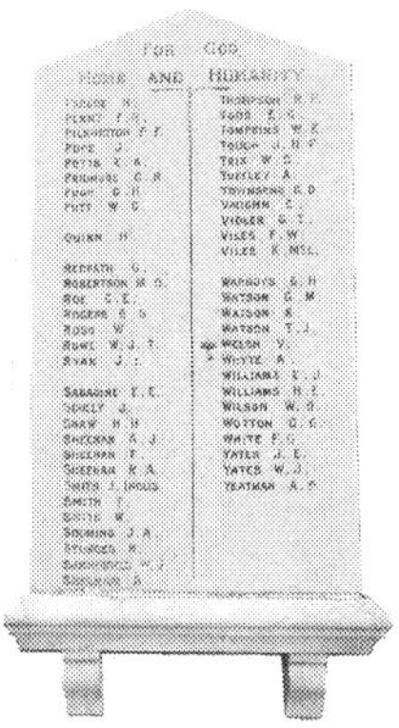

Victor Welsh was born on the 25th of October 1898 in Townsville, Queensland. He was the youngest child of four born to Samuel Welsh and his wife Eliza (née Hutchings). Samuel Welsh was a fireman, and the family lived in a property on Humphry St Townsville. Victor Welsh grew up in Townsville, and he attended the West End State School. He also served in the Senior Cadets for four years. After his school years were finished, Victor Welsh worked as a labourer.

On the 6th of November 1916, Victor Welsh applied to join the Australian Imperial Force (AIF). He declared that he was 18 years old, single, worked as a labourer and had experience in the Senior Cadets. His enlistment medical examination noted that he was 5'5¼" (166 cm) tall, weighed 8st 5lbs (53 kg), had a sallow complexion, slate grey eyes, and very dark brown hair. He identified as Church of England and nominated his mother as his next of kin, giving her address as Humphrey Street, Townsville.

Victor Welsh was accepted into the AIF, allocated service number 3202, assigned to the infantry, and posted to the 41st Infantry Battalion, 7th Reinforcements. When Victor enlisted, the 41st Battalion was training in England prior to crossing to Francc to fight on the Western Front. For the next three months, Private Victor Welsh and the rest of the Reinforcements completed basic training in Australia before they embarked from Sydney for England on the 7th of February 1917. They travelled as part of Convoy No. 29 aboard HMAT A18 *Wiltshire*, which carried 1,746 troops. The entire convoy of eight ships carried a total of 10,679 troops on the voyage. Nine weeks later, Private Welsh and the contingent of Reinforcements disembarked at Devonport in England. From there, they were soon transported to Dorrington Camp for further training.

It was during his time in Dorrington that Private Welsh went Absent Without Leave (AWOL) for seven days. On his return, he was charged, found guilty, and awarded 14 days Field Punishment No.2, which consisted of heavy labouring duties, possibly being restrained in handcuffs or fetters. He also forfeited 21 days' pay. On the 4th of March 1918, the Reinforcements crossed to France and were transferred from the 42nd Battalion Reinforcements to the 52nd Infantry Battalion. The 52nd Battalion was deployed in the vicinity of Locre in Belgium when Private Victor Welsh was one of 85 reinforcements taken on strength on the 9th of March 1918.

At the end of March 1918, the 51st, as part of the 4th Division, deployed to the front line in the vicinity of Buire-Dernancourt to defend positions on the River Ancre. The Battalion assisted in the repulse of a large German attack on the 5th of April 1918, launching a critical counterattack late in the day. It was during this attack that Private Victor Welsh was recorded as Wounded in Action and missing. In a statement made in November 1918 to the Australian Red Cross, 2184 Private George Hornung said, *"He was in B. Company, 6th Platoon. He was 5ft.9ins. medium and 21. At Dernancourt on the 5th of April 1918 at 9 A.M., we were holding the line against a violent enemy attack. Welsh was hit in the back with machine gun bullets. I bound him up, but when we were driven back by Fritz, we would not remove Welsh. The last I saw of him was when he was lying in the trench, badly wounded but conscious."* It was not until a Court of Enquiry was held in April 1919 that Private Victor Welsh was declared to have been Killed in Action on the 5th of April 1918. He has no known grave.

Victor Welsh was 19 years, five months old at the time of his death and had served with the AIF for one year, five months. Victor Welsh's name is inscribed on The Australian War Memorial Roll of Honour (Panel 156), The Australian National Memorial, Villers-Bretonneux, France, The Townsville West State School Memorial Gate, The City of Townsville WW1 Honour Roll, and The Townsville Cenotaph. Welsh Street Rosslea, Townsville, is named in his memory.

3202 Private Victor Welsh was awarded the British War Medal (no: 56276) and the Victory Medal (no: 55396). A memorial scroll and a memorial plaque were passed to his father in 1922.

1610 CORPORAL PERCIVAL FREDERICK PILKINGTON (1892-1918)

47th Infantry Battalion

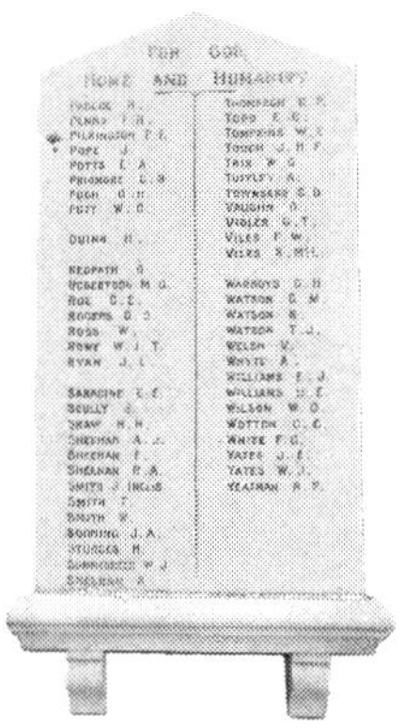

Percival (Percy) Frederick Pilkington was born on the 26th of November 1892 in Charters Towers, Queensland. He was the youngest of eight children of Daniel and Martha Pilkington (née Salisbury). He had two brothers, one of whom died in 1892, and five sisters, one of whom passed away in 1902. His father worked as a gardener and then as a sawyer. Martha Pilkington died in 1894, the day after giving birth to her youngest child.

Raised by his father, Percy Pilkington attended Richmond Hill State School in Charters Towers. After completing school, he moved to Townsville and worked as an engine driver/fireman.

On the 28th of December 1914, not long after war was declared in September, Percy Pilkington applied to enlist in the Australian Imperial Force (AIF). He declared that he was single, 22 years old, and had no previous military service. His enlistment medical examination noted that he was 5'6" (168 cm) tall and weighed 9st 5lbs (59 kg). He had a fresh complexion, grey eyes, and dark hair. He identified as Church of England and nominated his father as his next of kin, giving his address as c/- A.H. Price, Fifth St, Townsville.

Accepted into the AIF, Percy Pilkington was allocated service number 1610, assigned to the Infantry, and initially posted to the 9th Battalion, 3rd Reinforcements. The 9th Battalion was one of the initial infantry units raised for the AIF in the First World War. It held the distinction of being the first battalion recruited in Queensland and, in conjunction with the 10th, 11th, and 12th Battalions, formed the 3rd Brigade. When Percy Pilkington enlisted, the 9th Battalion was training in Egypt.

Private Percy Pilkington and his contingent of reinforcements completed six weeks of basic training in Australia before embarking from Brisbane for Egypt on 13 February 1915. They sailed as part of Convoy No. 3 aboard HMAT A49 *Seang Choon*, which carried 1,024 troops. The entire convoy consisted of 21 ships, transporting 8,828 troops and 6,027 horses.

Upon disembarking in Egypt, the reinforcements underwent a short training period before being deployed to the Gallipoli Peninsula. Private Pilkington joined the 9th Battalion at Gallipoli on the 6th of May 1915, where the battalion faced heavy shelling from Orange Grove and Kabe Tepe while defending the forward trenches.

For the next seven months, Private Pilkington fought with the 9th Battalion until November 1915, when he contracted severe influenza and was evacuated to a hospital in Heliopolis, near Cairo. He rejoined the 9th Battalion at Tel-el-Kebir on the 21st of January 1916, following the general withdrawal from Gallipoli in December 1915.

Private Pilkington was readmitted to the hospital in March 1916 due to a skin condition and remained there for a few weeks. In April, he was transferred to the 47th Battalion, formed in Egypt on 24 February 1916 as part of the AIF's "doubling."

The 47th Battalion left Egypt in early June 1916 to join the British Expeditionary Force on the Western Front in France. They arrived in France on the 9th of June 1916 and entered the trenches on the 3rd of July. The Battalion fought in its first major engagement at Pozières, initially providing labour during the 2nd Division's attack on the 4th of August. Later, they defended the captured territory with their own division.

Private Percival Pilkington was wounded by a gunshot to the face on the 31st of August 1916. The wound required him to be evacuated back to England, where he was admitted to the 2nd General Hospital, Leeds, followed by transfer to the 1st Australian Auxiliary Hospital in Harefield. He was discharged from the hospital on the 21st of September 1916 and transferred to Australia's No. 1 Command Depot, Perham Downs.

Private Pilkington was granted 14 days furlough in England from the 23rd of September 1916. He returned on the 27th of October 1916 and was charged with being Absent Without Leave (AWOL) from the 7th to the 25th of October 1916. Found guilty, he was jailed for 14 days and forfeited 33 days' pay.

After his release from military prison, he remained at Perham Downs until the end of November 1916, when he was admitted to the hospital on the 1st of December 1916 suffering from an infection. He underwent treatment for 21 days and returned to Perham Downs on the 21st of December 1916.

One month later, he was declared an illegal absentee after going AWOL from the 20th of January 1917 until he was apprehended at Warwick Square, London, on the 9th of March 1917. Private Pilkington was court-martialled, found guilty, imprisoned for 70 days, and forfeited 160 days' pay. He was released and sent back to France on the 29th of May 1917.

When he was released from military prison, Private Percy Pilkington had served a total of two years and five months in the AIF, with a little over seven months (214 days) without pay. After completing three weeks of refresher training in Étaples, Private Pilkington rejoined the 47th Battalion on the 21st of June 1917 near the village of Doulieu in northern France. The Battalion rotated between frontline and rear areas during the winter of 1917-18.

Private Percy Pilkington was promoted to Lance Corporal in November 1917 and promoted again to Corporal at the end of December 1917. He was granted two weeks' leave in England in January 1918, and this time he returned, on time, in February. In March 1918, Corporal Percy Pilkington was detached for three weeks to the 12th Australian Infantry Brigade School.

In the spring of 1918, the 47th Battalion played a role in halting the German Spring Offensive. It effectively repelled attacks in the vicinity of Dernancourt,

France, from late March to early April. It was during this action that Corporal Percival Pilkington was Wounded in Action for the second time on the 27th of March 1918. He was hit by a high-explosive shell and suffered serious wounds to both legs, his left thigh, and his left forearm. Corporal Pilkington was evacuated to the 8th Stationary Hospital at Wimereux. He survived until the 14th of April 1918 when he succumbed and Died of Wounds.

Corporal Percival Pilkington was 25 years old and had served with the AIF for three years at the time of his death. He was laid to rest at Wimereux Communal Cemetery (Plot X, Row D, Grave No. 2A), France. His death was notified in the 395th and 396th lists of Australian Casualties released on the 5th of May 1918.

Percival Frederick Pilkington's name is inscribed on The Australian War Memorial Roll of Honour (panel 144), The City of Townsville WW1 Honour Roll, The Townsville Cenotaph, and the Richmond Hill Teachers and Ex-Pupils Great War Honour Roll. Pilkington St, Garbutt, Townsville, is named in his memory.

1610 Corporal Percival Frederick Pilkington was awarded the 1914/15 Star (no: 2639), the British War Medal (no: 8533), and the Victory Medal (no: 8493). A memorial scroll and a memorial plaque were passed to his father in 1921 and 1923, respectively.

671A PRIVATE BERTRAM EVERARD ARMIT (1890 – 1918)

7th Machine Gun Company

Bertram (Bert) Everard Armit was born to Richard and Johanna Armit (née Hannigan) of Stanley Street, Townsville, on the 22nd of July 1890 in Camboon, Queensland. Bert Armit had six brothers and two sisters. He was the fourth son of nine surviving children (one child did not survive). Richard Armit was an inspector of telegraph/electric engineer, and the Armit family lived in a property on Stanley St, North Ward, Townsville.

Bert Armit attended Christian Brothers in Brisbane, and after his school years were finished, worked as a merchandise clerk with Burns Philp and Co. The four oldest boys in the Armit family all enlisted in the AIF. Richard in September 1914, Ernest in April 1915, Bert in July 1916, and Harold in December 1916.

Bert Armit applied at Townsville to join the AIF on the 13th of July 1916. He declared that he was single, was employed as a merchandise clerk, and one month short of his 26th birthday. He also declared that he had previously served in the Navy for two years and the 15th Light Horse for three years. His enlistment medical noted that he was 5'7¼" (171 cm) tall, weighed 9 stone 4lbs (59kg), had a fresh complexion, brown eyes, and dark brown hair. Bert Armit identified as Roman Catholic and nominated his mother as his next of kin, giving her address as Stanley St. North Ward, Townsville.

Accepted into the AIF, Bert Armit was allocated service number 67, assigned to the Infantry, and posted to the 2nd Machine Gun Battalion (7th Machine Gun Company, 12 Reinforcements). The 2nd was fighting on the Somme on the Western Front when Bert Armit enlisted. After nine months training in Australia, Private Bert Armit embarked with the 12th Reinforcements contingent from Melbourne bound for England on the 21st of June 1917. They joined Convoy No. 32 and sailed aboard HMAT A29 *Suevic*, which carried 1,567 troops. The entire convoy of seven ships carried a total of 6,406 troops on the voyage. Private Bert Armit was appointed as an Acting Sergeant on the 26th of June 1917.

On arrival in England, the Reinforcements were posted to the 7th Training Battalion located at Rollestone (near Amesbury on Salisbury Plain in Wiltshire). Acting Sergeant Bert Armit relinquished his acting rank and reverted to Private on arrival at the Training Battalion. In September 1917, Private Bert Armit was transferred to the 25th Infantry Battalion and was promoted to Acting Lance Corporal. He remained in Rollestone until the 19th of December 1917 when he crossed as a Private to France from Southampton as part of the 25th Battalion. Once landed in France, Private Bert Armit marched into the Australian Division Base Depot at Havre. From there, he was taken on strength of the 25th Battalion on the 23rd of December 1917 and allocated the letter "A" to his regimental number. The 25th Battalion wintered in Belgium until early 1918.

The commencement of the German spring offensive on March 21, 1918, marked a critical moment in the war effort. This offensive, a desperate attempt by the German Army to secure victory, saw the deployment of 63 divisions along a 70-mile (110 km) front. In response, Australian Divisions swiftly moved south to the Somme region to aid in repelling the enemy advance. On the 20th of March 1918, Private Bert Armit was transferred to the 2nd Machine Gun Battalion. This was one of five Machine Gun Battalions established during World War I. Formed in March 1918 as a component of the 2nd Division, the 2nd Machine Gun Battalion comprised four machine gun companies. These companies were the 5th, 6th, 7th, and the 22nd, and had previously operated independently. Private Bert Armit was assigned to the 7th Machine Gun Company (7th MG Coy) within the 2nd Machine Gun Battalion.

On the 17th of April 1918, the 7th MG Coy war diary noted that *"enemy artillery was fairly active during the day and night in an erratic and scattered way."* One soldier was wounded, and one was killed. The soldier killed was Private Bertram Everard Armit, recorded as Killed in Action on the 17th of April 1918 at Buire near Amiens. He was 27 years and eight months old and had served for 21 months in the AIF at the time of his death. He was laid to rest at the Warloy-Baillon Communal Cemetery Extension (Plot II, Row F,

Grave No. 7), France. His death was notified in the 400th list of Australian Casualties released on the 23rd of May 1918.

Bertram Everard Armit's name is inscribed on Australian War Memorial Roll of Honour (panel 175), The City of Townsville WW1 Honour Roll, and The Townsville Cenotaph. Armit Street, Aitkenvale, Townsville, is named in his memory.

671A Private Bertrand Everard Armit served in WW1 on the Western Front. He was awarded the British War Medal (no:57966) and the Victory Medal (no:56966). These medals, as well as a memorial scroll and a memorial plaque, were passed to Bert's mother at the request of his father. Johanna Armit was granted a pension of two pounds (£2/-/-) per fortnight for the loss of her son. This amount was nearly half of the wage paid to AIF soldiers in 1918. Bert Armit's three brothers all returned to Australia at the end of the war.

The following notice was published in the *Townsville Daily Bulletin* on the 4th of May 1918:

ARMIT. — Killed in Action in France on 17th April, Bertram Everard Armit, of 7th Australian Machine Gun Company, and youngest of four brothers serving with the Australian Imperial Forces.

"Fortis in Arduis."

970 DRIVER HAROLD HERBERT SHAW (1891 – 1918)

5th Light Horse Regiment/14th Field Artillery Brigade

Harold Herbert Shaw was born on the 5th of December 1891 in Hughenden. He was the second child and first son born to Herbert and Marie Shaw (née Gielis). He had two brothers and three sisters. Herbert Shaw was an overseer of a property in Hughenden, and it was here that Harold Shaw grew up.

Later in life, Harold Shaw took a job as a labourer and shed hand in Hughenden, where he lived at a property on Uhr Street. On the 21st of December 1914, he made the long trip to Townsville and applied to join the Australian Imperial Force (AIF). He declared that he was 23 years old, single, and had no previous military experience. His medical examination noted that he was 5'5½ " (166 cm) tall, weighed 8st 13½lbs (57kg), had a fair complexion, blue eyes, and fair hair. He identified as Roman Catholic and nominated his father as his next of kin, giving his address simply as Hughenden.

Accepted into the AIF, Harold Shaw was allocated service number 970, assigned to the infantry and posted to the 5th Light Horse Regiment, 5th Reinforcements. The 5th Light Horse Regiment, formed in September 1914 in Brisbane, was composed entirely of Queensland enlistees and was a unit of the 2nd Light Horse Brigade.

When Harold Shaw enlisted, the 5th Light Horse had departed from Sydney bound for Egypt. While initially deemed unsuitable for the Gallipoli operations, the light horse troops were later deployed without their horses to provide reinforcement to the infantry. After completing basic training in Australia, Trooper Harold Shaw and the other reinforcements embarked from Newcastle NSW, bound for Egypt, on the 17th of May 1915. They sailed as part of Convoy No.5 aboard HMAT A57 *Malakuta*, which carried 109 troops. The entire convoy of seven ships transported a total of 3,174 troops and 2,018 horses on the voyage.

After a period of training in Egypt, Trooper Shaw and his fellow reinforcements joined the 5th Light Horse on the Gallipoli Peninsula at the end of September 1915. The 5th Light Horse played a defensive role for most of the rest of the campaign but did engage in several minor attacks. Trooper Shaw left Gallipoli bound for Alexandria on the 18th of December, and the entire 5th Light Horse left the Peninsula by the 20th of December 1915. Soon after arrival back in Egypt, Trooper Shaw was hospitalized with an illness for some weeks. The exact nature of his illness is not recorded.

Trooper Harold Shaw was transferred out of the 5th Light Horse in late March 1916 and taken on the strength of the newly minted 14th Field Artillery Brigade. He was posted as a Driver to the 55th Battery located at Tel-el-kebir. A Driver (Dvr) was a military rank used in the AIF, equivalent to the rank of Private. This rank was used in the Artillery for the men who drove the teams of horses that pulled the guns and the ammunition resupply wagons. It was a hazardous job, and Drivers and their teams were high-value targets for enemy artillery. In the words of one Battery Commander in the AIF, *"even the animals came to know when a shell was coming close, and if, when halted, the horses heard the whine of an approaching salvo, they would tremble and sidle closer to their drivers."*

The 14th Field Artillery Brigade left Egypt in June 1916 and disembarked at Marseilles in France on the 30th of June 1916. From there, the Brigade moved into positions near Blaringhem in Northern France. In July 1916, the 5th Division, including the 14th Field Artillery Brigade, was committed to the Battle of Fromelles. Fromelles was the first major battle fought by Australian troops on the Western Front.

The engagement stands as a grim milestone in Australian military history. The AIF suffered a toll surpassing that of the entire Gallipoli campaign, with casualties exceeding 5,500. Among them, close to 2,000 soldiers succumbed to fatal wounds or were killed in action, while approximately 400 were taken captive. This represents the most substantial loss suffered by a single division within a 24-hour span throughout the entirety of the First World War.

As a result of its losses, the 5th Division's effectiveness was reduced, and it was not considered fit for offensive action for many months. Once reinforced and rested, the 5th Division, along with Driver Harold Shaw, took part in several major actions over the next two years. The Division took part in the Battle of Bullecourt in April 1917 and the Battle of Passchendaele (a.k.a. the Third Battle of Ypres) in July 1917. Both battles were major engagements with 3,000 Australians killed or wounded at Bullecourt and 38,000 Australians either killed, wounded, or missing at Passchendaele. Driver Harold Shaw would have been heavily involved in the dangerous task of bringing ammunition forward for these and the other actions that the Australians were committed to.

In early 1918, the Germans launched the Spring Offensive, also known as the Kaiserschlacht. It was a massive series of offensives aimed at achieving a decisive victory before the arrival of American troops. As the offensive drove the Allies back, the 5th Division was transferred south from Belgium to the Somme, with its infantry battalions taking control of positions around Villers-Bretonneux, where they engaged in heavy fighting during the Second Battle of Villers-Bretonneux, which took place from the 24th to 27th April 1918.

The AIF was tasked with retaking the town of Villers-Bretonneux, which had been captured by German forces earlier in the month. The Australians launched a counterattack on the 24th of April 1918 and, after a fierce battle, were able to retake the town and restore the original front line. The battle was considered a significant victory for the Allies and a turning point in the war. The AIF suffered heavy casualties during the battle, with over 3,000 men killed, wounded, or captured. It was during this battle that Driver Harold Shaw was Killed in Action near Corbie, on the 24th of April 1918.

He was 26 years and five months old at the time of his death and had been a member of the AIF for a total time of three years and five months. Driver Harold Herbert Shaw was laid to rest at Bonnay Communal Cemetery Extension (Row B, Grave No. 25), France. His death was notified in the 402nd list of Australian Casualties released on the 22nd of May 1918.

Harold Herbert Shaw's name is inscribed on The Australian War Memorial Roll of Honour (panel No. 18), the City of Townsville Honour Roll, and the City of Townsville Cenotaph. Shaw Street, Nelly Bay, Townsville, is named in his memory.

970 Driver Harold Herbert Shaw was awarded the 1914/15 Star (no: 22879), the British War Medal (no: 12619), and the Victory Medal (no: 12559). Harold's father died in 1922, and a memorial scroll, and a memorial plaque were eventually passed to his widowed mother in 1923.

The following notice was published in *Townsville Daily Bulletin* on 24th April 1919:

Family Notices

SHAW. — In memory of my pal, the late Driver Harold Herbert (Hal.) Shaw, (970), 14th A. F. A. Brigade, A. I. F., who was Killed in Action at Villiers Bretonneux

"True till Death."

Inserted by 948 Sapper H. Jones, 6th Div. Sig. Coy., A. I. F.

6851 PRIVATE JOHN HENRY FREDERICK TOUCH (1892 – 1917)

52nd Infantry Battalion

John Henry Frederick Touch was born in about February 1892 in Vancouver, Canada, before emigrating to Australia. In 1916 John Touch worked as a cook and lived in a property on the Strand in Townsville with his wife Gevennie (Gwinnie).

On the 26th of March 1917, John Touch applied to join the Australian Imperial Force (AIF) at Townsville, Queensland. He declared that he was 25 years old, married with one child, qualified as a chef and had no previous military experience. His medical examination noted that he was 5'10" (178 cm) tall, weighed 140lbs (64kg), had a fair complexion, brown eyes, and brown hair. He identified as Presbyterian and he nominated his wife as his next of kin giving her address as Manning St, Milton, Brisbane.

Accepted into the AIF, John Touch was allocated service number 6851, assigned to the Infantry and posted to the 26th Battalion, 20th Reinforcements. When John Touch enlisted, the 26th Battalion was in France on the Western Front, after having fought on Gallipoli. John Touch commenced his military training at Enoggera near Brisbane.

After three months of training, Private Touch embarked with the rest of the reinforcements for England from Sydney on the 14th of June 1917. They joined Convoy No. 32 and sailed aboard HMAT A20 *Hororata*, which carried 1,744 troops. The entire convoy of eight ships carried a total of 7,150 troops on the voyage. The *Hororata* docked in Liverpool, England, at the end of August 1917.

Once in England, the 20th Reinforcements were deployed to Rollestone and then to Fovant on the Salisbury Plain, where training continued until January 1918. During his time at Fovant, Private Touch went Absent Without Leave for five days. He was fined 12 days' pay and additionally forfeited five days' pay for the period of his illegal absence.

Training completed in England, Private John Touch and the rest of the 20th Reinforcements were shipped to No.1 Overflow Camp at Calais, France, on the 1st of April 1918. Three days later, Private Touch was transferred to the 52nd Infantry Battalion near Dernancourt and taken on strength.

On the 5th of April 1918, the 52nd Battalion assisted in the repulse of the largest German attack mounted against Australian troops during the war. The enemy threat persisted through April, and at about 2200hr (10:00 pm) on the 24th of April 1918, the 52nd began its advance to take Villers-Bretonneux from the Germans. It was during this battle that Private John Touch was Killed in Action on the 24th of April 1918, twenty days after joining the 52nd Battalion. He was 30 years old and had served with the AIF for one year, five months. He has no known grave. His death was notified in the 404th list of Australian Casualties released on the 27th of May 1918.

John Henry Frederick Touch's name is inscribed on The Australian War Memorial Roll of Honour (panel 15), The Australian National Memorial, Villers-Bretonneux, France, The City of Townsville WW1 Honour Roll, and The Townsville Cenotaph. Touch St Rosslea, Townsville, is named in his memory.

6851 Private John Henry Frederick Touch served on the Western Front in The Great War. He was awarded the British War Medal (no:57669) and the Victory Medal (no:56709). A memorial scroll and a memorial plaque were passed to his wife in 1922.

5338 SIGNALLER ADAM CARDNO ALEXANDER (1892 - 1918)

51st Infantry Battalion

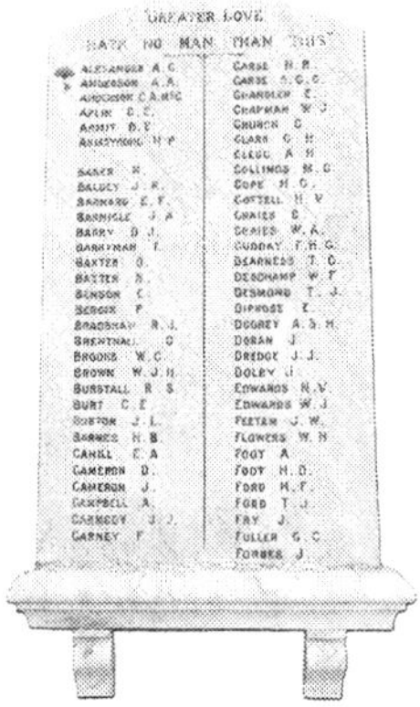

Adam Cardno Alexander was born on the 3rd of February 1892 in Brisbane, Queensland. He was the first-born child of five and the only son of Alexander and Amelia Alexander (née Cardno). Alexander Alexander was a shipwright, and the family lived in premises on Eyre Street in North Ward. Adam Alexander grew up in Townsville and attended Central State School. After his school years were finished, he worked as a compositor printer at D.W. Hastings Printer & Stationer in Flinders St Townsville.

On the 7th of February 1916, at the age of 24, Adam Alexander gave up his job and applied at Townsville to join the AIF. He declared that he was single, employed as a compositor, and had two and a half years' service in "D" Company of the Senior Cadets. His enlistment medical noted that he was 5'4" (162 cm) tall, weighed 8st (51kg), had a fair complexion, green/brown eyes, and brown hair. He identified as Presbyterian and nominated his father as his next of kin, giving his address as Stanley St, Townsville.

Accepted into the AIF, Adam Alexander was allocated service number 5338, assigned to the Infantry and initially posted to the 12th Battalion. Following 10 weeks of basic training in Australia, Private Adam Alexander and his fellow reinforcements embarked from Sydney, bound for Egypt on the 20th of April 1916. They sailed as part of Convoy No. 2 aboard SS *Hawkes Bay*, which carried 1,000 troops. The entire convoy of fifteen ships carried a total of 9,213 troops and 5,590 horses on the voyage. Private Alexander disembarked in Suez in late May 1916.

After three months in Egypt, he was shipped to England with other reinforcements and marched into the 3rd Training Battalion, Perham Downs Salisbury, on the 9th of August 1916. More training followed in England, and once completed, the reinforcements crossed to France on the 21st of August 1916 to reinforce the 12th Battalion. Once landed in France, the contingent of reinforcements, including Private Adam Alexander, were transported to the 1st Australian Division Base Depot (1st ADBD) at Étaples for further training before moving to the front line.

It was while he was in Étaples that Private Adam Alexander developed tonsillitis and was admitted to the 26th General Hospital on the 21st of September for treatment. He rejoined the 1st ADBD on the 7th of October. Two weeks after his return, Private Alexander was taken on strength of the 51st Infantry Battalion on the 21st of October 1916. The 51st Battalion was established in Egypt in early March 1916 as part of the "doubling" of the AIF. Roughly half of its troops had previously served in Gallipoli as part of the 11th Battalion, while the remaining half consisted of new reinforcements arriving from Australia.

When Private Alexander joined the Battalion, the 51st was deployed in Belgium after fighting at Pozières and Mouquet Farm during August and September. Following the battle of Mouquet Farm, the battalion spent the remainder of the year rotating between front-line duty and engaging in training and labouring behind the lines. This routine persisted throughout the harsh winter of 1916-17.

During his time with the 51st, Private Adam Alexander became a signaller, and he changed from Private to Signaller Alexander. On the morning of the 2nd of April 1917, the 51st Battalion (along with the 50th Battalion) attacked Noreuil, an outpost village near the Hindenburg Line. The attack was a success, but the Battalion suffered 239 casualties during the assault and subsequent defence of the village. Signaller Adam Alexander was one of the survivors of this attack and the other actions that the 51st Battalion was committed to. He was rewarded with three weeks leave in England in November 1917.

The next major engagement of the 51st Battalion occurred in April 1918. During the German Spring Offensive on the Western Front, German forces, utilizing infantry and tanks, captured Villers-Bretonneux, near Amiens, from weary British defenders on the 24th of April 1918. The 51st Battalion (as a component of the 13th Brigade) was deployed and engaged in the counter-attack, which effectively regained control of the town. During the action, the 51st Battalion lost 389 men in just two days. One of those casualties was Signaller Adam Alexander who was Killed in Action on the 25th of April 1918. He was 26 years old and had served in the AIF for two years, two months. He had been a member of the 51st Battalion for 18 months. He has no known grave, and his death was notified in the 193rd list of Australian Casualties released on the 14th of August 1918.

Adam Cardno Alexander's name is inscribed on The Australian War Memorial Roll of Honour (Panel 152), The Villers-Bretonneux Memorial (Australian National Memorial-France), The Townsville Roll of Honour, and The Townsville Cenotaph.

5338 Signaller Adam Cardno Alexander served in Egypt and on the Western Front in WW1. He was awarded the British War Medal (no:51163) and the Victory Medal (no:50483). These medals, as well as a memorial scroll and a memorial plaque, were passed to his father.

The following tribute was published in the *Townsville Daily Bulletin* on the 26th of April 1920:

ALEXANDER. — In loving memory of our beloved son and brother, Signaller A. C. Alexander, Killed in Action at the retaking of Villers Bretonneux, April 25th, 1918.

Too far away thy grave to see,
But not too far to think of thee.
And long as life and memory last,
We will remember thee.

(Inserted by his loving father and brother, A. Alexander and C. T. Alexander, Eyre Street, North Ward.)

MAJOR WILLIAM ALEXANDER CRAIES (1886 - 1918)

52nd Infantry Battalion

William Alexander Craies was born on the 19th of May 1886 in Townsville, Queensland. He was the oldest of eight children born to Walter and Emma Craies (née Wright). He had three brothers and four sisters. Walter Craies was an accountant, and the Craies family lived in a property on Ahearne St and later Alexandra St, North Ward, in Townsville. William Craies grew up in Townsville and attended Belgian Gardens State School and Sydney Military.

After his school years were completed, he gained employment as a clerk with Burns Philp & Co Ltd. He was eventually appointed to the position of business manager. Outside of his working life, William Craies also enlisted as a citizen soldier in the Australian Garrison Artillery, achieving the rank of Captain and spending six months with the garrison on Thursday Island. He was still serving in the Garrison Artillery when he applied at Townsville on the 1st of March 1915 to join the AIF. He declared that he was single and 28 years and nine months old.

William Craies was accepted into the AIF as an Officer (honorary Major) for the first time on the 1st of March 1915. His enlistment documents record that he was 5'7" (170 cm) tall and weighed 11st 8lbs (73kg). He identified as Church of England and nominated his father as his next of kin. Major William Craies was assigned to the Infantry and posted to the 25th Battalion, which was raised at Enoggera in Queensland in March 1915, as part of the 7th

Brigade. On the 29th of June 1915, Major Craies sailed with the 25th aboard HMAT A60 *Aeneas*, which carried 1,720 troops on the voyage. Their ship was one of four that made up Convoy No.7. The entire convoy transported 5,052 men and 20 horses. The 25th trained in Egypt during August, and in September was deployed to Gallipoli. Major Craies stayed on Gallipoli until the general evacuation back to Egypt in December 1915.

While stationed in Egypt, the AIF gained a reputation as highly undisciplined. Higher command determined to address the problem by sending ill-disciplined men and officers back to Australia. Major William Craies may have been one of those selected, and on the 3rd of March 1916, his appointment was terminated, and he was sent back to Australia the same month *"for Disciplinary Reasons Unsuitability & inefficiency."* He was discharged and returned to civilian life and his previous job.

Nine months later and undaunted by his previous experience in the AIF, William Craies successfully re-enlisted on the 1st of December 1916. This time he was appointed as a Lieutenant with the 52nd Battalion, 9th Reinforcements. The 52nd was raised in Egypt in 1916, with half the battalion comprising South and Western Australian and Tasmanian veterans from the 12th Battalion, and the other half largely comprised Queensland recruits who had recently arrived from Australia. Lieutenant Craies and his fellow reinforcements embarked from Sydney, bound for England, on the 24th of January 1917. They joined Convoy No. 29 and sailed aboard HMAT A33 *Ayrshire*, which carried 948 troops. The entire convoy of eight ships carried a total of 10,753 troops on the voyage.

Ayrshire docked at Devonport, England, on the 12th of April 1917. Once landed, the reinforcements were transported to the 13th Training Battalion at Codford near the Salisbury Plains in England. Ten weeks of further training in Codford followed before Lieutenant Craies and his fellow reinforcements crossed to France on the 26th of June 1917. A short period of training at Havre had to be completed before joining the 52nd Battalion on the 4th of July 1917. The Battalion was located at De Seule in the province of West-Vlaanderen, Belgium, and was preparing to enter the front line.

A month after joining the 52nd, Lieutenant William Craies was promoted to Captain and fought with the Battalion at the Battle of Polygon Wood, which took place from the 26th of September to the 3rd of October 1917. The Australian forces advanced behind creeping artillery barrages, intending to cover 1,000 – 1,500 yards (900m to 1400 meters) while securing reverse slopes for observation. Prompt preparations effectively countered German counterattacks by consolidating captured ground with extensive defences. This offensive resulted in substantial losses for the German 4th Army, hindering their control over Broodseinde ridge. Favourable weather conditions helped the allied advance, and subsequent German counterattacks proved unsuccessful. In November 1917, Captain Craies was granted three weeks leave in England. He returned to the battalion in early December 1917.

In early 1918, after the collapse of Russia, the Germans initiated the Spring Offensive which was a series of powerful attacks aimed to break the stalemate of trench warfare and secure victory on the Western Front. In response to the Allied retreat towards Paris, the Australian 4th Division was quickly redeployed from their winter position in Belgium. They were tasked with establishing a hurried defensive line to the northwest of the Ancre River, situated between Buire and Dernancourt. On the 5th of April 1918, the 52nd Battalion was in the front line in the Buire-Dernancourt Sector when the Germans initiated a significant attack, considered one of the most formidable faced by the Australians. By midday, the 12th Brigade had been pushed back from the railway to the top of the slope. Subsequently, at 1600 hr (4:00 pm), the 13th Brigade also faced a retreat from the railway. However, at 1700 hr (5:00 pm), the 13th Brigade executed a counterattack. Captain Craies played a vital role in regrouping the remaining men to ensure the battalion's reconstruction, effectively leading them back to the railway. His performance in this role was outstanding, leading to special mention in the despatches of Field Marshal Sir Douglas Haig on the 7th of April 1918. On the 24th of April 1918, the 52nd Battalion was committed to the allies' counterattack at Villers-Bretonneux, launched at 2200hr (10:00 p.m.). Honorary Major Craies, now in charge, decided to consolidate his Company's position and dig in upon reaching a locality just south of Villers-Bretonneux, around midnight. He was the only Company Commander left out of the four who had commenced the attack two hours before. Lieutenant Colonel J.L. Whitham, his commanding officer, made the following observations in a letter to AIF Headquarters dated the 25th of May 1915: *"He, therefore, had large responsibilities, which he accepted, and his work throughout the night and during the whole of the 25th of April was splendid. I cannot speak too highly of his bravery and fearlessness, and it was with very deep regret that I heard on that Thursday evening (25th April) that he had been dangerously wounded by a shot from an enemy sniper."*

Major William Craies was shot in the stomach on the 25th of April, and, despite the best efforts of his unit, he Died of Wounds on the same day. He was one month short of his 32nd birthday at the time of his death and had served with the AIF for a total of two years, four months.

Major William Craies was buried in the field near where he fell. In 1920, Major William Craies' body was reinterred and laid to rest in the Adelaide Cemetery at Villers-Bretonneux and later reinterred again and laid to rest at Villers-Bretonneux Military Cemetery, France. Major William Craies' death was notified in the 402nd list of Australian Casualties released on the 22nd of May 1918.

For his work in the counterattack, Major Craies was again (posthumously) Mentioned in Despatches. William Craies' name is inscribed on The Australian War Memorial Roll of Honour (panel 155), The Belgian Gardens State School Honour Board, The City of Townsville WW1 Honour Roll, and The Townsville Cenotaph. Craies Street, Belgian Gardens, Townsville, is named in his memory.

Major William Alexander Craies was awarded the 1914/15 Star (no:26187), the British War Medal (no:16595), and the Victory Medal with a large oak leaf (no:16528). A memorial scroll and a memorial plaque were passed to his father in 1921 and 1922, respectively. Emma Craies was granted a pension of one pound (£1/-/-) per week for the loss of her son This amount was only half of the wage paid to an AIF Private soldier in 1918.

The following notice was published in *The Brisbane Courier* (Qld.) on the 25th of April 1919:

CRAIES. - In fond and loving memory of Major William Alexander Craies, one of the most beloved officers who fell in the battle of Villers Bretonneux, on 25th April 1918, Anzac Day.

His victory, Liberty, the greatest possession of mankind, and worthy of the greatest sacrifice.

3381 LANCE CORPORAL LEWIS WILLIAM FREDERICK HOOPER (1897 - 1918)

49th Infantry Battalion

Lewis William Frederick Hooper was born on the 3rd of February 1897, in Townsville, Queensland. He was the oldest of seven children born to David Hooper and his wife, Elizabeth (née Bargent). David Porter worked as a cab proprietor and, later, as a porter. The family lived in premises on Alexandra St, North Ward, Townsville. Lewis had four brothers and two sisters. One of his sisters died, aged five years, in 1909.

Lewis Hooper grew up in Townsville and attended Central State School & Technical College. After completing his school years, he worked as a clerk with the Railways Office. On the 16th of August 1915, Lewis Hooper left the Railways and applied at Townsville to join the AIF. He declared that he was single, 18½ years old, and working as a clerk. He also declared four years of previous service with the Senior Cadets, where he was gazetted as a 2nd Lieutenant in June 1915. As he was less than 21 years of age, Lewis Hooper required his parents' written permission to be posted to active service overseas. The permission document was signed by both parents.

Lewis Hooper's enlistment medical noted that he was 5'7¼" (171 cm) tall, weighed 8st 7lbs (54kg), had a fair complexion, blue eyes, and light brown hair. He identified as Church of England, and he nominated his father as his next of kin. Accepted into the AIF, Lewis Hooper was allocated service number 3381, assigned to the Infantry, and posted to the 9th Battalion, 11th Reinforcements. The 9th Battalion was fighting at Gallipoli when Lewis Hooper enlisted.

From August to October 1915, Lewis Hooper and his fellow reinforcements underwent recruit training in Australia. On the 21st of October 1915, the reinforcements and Private Lewis Hooper embarked for Egypt from Brisbane.

They joined Convoy No. 13 and sailed aboard A48 *Seang Bee*, which carried 1,013 troops. The entire convoy of 11 ships transported a total of 11,646 troops and 400 horses on the voyage. After disembarking in Egypt, the reinforcements were moved to Zeitoun Camp.

On the 29th of February 1916, Private Lewis Hooper was transferred from the 9th Battalion to the 49th Battalion. Formed just two days prior on the 27th of February 1916, the 49th Battalion was established as part of the "doubling" initiative within the AIF. Its ranks comprised approximately half of former 9th Battalion soldiers who had previously served in Gallipoli, while the remainder, including Private Hooper, were fresh recruits from Australia. The 49th's composition closely resembled that of the 9th Battalion, boasting a significant contingent of soldiers from Queensland.

In June 1916, the 49th Battalion sailed from Alexandria bound for France to join the British Expeditionary Force on the Western Front. After disembarking in France on the 12th of June 1916, the 49th Battalion was quickly committed to action and occupied the Western Front trenches for the first time on the 21st of June. The battalion's initial major engagement occurred at Mouquet Farm in August 1916, during which it incurred significant casualties, especially during the assault on the 3rd of September. For the remainder of that year, the battalion maintained a cycle of frontline service, interspersed with periods of training and labouring behind the lines. This routine persisted throughout the harsh winter of 1916 - 17.

In January 1917, Private Hooper was granted three weeks leave in England, returning to the field in late February as the 49th participated in the advance that followed the German retreat to the Hindenburg Line. In May 1917, Lewis Hooper was promoted to Lance Corporal (LCpl), and in June, he was admitted sick to hospital. LCpl Hooper was to spend a total of 53 days in medical care in Le Havre. He re-joined the 49th in the Ypres sector in Belgium in late August 1917, in time to participate in the battle of Polygon Wood, which took place from the 26th of September to the 3rd of October 1917. Australian forces advanced under creeping artillery cover, aiming to secure reverse slopes for observation. Quick responses countered German attacks, strengthening captured ground. This offensive significantly weakened the

German 4th Army's hold on Broodseinde Ridge. Favourable weather aided the Allied advance, while German counterattacks failed.

At the end of December 1917, LCpl Lewis Hooper was detached to Headquarters in London where he was selected for a commission in the Indian Army. He did not proceed with the appointment for reasons which are not recorded. Instead, he returned to the 49th and was sent off to the 4th Divisional Signals School where he trained until the end of March 1918. By this time, the 49th, as part of the 4th Division, had taken up positions around Dernancourt where they took part in the Second Battle of Dernancourt on the 5th of April 1918. The Battalion suffered heavily in this action, losing 14 officers and 207 other ranks. Twenty days later, on Anzac Day 1918, the 49th took part in the Second Battle of Villers-Bretonneux. The Battalion launched an attack that successfully recaptured the town which had been lost the previous day to the Germans. It was during this battle that LCpl Lewis Hooper was Killed in Action on the 25th of April 1918. He was 21 years old and had been a member of the AIF for two years, eight months.

LCpl Lewis Hooper was buried in the field near Villers-Bretonneux. His body was later reinterred and laid to rest at Ribemont Communal Cemetery Extension (Plot I, Row N, Grave No. 6), France. His death was notified in the 402nd casualty list issued on the 22nd of May 1918.

Lewis William Frederick Hooper's name is inscribed on the Australian War Memorial Roll of Honour (panel 148), the City of Townsville WW1 Honour Roll, and the Townsville Cenotaph. Hooper Street, Belgian Gardens Townsville, was named in his memory in October 1919.

3381 Lance Corporal Lewis William Frederick Hooper served in WW1 in Egypt and on the Western Front. He was awarded the 1914/15 Star (no:1828), the British War Medal (no:21620), and the Victory Medal (no:21544). A memorial scroll and a memorial plaque were passed to his father in 1921 and 1922, respectively.

The following notice was published in the *Townsville Daily Bulletin* on the 25th of April 1919:

HOOPER—In loving memory of our eldest son and brother L/c H.Q. Signaller Lewis W. F. Hooper, 49th Battalion. Killed in Action, at Villers Bretonneux, April 25th, 1918, aged 21 years.

'Tis just twelve months ago today
(So the Base Records tell)
Since the eldest son of our family,
Was slain by a Hunnish shell.
Just a cross made by his comrades,
On a lonely hillside grave,
Marks the resting place of our hero
One of Australia's brave.
High explosive shell and shrapnel,
Burst no longer o'er his head,
But lilies white and fragrant,
Grow o'er the honored dead.

Inserted by his sorrowing parents, Brothers and sister David W. and Elizabeth Hooper, Archie, Joseph, Henry, Francis and Daisy Hooper, also his Grandmother Ann Bargent, D. W. and E. Hooper, brothers and sister.

3094 PRIVATE WILLIAM CRISPIN PUTT (1895 – 1918)

49th Infantry Battalion

William Crispin Putt was born on the 2nd of September 1895 in Townsville. He was the first-born of five children, all boys, of William and Bertha Putt (née Lawes). William Snr worked as a seaman, and the family lived at a property on the corner of Bell & Archer St, South Townsville. When William Putt was 14 years old in 1909, his mother died. William Putt snr. married Amelia Wells in 1911.

Growing up in Townsville, William jnr. attended the South Townsville State School. With his school years finished, he worked as a labourer. He also joined the Senior Cadets, served for two years, and then joined the Citizen Military Forces for one year.

On the 2nd of October 1915, William Putt applied, at Townsville, to enlist in the Australian Imperial Force (AIF). He declared that he was single and 20 years old. He also declared his previous military service in the Cadets and the Citizen Forces. His medical examination noted that he was 5'6" (168 cm) tall, weighed 10st 7lbs (67kg), had a fair complexion, grey eyes, and fair hair. He identified as Roman Catholic and nominated his father as his next of kin.

Accepted into the AIF, William Putt was allocated service number 3094, assigned to the Infantry and posted to the 9th Battalion, 10th Reinforcements. The 9th Battalion was fighting at Gallipoli when Putt enlisted. Having previous military experience meant that William Putt was already trained to be deployed overseas.

On the 5th of October 1915, three days after he enlisted, Private William Putt and 673 other troops embarked for Egypt from Brisbane. The contingent joined Convoy No.12 and sailed aboard HMAT A69 *Warilda*. The entire convoy of five ships transported a total of 5,541 troops on the voyage. The voyage and land travel took another three and a half months, and it was not until the 21st of January 1916 that Private Putt joined the 9th Battalion at Tel-el-Kebir in Egypt.

Two weeks later Private Putt was transferred to the 49th Infantry Battalion. Formed on the 27th of February 1916, the 49th Battalion was established as part of the "doubling" initiative within the AIF. Its ranks comprised approximately half of former 9th Battalion soldiers who had previously served in Gallipoli, while the remainder, including Private Putt, were fresh reinforcements from Australia. The 49th's composition closely resembled that of the 9th Battalion, boasting a significant contingent of soldiers from Queensland.

In June 1916, the 49th Battalion sailed from Alexandria bound for France to join the British Expeditionary Force (BEF) on the Western Front. After disembarking in France on the 12th of June 1916, the 49th Battalion was quickly committed to action and occupied the Western Front trenches for the first time on the 21st of June. The battalion's initial major engagement occurred at Mouquet Farm in August 1916.

Private William Putt received a gunshot wound on 13th August to the shoulders and back. He was evacuated to hospital in Rouen and, from there, back to England to the Lord Derby War Hospital. He was in hospital until early December 1916 when he returned to France for training prior to re-joining the 49th Battalion. As luck would have it, Private Putt contracted influenza after four days and was re-admitted to hospital, in France this time. He did not re-join his Battalion until 16th January 1917.

In 1917, the 49th Battalion participated in key battles: Second Bullecourt in May, Menin Road in June, Polygon Wood in September, and Passchendaele in October. They suffered heavy casualties and were periodically withdrawn for recuperation and training.

In July 1917, Private William Putt was court-martialled in France for disobeying a lawful command to "carry his rifle on his right shoulder." He was found guilty and sentenced to 90 days Field Punishment No.2. There were two categories of field punishment. Field punishment No.1 consisted of heavy labouring duties, possibly being restrained in handcuffs or fetters, and being tied to a post or wheel. Field punishment No.2 was the same as No.1 except that the offender was not attached to a fixed object.

On the 16th of July 1917, the 49th Battalion was holding the front line near Deseule. It was during this action that Private William Putt was wounded by gunshot for a second time. He was severely wounded in the face and arms and was evacuated back to England and admitted to the 2nd Southern General Hospital in Bristol. After three weeks, he was transferred to No. 3 Australian Auxiliary Hospital, Dartford. From here he was eventually transferred to convalescence units in England. He was shipped back to France in early December 1917 and marched into the training Depot.

Private Putt rejoined the 49th Battalion on the 15th of December 1917. The battalion was near Templeux-la-Fosse, 30 miles (48 km) east of Amiens. In early 1918, after Russia's collapse, the Germans launched the Spring Offensive. In late March, the 4th Division deployed to defend positions around Dernancourt on the River Ancre.

On the 25th of April 1918, the 49th Battalion was heavily engaged with the enemy near Laneuville as part of the Second Battle of Villers-Bretonneux. It was during this battle that Private William Putt was Killed in Action. He was 22 years old and had been a member of the AIF for two years and seven months. He has no known grave, and his death was notified in the 402nd list of Australian Casualties released on the 22nd of May 1918.

Private William Crispin Putt's name is inscribed on The Australian War Memorial Roll of Honour (panel 149), The Australian National Memorial, Villers-Bretonneux, France, The 49th Battalion Roll of Honour, The City of Townsville WW1 Honour Roll, and The Townsville Cenotaph. Putt Street, Railway Estate, Townsville, is named in his memory.

3094 Private William Crispin Putt served in WW1 in Egypt and on the Western Front. He was awarded the 1914/15 Star (no:2143), the British War Medal (no:21575), and the Victory Medal (no:21499). These medals, as well as a memorial scroll and a memorial plaque, were passed to his father in 1921 and 1922, respectively.

The following tribute was printed in the *Townsville Daily Bulletin* on 25th April 1922:

PUTT. —In loving memory of our dear son and brother, William Crispin, who was Killed in Action at Villiers Bretonneux, on the 25th of April 1918.

He fell to uphold his country's honor, In the strength of manhood's might.

Inserted by his loving mother, father, and brother.

497 PRIVATE JAMES FORBES (1874 – 1918)

Australian Naval & Military Expeditionary Force

James Forbes was born in September 1874 at Alford, Scotland. He arrived in Townsville on the 10th of August 1900 and worked as a labourer. On the 26th of June 1909, he married Rose Ann Stone in Queensland, and together they had three children and lived in a property on Bell St, South Townsville.

James Forbes applied to enlist in the 3rd Australian Naval and Military Expeditionary Force (AN&MEF) on the 22nd of December 1914. He declared that he was 40 years old and a labourer without previous military service. His medical examination noted that he was 5'9¼ " (170 cm) tall, weighed 10st 11lbs (68kg), had a brown complexion, blue eyes, and brown hair. He identified as Presbyterian and nominated his wife as his next of kin and assigned 3/5ths of his pay to her.

Accepted into the AN&MEF, James Forbes was allocated service number 497 and assigned to the 3rd Battalion, AN&MEF, which was known as the Tropical Force because it had been specially enlisted for service in the tropics.

The Australian Naval and Military Expeditionary Force (AN&MEF) comprised around 2,000 volunteers, swiftly assembled in Australia in the early stages of World War I. Their mission: *to seize and dismantle German wireless stations situated in German New Guinea, located in the south-west Pacific.* These stations, used by Vizeadmiral (Vice Admiral) Maximilian von Spee's East Asia Squadron of the Imperial German Navy, posed a threat to merchant shipping in the vicinity. Despite dogged opposition, resulting in the initial casualties of the Great War, the AN&MEF prevailed over the German Forces. Terms were formally agreed upon on the 17th of September 1914, marking the cessation of military hostilities. On the 21st of September 1914, the remaining 40 German soldiers and 110 natives surrendered. The occupation of the German colony at Madang on Kaiser-Wilhelmsland (the New Guinea mainland) occurred on the 24th of September 1914. With the successful capture of the German-held areas, the AN&MEF assumed responsibility for occupation duties for the remainder of the war. The size of the garrison at this time was set at a total of 600 men.

Private James Forbes left Sydney on the 21st of January 1915 aboard H.M.A.T.S.S. *Eastern*, bound for New Guinea to join the 3rd Battalion. He was posted to Rabaul and was employed as the Garrison Cook for a short period. He returned to Australia twice on furlough. The first time from the 6th of December 1915 until the 10th of February 1916, and the second time from the 1st of January 1917 until the 31st of March 1917.

In November 1917, Private James Forbes was promoted to Acting Corporal, and he was employed in the Regimental Institute. He stayed in this role until he became ill and died on the 30th of April 1918 from the effects of septicaemia. Acting Corporal James Forbes was 43 years, eight months old and had served in the AN&MEF for three years, eight months at the time of his death. He was laid to rest at Bita Paka War Cemetery, Rabaul, Papua New Guinea.

James Forbes' name is inscribed on The Australian War Memorial Roll of Honour (panel 188), The City of Townsville WW1 Honour Roll, and The Townsville Cenotaph. Forbes Street, Cluden, Townsville, is named in his memory.

497 Private James Forbes served in WW1 in New Guinea. He was awarded the British War Medal (no:5239). A memorial scroll and a memorial plaque were passed to his widow in 1921 and 1925, respectively.

The following notice was published in the *Townsville Daily Bulletin* (Qld) on the 9th of May 1918:

FORBES. - On April 30th, at Rabaul, PRIVATE James Forbes, beloved husband of Rose Ann Forbes, Tully Street, South Townsville. Deeply mourned.

4th ENGINEER WILLIAM MINEHANE (1894 – 1918)

Merchant Marine

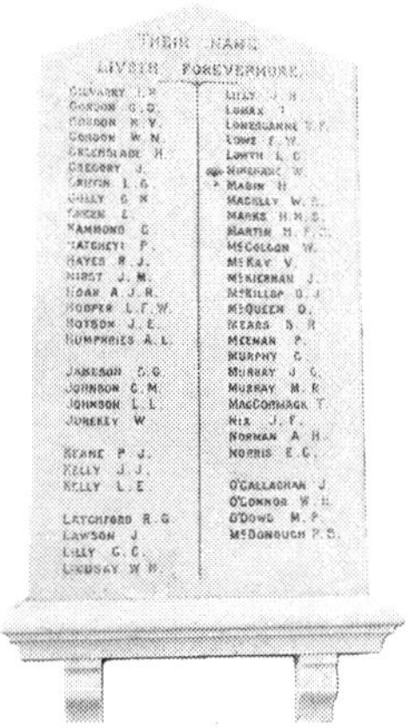

William Minehane (aka William Minchan) was born on the 25th of March 1894, in Townsville, Queensland. He was the fourth-born child and second oldest son of Jack and Bridget Minehane (née Cully). He had three brothers and six surviving sisters. Jack Minehane was a storeman, and the family lived in Bell Street, South Townsville, in 1903.

After his school years were completed, young William Minehane joined the merchant marine, and by 1909, he was a Master Mariner living in a property on Second St, South Townsville. His older brother, James, was also a seaman at that time.

During WW1, William Minehane was employed by Australian Steamships Ltd. On the 1st of May 1918, he was working on the SS *Era*, belonging to Howard Smith & Co., when the ship was torpedoed and sunk by the German submarine SM U-32 in the Mediterranean Sea off Cape Ténès, Algeria (36°45°N 1°56°E). Twelve crew were lost, one of which was William Minehane. He was 24 years old, and his body was not recovered.

William Minehane's name is recorded in The Australian War Memorial Commemorative Roll. His name is inscribed on The Tower Hill Memorial, London, England, and The Townsville Cenotaph. Bridget Minehane received £200 compensation for the death of her son. This amount was approximately equal to one year's salary.

The following article was published in *The Australasian* (Melbourne) on the 15th of June 1918:

THE ERA TORPEDOED.

The *Commonwealth* Shipping Board has received a cable message from the British Admiralty, stating that the cargo steamer Era, 2,379 tons, has been torpedoed. The Era was built in 1888 at West Hartlepool. Since leaving the slip, she had been the property of Howard Smith Ltd. and was employed for many years on the coal and general cargo service between Melbourne and Sydney and Newcastle. The Era had a carrying capacity of about 3,000 tons. She was withdrawn from the interstate trade on about January 18, after landing coal cargoes at Adelaide and Wallaroo.

It is believed that the Era was lost in British waters, and that the crew consisted of men who had signed on at Sydney. At least one stoker was drowned, and seven or eight other members of the crew are missing.

6026 PRIVATE WYLIE NORMAN GORDON (1896 – 1918)

9th Infantry Battalion

Wylie Norman Gordon was born in Queensland on the 1st of January 1896 to James and Catherine Gordon (née Armstrong). Wylie was the oldest son and second born of eight children, three of whom died as infants. The family lived in a property on Stagpole St, West End in Townsville, and Wylie Gordon attended the Belgian Gardens State School (Townsville North School).

After his school years were completed, Wylie Gordon worked as a stockman. He also served for four years in the Cadets and was still serving on the 5th of February 1916 when he applied to enlist in the Australian Imperial Force (AIF). He declared his Cadet service and that he was single, 20 years old, and working as a stockman. His enlistment medical examination noted that he was 5'3½" (161 cm) tall, weighed 8st 6lbs (52 kg), had a fair complexion, grey eyes, and light brown hair. He identified as Church of England and nominated his father as his next of kin.

Accepted into the AIF, Wylie Gordon was allocated service number 6026, assigned to the Infantry and posted to the 9th Battalion, 19th Reinforcements. The 9th Battalion was deployed in Egypt when he enlisted. The battalion had played a significant role at Gallipoli and was resting and rebuilding in preparation for the planned move to the Western Front in France when Wylie Gordon enlisted.

From February to August 1916, Private Gordon underwent recruit training in Australia at the 11th Depot Battalion at Enoggera in Brisbane. Reinforcements were given basic training in Australia, and advanced skills were taught in training units in England. On the 8th of August 1916, Private Gordon and his fellow reinforcements embarked from Brisbane, bound for England. They joined Convoy No. 23 and sailed aboard HMAT A50 *Itonus*, which carried 780 troops. The entire convoy of 12 ships carried a total of 11,343 troops on the voyage.

The reinforcements disembarked at Plymouth, England on the 18th of October 1916. From there, they were transported to the 3rd Training Battalion at Fovant near Salisbury. Private Wylie Gordon fell ill during his time in England and was admitted to the hospital twice during his training. These stays in the hospital delayed his movement to France. It was not until the 11th of April 1917 that he crossed to the 1st Australian Divisional Base Depot (1st ADBD) at Étaples, France. Nine days later, on the 20th of April 1917, he joined the 9th Battalion in the field in France.

The 9th Battalion was deployed in the vicinity of Morchies, about 17 miles (27 km) southeast of Arras. The Battalion followed the German retreat to the Hindenburg Line, participating in fighting at Lagnicourt in April 1917 and at Bullecourt in May 1917. Private Wylie Gordon was a rifleman in the 9th when these battles occurred. Returning to Belgium in September, the battalion fought in major battles including Menin Road, Broodseinde Ridge, Poelcappelle, and Passchendaele during the Third Ypres campaign.

Private Wylie Gordon had seen plenty of action over the previous months, and he was granted leave in England for a couple of weeks in January-February 1918. On his return, he was deployed back to the front line in France. On the 16th of March 1918, Private Wylie Gordon was posted as sick and was admitted to the hospital suffering from trench foot. Although he did not know it, Private Gordon was destined to never re-join the 9th Battalion again.

Following a month of convalescence in Boulogne, he returned to the 1st ADBD in Havre to prepare for return to his unit. Four days after arrival at the depot, Private Gordon again reported sick and was admitted to the hospital suffering from pleurisy. At 1:45 p.m. on the next day, the 3rd of May 1918, Private Wylie Gordon died of lobar pneumonia. His death was attributed to *"exposure in connection with military service."* Private Wylie Gordon was 22 years old at the time of his death and had served in the AIF for two years and three months. He was laid to rest at Ste Marie Cemetery (Plot III, Row G, Grave No. 5), Le Havre, France. His death was notified in the 402nd list of Australian Casualties released on the 22nd of May 1918.

Private Wylie Gordon served on the Western Front in WWI. His name is inscribed on The Australian War Memorial Roll of Honour (panel 56), The City of Townsville WWI Honour Roll, and The Townsville Cenotaph. Gorden Street, Garbutt Townsville, is named in his and G.D. and V.G. Gordon's memory (n.b. The spelling of the street name, i.e., Gorden instead of Gordon, has been attributed to an error).

6026 Private Wylie Norman Gordon was awarded the British War Medal (no: 44537) and the Victory Medal (no: 44034). A memorial scroll and a memorial plaque were passed to his father in 1922.

The following notice was published in the *Townsville Daily Bulletin* (Qld) on 3rd May 1920:

GORDON. —In sad but loving memory of our dear son and brother, Private Wylie Norman Gordon, who died of pneumonia, in France, May 3, 1918.

> This day recalls sad memories,
> Of our dear son and brother at rest.
> And those who think of him to-day,
> Are those who loved him best.

Inserted by his sorrowing parents, sisters, and brothers, Mr. and Mrs. J. Gordon and family.

Similar Notices were published in 1921, 1931, 1933, and 1934.

6243 PRIVATE JOHN DOLBY (1879 – 1918)

9th Infantry Battalion

John Dolby was born at Sheffield, Yorkshire, England, in March 1879, to Thomas & Mary Dolby (née Asher). He was the youngest of four children. The Dolby family arrived in Australia in 1887 when young John was eight years old. By 1903, the family was living at a property at Stewarts Creek in Townsville, and Thomas Dolby was working as a labourer. Mary Dolby died on the 18th of August 1915.

After his school years were completed, John Dolby worked as a labourer at His Majesty's Penal Establishment, Stewart's Creek until the 10th of April 1916 when he applied at Townsville to enlist in the Australian Imperial Force (AIF). He declared he was single, 37 years old, working as a labourer, and had no previous military experience. John Dolby's enlistment medical noted that he was 5'11½" (181 cm) tall, weighed 10st 13lbs (69kg), had a dark complexion, brown eyes, and black hair. He identified as Church of England and nominated his father as his next of kin.

Accepted into the AIF, John Dolby was allocated service number 6243, assigned to the Infantry, and posted to the 9th Battalion, 20th Reinforcements. From April to September 1916, John Dolby underwent recruit training in Australia at the 11th Depot Battalion at Enoggera in Brisbane. Reinforcements were given basic training in Australia, and advanced skills were taught in training units in England.

On the 7th of September 1916, Private John Dolby embarked with his fellow Reinforcements from Brisbane, bound for England. They joined Convoy No. 24 and sailed aboard HMAT A46 *Clan McGillivray* which boarded 924 troops in Brisbane and another 159 in Fremantle. The entire convoy of 13 ships carried a total of 12,785 troops on the voyage. Eight weeks later, the Reinforcements disembarked at Plymouth, England, on the 2nd of November 1916. From there, they were moved to the 3rd Training Battalion located at Fovant near Salisbury.

Following six weeks of further training, Private John Dolby and the Draft of Reinforcements crossed to France via Folkestone on the Golden Eagle on the 27th of December 1916. Once landed in France, they marched into the 1st Australian Divisional Base Depot (1st ADBD) at Étaples, on the 17th of December 1916. About two weeks later, Private John Dolby joined 9th Battalion in the field on the 1st of January 1917, just in time to spend the bitterly cold winter of 1916-1917 in the trenches.

During early 1917, the 9th Battalion remained deployed on the Somme, following up on the German withdrawal to the Hindenburg line. Private John Dolby saw considerable action as the battalion engaged in combat at Lagnicourt in April and Bullecourt in May. In September, the battalion relocated to Belgium, where it played significant roles in pivotal battles such as Menin Road, Broodseinde Ridge, Poelcappelle, and Passchendaele, all part of the Third Ypres campaign.

Private John Dolby was Wounded in Action by shrapnel in Belgium on the 20th of September 1917. After evacuation to the 5th General Hospital at Rouen for treatment, and a period of convalescence and training, he re-joined the 9th Battalion on the 6th of November 1917. The 9th Battalion remained in Flanders in the Messines sector of the front into early 1918.

In mid-April 1918, the 9th Battalion, with other units of the 1st Australian Division, was rushed back to northern France to help contain an enemy breakthrough aimed at the vital town of Hazebrouck. Private John Dolby was Wounded in Action for a second time on the 24th of April 1918. This time, with gunshot wounds to the head. He was evacuated and finally ended up in the Guilford War Hospital England where he succumbed to his wounds and Died of Wounds received in action on the 10th of May 1918. He was 39 years old and had served in the AIF for two years, one month at the time of his death. Private John Dolby was laid to rest at Guildford (Stoke) Cemetery (Row G, Grave No: 373), Surrey, England. His death was notified in the 405th list of Australian Casualties released on the 30th of May 1918.

6243 Private John Dolby served on the Western Front in WW1. His name is inscribed on The Australian War Memorial Roll of Honour (panel 55), The

City of Townsville WW1 Honour Roll, and The Townsville Cenotaph. Dolby Street, Cluden, Townsville is named in his memory.

6243 Private John Dolby was awarded the British War Medal (no:46699) and the Victory Medal (no:46198). A memorial scroll and a memorial plaque were forwarded to his father in 1921 and 1922 respectively, even though his father had died at Townsville on the 7th of March 1918. John's brother William took possession of John's awards.

The following notice was published in the *Townsville Daily Bulletin* (Qld) on 8th June 1918:

DOLBY. - Officially reported, died of wounds, at Guildford War Hospital, May 10th, 1918, 6243 Pte John Dolby 9th Battalion, third son of the late Thomas and Mary Dolby, Stewart's Creek.

> We have pictured you safe returning,
> And longed to clasp your hand,
> But God has postponed our meeting,
> It will be in a better land.
> Inserted by his loving brothers and sisters-in-laws.

5244 PRIVATE CLARENCE EDGAR BURT (1891 – 1918)

31st Infantry Battalion

Clarence Edgar Burt was the youngest of five children, born to John and Jane Burt (née Pearce), on the 26th of October 1891, in Collingwood, Melbourne. The Burt family moved to Townsville by the 1900s and lived at Denham St Stanton Hill. John Burt worked as a dredge master. Clarence Burt grew up in Townsville and attended the Central State School. Once his school years were completed, he followed his father's example and gained employment as a seaman. He also joined the Royal Australian Naval Reserve.

On the 15th of November 1917, Clarence Burt applied at Townsville to enlist in the Australian Imperial Force (AIF). He declared that he was single, 26 years old, and a discharged seaman with four years of previous service in the Royal Australian Naval Reserve. Clarence Burt's medical assessment noted that he was 5'7" (170 cm) tall, weighed 176 lbs (75 kg), had a fresh complexion, blue eyes, and brown hair. He identified as Methodist and listed his father as his next of kin, giving his address as the Imperial Hotel, Townsville.

Accepted into the AIF, Clarence Burt was allocated service number 5244, assigned to the Infantry, and posted to the 31st Battalion, 15th Reinforcements. When Clarence Burt enlisted, the 31st was deployed in France after being involved in major fighting at Polygon Wood in September 1917. For about three months after he enlisted, Clarence Burt underwent the required training to turn himself from a sailor of four years' service into an infantryman. That done, he and his fellow reinforcements embarked from Sydney, bound for England via Egypt, on the 2nd of March 1918.

Private Clarence Burt's contingent joined Convoy No.37 and sailed aboard HMT S.S. *Ormonde*, which boarded 528 troops in Brisbane, 363 in Melbourne, and 241 in Fremantle. The entire convoy of three ships transported a total of 4,021 on the voyage. They arrived at Suez in April 1918. It was only to be a short stay in Suez for, on the 4th of May, Private Burt embarked on the same ship, this time bound for England. It was during this voyage that Private Burt became seriously ill, and he was transferred to the British 2nd General Hospital in Le Havre, France on the14th of May 1918.

He survived his illness for four days after admission but died on the 18th of May 1918. His death was recorded as *"Died of disease (Broncho, Pneumonia, and Pericarditis). Disease was aggravated by privation and exposure whilst on Military Service."*

Private Clarence Burt was 26½ years old and had served with the AIF for six months on the date of his death. He was buried at the Cimetière Sainte-March in Le Havre; France and his death was notified in the 408th list of Australian Casualties released on 21st June 1918.

Clarence Burt's name is inscribed on The Australian War Memorial Roll of Honour (panel 118), The City of Townsville WW1 Honour Roll, and The Townsville Cenotaph. Burt Street, Aitkenvale, Townsville is named in his memory.

5244 Private Clarence Edgar Burt was awarded the British War Medal (no:59605). A memorial scroll and a memorial plaque were passed to his father in 1922. Jane Burt was granted a war pension of one pound (£1/-/-) per fortnight for the loss of her son. This amount was about one quarter of the wage paid to AIF soldiers in 1918.

The following tribute was published in the *Brisbane Courier* on 19th May 1919:

ROLL OF HONOUR

BURT. —In fondest and loving memory of our dear brother, Pte Clarence Edgar Burt, 16/31st Battalion, died of illness, Havre, France, May 18, 1918. His King and Country called him,

> The call was not in vain:
> And on a Queensland roll of honour
> You will find his name.

Inserted by his loving brother and sister, Harold and Gladyce Burt.

6349 LANCE CORPORAL SYDNEY RUSSELL MEARS (1880 – 1918)

26th Infantry Battalion

Sydney (Syd) Russell Mears was born on the 12th of June 1880 in Sydney, New South Wales, the oldest child and only son of Henry and Mary (aka Minnie) Mears (née Russell). He had three younger sisters. Syd Mears grew up in NSW before his work took him to Townsville. In 1913, he lived at the Queens Hotel in Townsville and worked as the manager of the Australian Estates & Mortgage Company Ltd in Denham Street.

On the 2nd of September 1916, Syd Mears applied at Townsville to enlist in the Australian Imperial Force (AIF). He was 36 years, two months old and declared that he was single, employed as a mercantile manager, and had no previous military experience. His enlistment medical assessment noted that he was 5'6½" (169 cm) tall, weighed 9st 6lbs (60 kg), had a dark complexion, grey eyes, and brown hair. He identified as Church of England and nominated his mother as his next of kin, initially giving her address as *"Acadia"* Virgil St, Bronte, NSW; however, this was later changed to Ocean St, Woollahra, Sydney.

Accepted into the AIF, Syd Mears was allocated service number 6349, assigned to the Infantry, and posted to the 26th Battalion, 18th Reinforcements. The 26th was in the Somme Valley when Syd Mears joined the AIF. After three months of basic training, Private Mears and his fellow reinforcements departed Sydney for England on the 23rd of December 1916 as part of Convoy No. 27. They sailed aboard HMAT A64 *Demosthenes*, which carried 1,240 troops. The 13-ship convoy carried a total of 15,278 troops and 311 passengers.

Fourteen weeks later, the *Demosthenes* docked at Plymouth, England. The reinforcements disembarked and were transferred to the 7th Training Battalion located at Rollestone (near Amesbury on Salisbury Plain in Wiltshire). Three months later, Private Syd Mears crossed to France and, following a further two weeks of training at Le Havre, was taken on strength of the 26th Battalion on the 9th of July 1917. The battalion was undertaking a period of training near Bapaume, France, when he joined the unit.

Later that year, the AIF shifted operations to Belgium. Private Mears fought in the Battle of Menin Road on the 20th of September 1917 and helped capture Broodseinde Ridge on the 4th of October 1917. He was promoted to Lance Corporal in late November 1917 and soon after attended the Divisional Bomb School. There, he learned grenade mechanics and tactics, becoming proficient with all types, including enemy grenades. After qualifying, he re-joined the battalion in December 1917, as they took positions on the front line near the Douve River.

Private Syd Mears continued to fight with the Battalion until March 1918, when he was granted three weeks' leave in England. Like most AIF battalions, the 26th fought to turn back the German spring offensive in April 1918. In May 1918, they were in France near Buire, which is about eight kilometres from Albert, Somme. On the 25th of May 1918, Private Syd Mears was Killed in Action.

In a statement made in August 1918, 5422 Signaller R.C. Tilden described the moment Syd was killed: *"L/Cpl Mears was killed while having his tea in a sunken road at the left of Buire. He was sitting on top of the dugout while I was inside it when a 5.9 exploded, killing Mears and wounding two others. I do not know where he was buried."*

Private Syd Mears was one month short of his 38th birthday when he was killed. He had served with the AIF for 21 months. His body was recovered, and he was laid to rest in the Ribemont Communal Cemetery (Row A, Grave No. 5), France. His death was notified in the 411th list of Australian casualties released on the 24th of June 1918.

Sydney Russell Mears' name is inscribed on The Australian War Memorial Roll of Honour (panel 108), The City of Townsville WW1 Honour Roll, and The Townsville Cenotaph. Mears Street, Mysterton, Townsville, is named in his memory.

6349 Lance Corporal Sydney Russell Mears was awarded the British War Medal (no:55191) and the Victory Medal (no:54352). A memorial scroll and a memorial plaque were passed to his mother in 1923.

4000 LCPL WILLIAM OVEREND WILSON D.C.M. D.M. (1893 – 1918)

31st Infantry Battalion

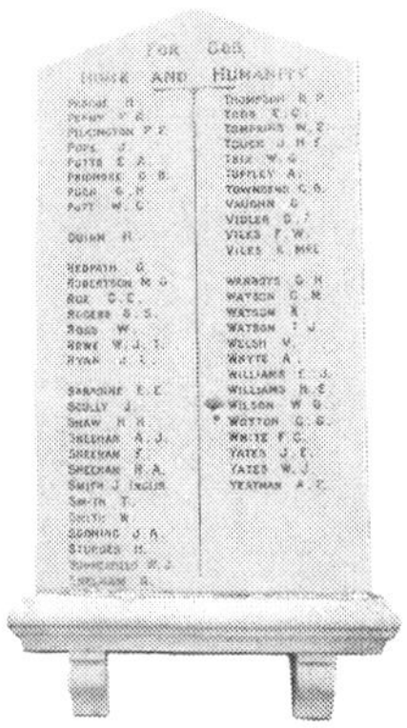

William Overend Wilson was born on the 23rd of August 1893 in Townsville, Queensland, to John and Marie Wilson (née Browne). He was the second child and first son among seven siblings, consisting of three brothers and three sisters. John Wilson was a station manager, and in 1903, the family lived at Canobie Station, a cattle station approximately 143 kilometres north of Cloncurry. Due to the isolation, William and his siblings were home-schooled.

In 1912, Canobie West Station was sold along with five other properties to the New Zealand and Australian Land Co. Ltd. By 1913, John Wilson had become a manager residing at a property on Echlin Street in Townsville. In 1915, William was employed as a stockman at Cork Station, located beside the Diamantina River, 160 km southwest of Winton, Queensland. He worked there until 1916 when he decided to enlist in the Light Horse, following his younger brothers, Eric and Arthur, who had enlisted in December 1914.

Traveling to Brisbane, he applied to join the Australian Imperial Force (AIF) on the 28th of April 1916. William declared that he was 22 years and eight months old, single, employed as a stockman, and with no previous military service. His medical assessment noted that he was 5'10½" (179 cm) tall, weighed 123 lbs (56 kg), had a dark complexion, grey eyes, and black hair. Identifying as Roman Catholic, he listed his mother as next of kin, providing her address as Flinders Street, West End, Townsville.

Accepted into the AIF, William Wilson was allocated service number 4000, assigned to the Infantry, and posted to the 31st Battalion, 10th Reinforcements. At the time of his enlistment, the 31st Battalion was in Egypt, preparing to proceed to France as part of the 8th Brigade.

After months of training in Australia, Private William Wilson embarked with fellow reinforcements on the 21st of October 1916 from Brisbane to England. They sailed as part of Convoy 25 aboard HMAT A36 *Boonah*, which carried 1,235 military personnel. The entire 12-ship convoy transported a total of 12,513 personnel and 1,800 horses on the voyage.

On the 10th of January 1917, the *Boonah* docked at Plymouth, England, and the reinforcements disembarked into the freezing winter, approximately 12 weeks after leaving Australia. From Plymouth, they were moved to the 8th Training Battalion at Codford, near Salisbury Plain, for further training.

Private William Wilson exhibited leadership qualities early on, resulting in his promotion to E.D.P. Cpl (Extra Duty Pay Corporal) and then to Lance Sergeant in the first week of February. In late March 1917, he was appointed Acting Sergeant. Six days later, on the 29th of March 1917, Acting Sergeant Wilson crossed to Étaples in France but reverted to Private upon arrival.

After a week of training, on the 2nd of April 1917, Private William Wilson joined the 31st Battalion in the field. The Battalion was stationed in the front line near Beaumetz on the Somme, supporting the 55th and 56th Infantry Battalions. In early 1917, as the German Army withdrew to the Hindenburg Line, the British front advanced, and the 31st Battalion participated in the follow-up operations.

In mid-June 1917, Private Wilson contracted Influenza and was evacuated to the 56th Casualty Clearing Station, where he was diagnosed with Trench Fever. He rejoined the 31st on the 22nd of July 1917. By September 1917, the Battalion was deployed in the Ypres sector in Belgium, playing a significant role in the Battle of Polygon Wood on the 26th of September 1917.

It was during this battle that Private William Overend Wilson was recommended for an award for his actions. He was subsequently awarded the Distinguished Conduct Medal. The Citation reads:

"At Polygon Wood, east of Ypres, during the period 25th-28th Sept 1917, this soldier displayed the greatest coolness, initiative, and deliberation. When an enemy MG in the "Pill Boxes" at Jerk House commenced a murderous enfilade fire on the first and second waves, he, notwithstanding that his officer and NCOs had been killed, saw the situation at a glance, and calling on his section to follow, gallantly led the way. By his vigorous example and action, the section surrounded the Pill Box, capturing and killing all the occupants in addition to seizing four machine guns, thereby saving the right flank from entire annihilation. At all times and under all circumstances, his cheerfulness and devotion to duty inspired his comrades. He acted as a guide to relieving troops and was always the first to volunteer for dangerous tasks."

Private William Wilson was awarded the Decoration Militaire (Belgium) for the same action, which was presented to him personally by the King of Belgium in March 1918. On the 1st of October 1917, he was promoted to Lance Corporal. Nine days later, the 31st Battalion was deployed in the Support Line in the Chateau Segard area near Dickebusch, Belgium.

At 1715 hours (5:15 p.m.), the Battalion faced enemy artillery bombardment, during which Lance Corporal Wilson was wounded in the left shoulder by shrapnel. He was evacuated to the 3rd Australian Field Ambulance and then to England, where he was admitted to Queen Mary's Hospital on the 20th of October 1917.

It took nearly 20 weeks for Lance Corporal Wilson to recover from his wound and rejoin the 31st Battalion. He returned on the 23rd of February 1918, finding the Battalion in a support area at the Rossignol Camp at Kemmel, Flanders, Belgium. The weather was freezing, and the Battalion was tasked with providing work parties.

On the 15th of March 1918, the 31st Battalion marched back to the frontline near Lumm Farm in the Wytschaete Sector, holding the left centre sector, the Wambeek Spur. In April and May 1918, the Battalion patrolled and fought in the Corbie Sector.

On the 25th of May 1918, at 2:15 a.m., a raiding party of 42 soldiers and two officers from the 31st Battalion attempted a raid on enemy trenches. Machine gun fire and grenades were directed at the raiding party, resulting in three killed and seven wounded. Among those Killed in Action was Lance Corporal William Overend Wilson D.C.M., D.M., who was 24 years and nine months old at the time of his death. He had served with the AIF for two years and one month. His death was notified in the 410th list of Australian Casualties released on the 19th of June 1918.

Lance Corporal Wilson was initially laid to rest in the field at Hamelet in Northern France, about eight kilometres from Amiens. His body was later reinterred at Adelaide Cemetery, Villers-Bretonneux (Plot III, Row H, Grave No. 15), France. His father requested that his headstone be engraved with the Latin family motto: "*PRO REGE, LEGE, ET GREGE,*" which translates to For The King, The Law, And The People.

William Wilson's name is inscribed on The Australian War Memorial Roll of Honour (Panel 119), Nanango War Memorial, The City of Townsville WW1 Honour Roll, and The Townsville Cenotaph. Wilson Street, West End, Townsville, is named in his memory.

4000 Lance Corporal William Overend Wilson D.C.M., D.M. was awarded the British War Medal (no: 50716) and the Victory Medal (no. 50079). During his service, he was awarded the Distinguished Conduct Medal and was personally presented the Decoration Militaire (Belgium) by the King of Belgium. A memorial scroll and a memorial plaque were passed to his father in 1921 and 1922, respectively. William Wilson's two younger brothers both survived the war and returned to Australia.

The following article was published in the *Townsville Daily Bulletin* on 15th June 1918:

THE GREAT WAR.

Mrs. J. Radcliffe Wilson, of Hermit Park, received news from Base Records on Saturday that her eldest son, Corporal William Overend Wilson, D.C.M., had been Killed in Action in France. The deceased, who joined up in Brisbane, had been two years on active service. Two other sons of the family have been three years on service, one having been wounded twice.

The following tribute was published in the *Townsville Daily Bulletin* on 24th May 1919:

Family Notices

WILSON — In fond remembrance of Lance-Corporal William Overend Wilson. D.C.M., and Medalle Militaire, of the 31st Battalion, A.I.F., killed in a raid on the German trenches, in France, on the 25th of May, 1918, eldest beloved son of John Radcliffe and Minnie Wilson of Townsville, in his 25th year.

> "Pro Rege, Lege, et Grege."
> for the king, the law, and the people.

3887 PRIVATE CHARLES BANKS PRIDMORE (1889 – 1918)

49th Infantry Battalion

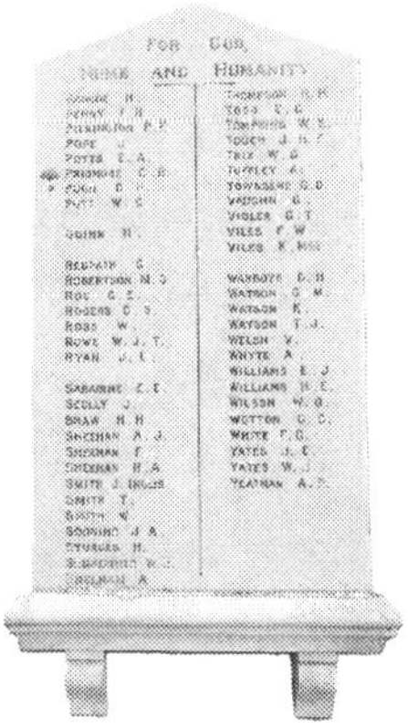

Charles Banks Pridmore was the second son born to Thomas and Caroline Pridmore (née Larsen) on the 17th of January 1889 in Georgetown, Queensland. Eventually, there were eight children in the family, and Charles Pridmore had two brothers and five sisters. Growing up in Townsville, he attended the Townsville West End and Belgian Gardens State Schools, and after his school years finished, he worked as a labourer and plumber.

Charles Pridmore's older brother, George, joined the AIF in September 1915. Three weeks later, Charles Pridmore followed his brother's example, and on the 4th of October 1915, he applied to enlist in the Australian Imperial Force (AIF) in Townsville. He declared that he was single, 26 years, eight months old, employed as a labourer, and without any previous military experience. His enlistment medical examination noted that he was 5'6½" (169 cm) tall, weighed 9st 1lbs (58 kg), had a dark complexion, hazel eyes, and dark brown hair. He identified as Church of England and nominated his mother as his next of kin, giving her address as c/- Mrs. Larsen, Jane St, West End, Townsville.

Accepted into the AIF, Charles Pridmore was allocated service number 3887, assigned to the Infantry, and posted to the 9th Battalion, 12th Reinforcements. The 9th Battalion was fighting at Gallipoli when he enlisted.

After three months of training in Australia, Private Pridmore embarked from Brisbane on the 30th of December 1915 with his fellow reinforcements. They sailed aboard HMAT A50 *Itonus* as part of Convoy No. 15, which carried 8,991 troops across 11 ships.

The journey, including land travel, took 10 weeks. On the 17th of March 1916, Private Pridmore joined the 9th Battalion at Zeitoun, Egypt.

Three weeks after arrival in Egypt, Private Pridmore was transferred to the 49th Infantry Battalion, which was raised in Egypt on 27th February 1916 as part of the "doubling" of the AIF. Approximately half of the 49th's soldiers had previously served in Gallipoli with the 9th Battalion, and the other half were fresh reinforcements from Australia. The unit's composition closely mirrored that of the 9th Battalion, with a significant proportion of Queensland men.

In June 1916, the 49th Battalion sailed from Alexandria to join the British Expeditionary Force on the Western Front. They arrived in France on the 12th of June and entered the trenches for the first time on the 21st of June.

Their first major engagement was at Mouquet Farm in August 1916, where the 49th suffered heavy losses. During the battle, Private Charles Pridmore sustained a head wound and was evacuated to the 26th General Hospital in Étaples, France. He was later sent to England for further treatment followed by a period of furlough.

While on furlough, he contracted an infection and was readmitted to hospital on the 9th of October 1916. He underwent 32 days of treatment and, once recovered, trained before rejoining the 49th Battalion on the 13th of February 1917.

Ten days after rejoining the unit, he was hospitalized with pneumonia at the 1st Australian General Hospital in Rouen, France. After five days, he was invalided back to England and admitted to the War Hospital, suffering from pleurisy. Some weeks later, he was discharged from the hospital, granted furlough, and ordered to report to Perham Downs, Wiltshire, on the 4th of April 1917. The period from early April until the middle of November 1917 saw Private Charles Pridmore serving in England in training units and the 69th Battalion at Hurdcott.

On the 13th of November 1917, Private Pridmore returned to the 49th Battalion in France. Another bleak winter of trench routine followed, and four months later, Private Pridmore contracted Trench Fever and was hospitalized at Ecault Convalescent Camp near Boulogne in France. He re-joined the 49th Battalion in the field on the 4th of May 1918. The 49th was located near the River Somme, close to Blangy-Tronville in May. In June, they were near Querrieu.

On the 12th of June 1918, the 49th was engaged in clearing debris at Sailly Le Sac. There are no reports of enemy action in the War Diary for the 12th of June; however, Private Charles Pridmore was wounded by gunshot in the abdomen and back. He was evacuated to the 6th Australian Field Ambulance but Died of Wounds received in action on the 13th of June 1918.

Private Charles Pridmore was 29 years, five months old, and had been a member of the AIF for two years, eight months. He was laid to rest at Corbie Communal Cemetery Extension (Plot 2, Row E, Grave No. 59), France. His death was notified in the 415th list of Australian Casualties released on the 7th of July 1918.

Charles Banks Pridmore's name is inscribed on The Australian War Memorial Roll of Honour (panel 149), The City of Townsville WW1 Honour Roll, the Townsville West State School Memorial Gate of Honour, and The Townsville Cenotaph. Pridmore Street, West End, Townsville, is named in his memory.

3887 Private Charles Banks Pridmore served in WW1 on the Western Front. He was awarded the British War Medal (no: 28517) and the Victory Medal (no: 28337). A memorial scroll and a memorial plaque were passed to his father in 1921 and 1922, respectively.

The following notice was published in the *Townsville Daily Bulletin* (Qld) on the 13th of June 1921:

PRIDMORE. —In loving memory of our dear son and brother, No. 3887, Private Charles Banks Pridmore, 49th Battalion, A.I.F., Died of Wounds at Corbie, France, on June 13th, 1918. R.I.P. (Inserted by his loving mother, sisters, and brothers.)

PRIDMORE. —In loving memory of our dear brother, No. 3887, Private Charles Banks Pridmore, 49th Battalion, A.I.F., died of wounds, June 13th, 1918. Always Remembered. (Inserted by his brother and sister-in-law, A. E. and I. M. Pridmore.)

6318 PRIVATE JACK MURRAY McKIERAN (1896 – 1918)

15th Infantry Battalion

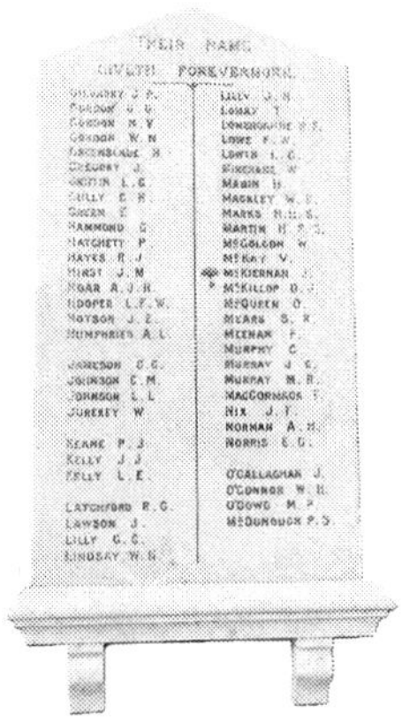

Jack Murray McKiernan was born on the 25th of October 1896 in Charters Towers, Queensland, the oldest child of James and Annie McKiernan (née McAllister). He had two brothers and one sister. Jack McKiernan was the grandson of Alderman Michael McKiernan (1832-1912), who was the Mayor of Townsville in 1897. The family lived in a property on Eyre St, Townsville in 1903 and later in Stanton Terrace.

Jack McKiernan grew up in Townsville and attended the Christian Brothers School. After his school years were completed, he followed in his father's footsteps and became a clerk with Messrs. Chapman and Company, a highly regarded department store of the time. Jack McKiernan also joined the Reserve Forces and served for five years.

On the 28th of February 1916, Jack McKiernan applied in Townsville to join the Australian Imperial Force (AIF). He declared that he was 19 years, four months old, a clerk, and that he had five years of previous military experience. His enlistment medical examination noted that he was 5'7½" (171 cm) tall, weighed 9st 2lbs (58 kg), had a dark complexion, brown/green eyes, and black hair. He identified as Roman Catholic, and nominated his father as his next of kin, giving his address as Fletcher Street, West End, Townsville.

Accepted into the AIF, Jack McKiernan was allocated service number 6318, assigned to the Infantry, and posted to the 15th Battalion, 20th Reinforcements. The 15th Battalion was originally raised in September 1914 and, when Jack McKiernan enlisted, the Battalion was deployed in Egypt after serving on the Gallipoli Peninsula.

After completing six months of basic training in Australia, Private Jack McKiernan embarked with his fellow reinforcements from Brisbane on the 7th of September 1916, bound for England. They sailed aboard HMAT A46 *Clan Macgillivray*, which boarded 1,075 troops and joined Convoy No. 24. The entire convoy of 13 ships transported a combined total of 12,785 passengers on the voyage.

The 20th Reinforcements disembarked at Plymouth, England, on the 2nd of November 1916. They spent the next six weeks training at Rollestone Camp on Salisbury Plain before they crossed to Étaples in France. Another 10 days of training were completed at the 4th Australian Divisional Base Depot. That finished, Private Jack McKiernan and the rest of the 20th Reinforcements were taken on strength of the 15th Battalion on the 6th of January 1917.

The 15th Battalion was stationed in the Ypres salient, facing harsh conditions and sickness in winter. They rotated through defensive and labouring duties before moving to Gueudecourt. On the 1st of February 1917, the 15th attacked Stormy Trench but was driven back with heavy losses due to uncut wire on the right of the attack. On the 2nd of February, German counterattacks began. An SOS signal from the 15th Battalion went unnoticed for 50 minutes. The battalion was repulsed with 144 casualties.

Private McKiernan survived this battle but received an accidental blast wound to his chest and back on the 21st of February 1917. The wounds were severe, and Private McKiernan was evacuated back to the 18th General Hospital located at Camiers near Étaples and from there back to Edmonton Military Hospital in England. A Commission of Enquiry was held into the accident in April and found that Private Jack McKiernan *"was in no way to blame."*

After treatment and recovery, Private Jack McKiernan returned to the 15th Battalion, located in the vicinity of Villers-Bretonneux, on the 19th of May 1918. He had been away from the battalion for about 15 months after the accident occurred. In June 1918, the Allies prepared for their own offensive, and American troops were assigned to the 15th Battalion for training. Lieutenant General John Monash, as the new commander of the Australian Corps, executed a combined arms attack to reduce the salient in front of Amiens around Hamel.

On the 4th of July 1918, Australian and American forces attacked Hamel. The 15th Battalion was tasked with attacking the German strong point "Pear Trench" in the battle. The battalion, along with an attached company of Americans, suffered heavy losses with nine officers and 231 other ranks killed out of 636 men committed. One of those casualties was Private Jack McKiernan, who was Killed in Action on the 4th of July 1918. He was 21 years, eight months old and had been a member of the AIF for two years, four months.

Private Jack McKiernan was buried in the field but his body was later reinterred and laid to rest at Villers-Bretonneux Military Cemetery (Plot III, Row B, Grave No. 4), Fouilloy, France. His death was notified in the 419th list of Australian casualties released on the 27th of July 1918.

Jack Murray McKiernan's name is inscribed on the Australian War Memorial Roll of Honour (Panel 76), the City of Townsville WW1 Honour Roll, and the Townsville Cenotaph. McKiernan Street, Belgian Gardens in Townsville, was named in his memory.

6318 Private Jack Murray McKiernan was awarded the British War Medal (no: 46661) and the Victory Medal (no: 46122). A memorial scroll and a memorial plaque were passed to his father in 1921 and 1922, respectively.

The following notice was published in *The Catholic Press* (Sydney, NSW) on 16th March 1916:

Another promising young man has left Townsville for Enoggera, in the person of Mr. Jack McKiernan, son of Mr. and Mrs. J. McKiernan. and grandson of the late Alderman McKiernan. Mr. McKiernan was an employee of Messrs. Chapman and Company and was presented with a luminous wristlet watch by the firm; he was also the recipient of a Gillette safety razor, a gold tip cigarette holder in a silver case, and a cheque to procure a bullet-proof wallet from the office staff and other employees of the firm. Messrs. Bird, Parker, Bragg, and Benson spoke in eulogistic terms of the departing soldier's good finalities, to which he suitably responded. In the afternoon Mr. McKiernan was entertained by Mrs. Chapman and family and presented with a Waterman's fountain pen.

270 PRIVATE MAX OSCAR ROBERTSON (1895-1918)

11th Light Trench Mortar Battery

Max Oscar Cassano was born on the 3rd of October 1895 in Innisfail, Queensland, to Christian and Emily Cassano (née James). He was the second child and second son. Max's father died in 1897 when Max was two. In 1899, his mother, Emily, remarried Andrew Robertson, a bootmaker, and Max took on the Robertson surname. The family lived in a property on Perkins St, Townsville and Max attended South Townsville State School. He later trained as a tinsmith.

On the 1st of November 1915, Max Robertson applied to enlist in the Australian Imperial Force (AIF) at Townsville. He declared that he was 22 years and one month old, single, and a qualified tinsmith. He also declared that he had four months of prior military experience with the 2nd Kennedy Infantry Regiment as part of the Expeditionary Force on Thursday Island.

His enlistment medical noted that he was 5'6" (168 cm) tall, weighed 8st 11lbs (56 kg), with a fair complexion, grey eyes, and dark brown hair. He identified as Presbyterian and listed his mother as his next of kin, providing her address on Perkins Street, South Townsville. He declared his father was deceased.

Accepted into the AIF, Max Robertson was allocated service number 270, assigned to the Infantry, and posted to the 41st Battalion. Raised in February

1916, the 41st Battalion was composed of recruits from Brisbane, northern Queensland, and the northern rivers district of NSW.

Private Max Robertson spent six months training in Australia with the newly formed battalion. Once trained and equipped, the 41st Battalion embarked from Sydney for England on the 18th of May 1916. They sailed aboard HMAT A64 *Demosthenes* which carried 1,612 troops as part of the 21st Convoy. The entire convoy of 18 ships, transported 25,290 troops. They sailed via Cape Town and the Cape Verde Islands, arriving in Plymouth, England, in July 1916.

Once disembarked, Private Max Robertson, with the rest of the 41st, moved to training grounds on Salisbury Plain at Larkhill. In November 1916, the battalion crossed from Portsmouth to Le Havre, France, to join the Western Front. They moved toward the border area near Armentières and entered the front line for the first time on Christmas Eve, 1916, at the Mushroom Salient near La Chapelle-d'Armentières. Tragically, four soldiers were killed by artillery fire on their first night at the front. During the bitterly cold winter of 1916-17, the battalion alternated between serving in the front line, training, and labouring in the rear areas.

On the 29th of December 1916, Private Max Robertson was transferred to the 11th Light Trench Mortar Battery. This battery was part of the 11th Brigade of the 3rd Division and was stationed close to the front line, placing the enemy within range. However, this positioning also meant that the trench mortars were vulnerable to enemy fire, rendering them unpopular among those serving in the nearby front line.

In early January 1917, Private Robertson contracted mumps and orchitis, which necessitated his hospitalization at Saint-Omer Hospital in Northern France. He returned to the Trench Mortar Battery in early February but was readmitted to the hospital the following day, still affected by mumps. Another three weeks passed before he rejoined his unit at the end of February 1917.

Private Robertson participated in the Battle of Messines in June 1917 and the Battle of Broodseinde in October 1917. In early 1918, following the collapse of Russia, German forces shifted focus to the Western Front. Private Robertson was part of the 3rd Division, which was quickly deployed to the Somme region to counter the German offensive in March 1918. The Division successfully halted the German advance at Morlancourt and Villers-Bretonneux. After the German offensive was exhausted, the Allies prepared for their own offensive in June 1918, involving small-scale advances known as "peaceful penetrations." Holding the high ground north of the Somme River, the 3rd Division participated in these operations around Sailly-le-Sec and achieved notable success.

In July 1918, the 3rd Division was part of the Australian Corps under Lieutenant General John Monash. Monash planned an attack on Hamel, selecting the 4th Division, with reinforcements from the 3rd Division (including the 11th Brigade and Private Max Robertson's unit), along with attached American Infantry, to lead the assault on the 4th of July 1918. During this attack, Private Max Robertson was Wounded in Action. Witness 3716 Private J.W.C. Swan reported, *"He was hit by a piece of shrapnel that ricocheted off a base plate he was carrying, wounding him in the chest and leg."* Private Robertson was evacuated to the 47th Casualty Clearing Station but Died of Wounds shortly after arrival.

At the time of his death, Max Robertson was 22 years, nine months old and had served with the AIF for two years, eight months. He was laid to rest at Crouy British Cemetery (Plot III, Row B, Grave No. 17), Crouy-Sur-Somme, France. His death was notified in the 419th list of Australian casualties released on the 27th of July 1918.

Max Oscar Robertson's name is inscribed on The Australian War Memorial Roll of Honour (panel 20), The City of Townsville WW1 Honour Roll, and The Townsville Cenotaph. Robertson Street, Railway Estate, Townsville is named in his memory.

270 Private Max Oscar Robertson was awarded the British War Medal (no:39911) and the Victory Medal (no:39527). A memorial scroll and a memorial plaque were passed to his mother in 1922.

806/2265 PRIVATE FRANCIS ROBERT PENNY (1895 – 1918)

5th Light Horse Regiment/15th Infantry Battalion

Francis Robert Penny was born on the 14th of June 1895 at Hughenden, Queensland. He was the youngest of nine surviving children born to Thomas and Jane Penny (née Gibson). He had five brothers and three sisters. Francis Penny's mother died about a month after he was born. His oldest sister, Elizabeth, was 19 years old when Francis was born, and he was raised by his father and his siblings. Thomas Penny was a butcher in Hughenden and became a grazier. Francis Penny grew up in Hughenden and attended the local Catholic School. After his school years were finished, he followed his father into the butcher's trade.

Australia entered WWI on the 4th of August 1914, joining Britain in the war effort against Germany and its allies. Five months later, on the 30th of December 1914, Francis Penny applied at Townsville to enlist in the Australian Imperial Force (AIF). He was 19 years and six months old with no previous military experience. His medical examination noted that he was 5'8¾" (175 cm) tall, weighed 10st 10lbs (68kg), had a fair complexion, grey eyes, and brown hair. He identified as Roman Catholic and nominated his father as his next of kin, giving his address simply as Hughenden.

Accepted into the AIF, Francis Penny was allocated service number 806, assigned to the infantry and posted to the 5th Light Horse Regiment, 3rd Reinforcements. The 5th Light Horse Regiment had been raised in Brisbane in September 1914, entirely from men who had enlisted in Queensland, and became part of the 2nd Light Horse Brigade. The 5th Light Horse had embarked from Sydney in December 1914 and was bound for Egypt when Francis Penny enlisted. The Regiment disembarked in Egypt on the 1st of February 1915.

Initially, the Light Horse Regiments were equipped as mounted infantry, but their horses were unsuitable for the rough terrain of the Gallipoli Peninsula.

Following heavy casualties, it was decided to deploy the Light Horse Regiments as unmounted infantry to reinforce the ANZAC forces. The 2nd Light Horse Brigade was among the reinforcements sent to Gallipoli, attached to the 1st Australian Division. After training in Australia and Egypt, Trooper Francis Penny joined the 5th Light Horse on the Gallipoli Peninsula at the end of July 1915.

Trooper Penny developed an infection about a month after arrival at ANZAC and was transferred to Lemnos and admitted for treatment. The next few months were to be spent in hospital as he recovered and then developed the heart condition recorded as mitral stenosis. The heart condition was serious enough for him to be put before a medical board which recommended that he be discharged as medically unfit. Accordingly, Trooper Francis Penny was returned to Australia, arriving in January 1916. On his return, he was granted a pension of 39 pounds (£39) p.a. This was 35% of the wage paid to a soldier in 1915.

Back in Australia, Francis Penny worked as an insurance clerk with F.A. Brodie & Co. in Townsville and lived in a property on Armstrong St with other family members. Immediately prior to his discharge from the Army, he stated that he now felt well again and that he would like his discharge cancelled. His request was denied and the Army proceeded to discharge him anyway.

Francis Penny's older brother, Henry, enlisted in September 1915. He was posted to the 15th Battalion and was Killed in Action in France on the 1st of February 1917 when the unit attacked at Stormy Trench, suffering heavy losses.

One month after his brother's death, Francis Penny enlisted for a second time on the 2nd of March 1917. He declared his previous service and the fact that he was medically discharged. He also nominated his sister Jane as his next of kin, giving her address as Armstrong Street, Hermit Park, Townsville. Francis Penny had obviously recovered from his heart problem as his enlistment medical examination made no mention of it. Accepted into the AIF for a second time, he was allocated service number 2265, re-assigned to the infantry, and posted to the 15th Battalion, 15th Reinforcements, the same unit in which his brother had served and died.

After six weeks of training in Australia, Private Francis Penny embarked with fellow reinforcements on the 11th of May 1917 from Melbourne, bound for England. They sailed with Convoy 31 aboard HMAT A9 *Shropshire*, which carried 1,674 troops. The entire convoy of 13 ships transported 7,802 troops, 1,878 sailors, and 2,077 civilians.

The 15th Reinforcements disembarked at Plymouth, England, on the 19th of July 1917, and spent six months training at Durrington near Salisbury Plain. Once training was complete, Private Francis Penny and his fellow reinforcements crossed to Le Havre, France, and spent another month in training.

Private Francis Penny joined the 15th Battalion on the 12th of February 1918 while stationed at a camp near Tournai, Belgium. At that time, the battalion was preparing for battles in March and April, playing a crucial role in halting the German spring offensive.

On the 4th of July 1918, Australian and American forces, including Private Penny, participated in an assault on the French town of Hamel. The 15th Battalion aimed to capture a heavily fortified German position known as the "Pear Trench."

Despite intense enemy fire, the battalion, supported by an American company, advanced on Pear Trench. They suffered heavy casualties, with nine officers and 231 other ranks killed or wounded out of 636 men engaged. This represented one of the highest losses incurred by an Australian battalion during the battle.

One of those casualties was Private Francis Penny who was Wounded in Action by a gunshot to the back. The wounds were severe, and he was evacuated within hours to the 5th Casualty Clearing Station for treatment. Unfortunately, he was unable to be saved, and he Died of Wounds on the 5th of July 1918. He was 23 years old and had been a member of the AIF for a total time of two years and eight months at the time of his death.

Private Francis Penny was laid to rest at Crouy British Cemetery (Plot III, Row B, Grave No. 23), Crouy-Sur-Somme, France. His death was notified in the 415th list of Australian casualties released on the 27th of July 1918.

Francis Penny's name is inscribed on The Australian War Memorial Roll of Honour (panel 77), The Hughenden War Memorial, and the City of Townsville Cenotaph.

806/2265 Private Francis Robert Penny was awarded the 1914/15 Star (no:15044), the British War Medal (no:8206), and the Victory Medal (no:8167). A memorial scroll and a memorial plaque were passed to his brother, James, in 1921 and 1922, respectively.

The following notice was published in *The Northern Herald* (Cairns, Qld.) on the 25th of July 1918:

Another Loss.

Mr. Penny, of the QN bank, Cairns, is advised that his brother, PRIVATE F. R. Penny, Died of Wounds on 5th July. Deceased, who was a single man 22 years of age, enlisted twice, first from Hughenden (being invalided home) and then, from Townsville - some eighteen months ago. An older brother was previously killed in action.

3613A PRIVATE ALFRED STANLEY HODGSON DOOREY (1896 – 1918)

25th Infantry Battalion

Alfred Stanley Hodgson (Stan) Doorey was born on the 25th of February 1896 in Townsville, Queensland. He was the only son of Stanley and Elizabeth Doorey (née Hodgson), and he had one surviving younger sister. Stanley Doorey was a signalman, and the family lived in a property on Griffith Street, Townsville, and then Main Street in Railway Estate in Townsville. Alfred Doorey grew up in Townsville and attended West End State School and the Ross Island School. After his school years were finished, he worked as a motor mechanic. He also spent two years in the Citizen Forces.

On the 2nd of September 1916, Alfred Doorey and Emily Pease married when he was 20 years old, and she was 26. They lived in Main Street, Railway Estate. The marriage did not turn out to be a harmonious one. Five months after his marriage, Alfred Doorey applied on the 8th of February 1917 to enlist in the Australian Imperial Force (AIF). He was 20 years, 11 months old, and he listed his profession as a motor mechanic. Alfred Doorey's enlistment medical examination noted that he was 5'2½" (156 cm) tall, weighed 110 lbs (50 kg), had a fair complexion, grey eyes, and brown hair. He identified as Church of England, and he nominated his sister, Ida, and not his wife as the sole beneficiary of his will and his next of kin. Accepted into the AIF, he was allocated service number 6864, assigned to the Infantry, and was posted to the 47th Battalion, 10th Reinforcements. The 47th Battalion was fighting on the Western Front when Alfred Doorey enlisted.

After completing recruit training at Camp Lytton in Brisbane, Private Alfred Doorey embarked with the rest of the 10th Reinforcements from Sydney aboard HMAT A28 *Miltiades* on the 2nd of August 1917, bound for England. Their ship travelled as part of Convoy 33 and carried 1019 men. The entire convoy of eight ships transported 6029 military personnel and 969 civilians on the voyage. The Reinforcements disembarked at Glasgow, Scotland, on the 2nd of October 1917 and were transported to the 12th Training Battalion, Codford, located near Wiltshire in the southwest of England. Three months further training followed before crossing to France on the 8th of January 1918. Private Doorey marched into the 47th Battalion on the 11th of January 1918. Four days later, he was re-posted to the 25th Battalion and allocated the new service number of 3613A.

The 25th Battalion suffered heavy casualties in 1918 during the battles in the German spring offensive and the push towards the defeat of the German Army. The battalion lost 1,026 killed and 2,821 wounded, the highest number of casualties among all Australian battalions in the war. Private Alfred Stanley Hodgson Doorey was one of those casualties. He was Killed in Action at Villers-Bretonneux on the 6th of July 1918. He was 22 years, four months old, and had served in the AIF for 22 months. His death was notified in the

420th list of Australian casualties released on 2nd August 1918. Lieutenant F. McComb made the following statement to Australian Red Cross Service investigators in 1919: *"I saw Doorey C Company killed instantly alongside of me at Villers-Bretonneux about 5 a.m. soon after we hopped over by machine gun bullet through throat. He was again hit by shell, but the bullet killed him. It was near the main road. I had a look at him and saw that he was dead. He was buried where he fell by Lieut Smith. He was my batman and was an absolutely game little chap. Doorey: – Small, dark, a married man."*

Alfred Stanley Hodgson Doorey served on the Western Front in WW1. His name is inscribed on The Australian War Memorial Roll of Honour (panel 104), The Australian National Memorial, Villers-Bretonneux, France, The Ross Island State School, The City of Townsville WW1 Honour Roll, and The Townsville Cenotaph. Doorey Street, Railway Estate, Townsville is named in his memory.

3613A Private Alfred Stanley Hodgson Doorey was awarded the British War Medal (no:58376) and the Victory Medal (no:57336). Following written representations to the Army, a memorial scroll, and a memorial plaque were passed to his father (instead of his wife) in 1921 and 1922, respectively.

The following tribute was published in the *Townsville Daily Bulletin* on the 25th of July 1918:

DOOREY. — Pte A. S. H. Doorey, 25th Batt., beloved only son of and Mrs. S. H. Doorey, Railway Estate, killed in action, July 4th, 1918, somewhere in France.

4115 PRIVATE DONALD CAMERON (1896 – 1918)

26th Infantry Battalion

Donald Cameron was born on the 2nd of December 1896 in Townsville, to Alexander and Margaret Cameron (née Ross). He was the sixth of nine children and the fifth son. The Cameron family emigrated from Scotland to Australia and North Queensland in 1887. They settled in Reid River, south of Townsville, and Alexander Cameron worked as a lengthman (someone who maintained a stretch of road). Donald Cameron grew up in Reid River and attended the Reid River Provisional School and the Ross Island School. After his school years were finished, he worked as a teamster and as a draper's assistant. He also served two years with the Militia and three years with the Senior Cadets.

On the 28th of September 1915, Donald Cameron applied to enlist in the Australian Imperial Force (AIF) at Brisbane. His older brother, James Cameron, had enlisted on the previous day. Donald Cameron initially declared that he was employed as a draper's assistant, but this was amended to read teamster. He was 18 years, nine months old; however, he stated that he was four months older at 19 years, one month. He declared his previous service in the Militia and Senior Cadets. His enlistment medical examination notes that he was 5'11" (180 cm) tall, weighed 9st 4lbs (59 kg), had a fresh complexion, brown eyes, and brown hair. He identified as Presbyterian, and he nominated his mother as his next of kin, giving her address as Logan Road, South Brisbane; however, this was changed to Reid River, Townsville.

Accepted into the AIF, Donald Cameron was allocated service number 4115, assigned to the Infantry, and posted to the 26th Battalion, 10th Reinforcements, as was his brother. The 26th Battalion was raised at Enoggera, Queensland, in April 1915 from recruits enlisted in Queensland and Tasmania. Along with the 25th, 27th, and 28th Battalions, it was one of the four infantry battalions in the 7th Brigade. The 26th Battalion was fighting on the Gallipoli Peninsula when Donald Cameron and his brother enlisted. Six months of basic training followed enlistment, and after that was completed, Private Donald Cameron embarked from Brisbane with his brother and the other Reinforcements, as part of Convoy No. 18, bound for England. They sailed with 958 passengers aboard HMAT A73 *Commonwealth*. The entire convoy of 13 ships transported 10,732 troops and 3,840 horses on the voyage.

Disembarking in England in June 1916, Private Donald Cameron and the rest of the 10th Reinforcements were transported to the 7th Training Battalion at Rollestone in Wiltshire, England. Advanced training followed at Rollestone until September 1916 when the Reinforcements crossed to Étaples in France and then to their unit on the 28th of September 1916. The 26th Battalion was deployed in northern France near Steenvoorde when the Reinforcements marched in. Private Donald Cameron was diagnosed with trench feet in October 1916 and was evacuated to England. While in England, he went Absent Without Leave from the 29th of December 1916 until the 3rd of January 1917, for his New Year's "holiday." Private Donald Cameron was charged, found guilty, and placed in detention for 120 hours. He lost a total of 13 days' pay. After six months of treatment, recovery, and punishment, he returned to the battalion in April 1917.

On the 3rd of May 1917, Private Donald Cameron's brother, Private James Cameron, was Killed in Action around Bullecourt. In June 1917, Private Cameron was readmitted to Hospital suffering from Trench Fever, a moderately serious disease transmitted by body lice. He was out of action for six weeks, rejoining the Battalion in August 1917. On the 31st of October 1917, Private Cameron was Wounded in Action and gassed by a mustard gas shell at Broodseinde Ridge, Belgium. He was evacuated to hospital in France and returned to his battalion five weeks later.

During the early months of 1918, the 26th Battalion was involved in fighting in the trenches in France and Belgium, including the Battle of Hazebrouck in April 1918. On the 14th of July 1918, the 26th Battalion was advancing at Villers-Bretonneux. It was during this operation that Private Donald Cameron was Wounded in Action with bomb (grenade) wounds to his legs and right arm. He was evacuated to the 5th Casualty Clearing Station. Tragically, he succumbed to his injuries there and Died from Wounds three days later, on the 17th of July 1918. Private Donald Cameron was 21 years, seven months old and had served with the AIF for two years, 10 months at the time of his death. He was laid to rest at Crouy British Cemetery, France. His death was notified in the 422nd list of Australian Casualties released on the 12th of August 1918.

2182 Private E. Arnold described the moment Donald Cameron was mortally wounded: *"He was badly wounded in a daylight raid on Syria trench to the right of Monument Wood, in front of Villers-Bretonneux at midday. Running along too soon after the bombing, he reached the traverse and got a bomb all to himself."*

Donald Cameron's name is inscribed on The Australian War Memorial Roll of Honour (panel 107), The City of Townsville WW1 Honour Roll, The Ross Island State School Honour Roll 1914-1918, and The Townsville Cenotaph. Cameron Street, Railway Estate, Townsville is named in his and his brother James' memory.

4115 Private Donald Cameron was awarded the British War Medal (no: 34710) and the Victory Medal (no: 34463). A memorial scroll and a memorial plaque were passed to his father in 1922.

759/759B PRIVATE ALBERT LINCOLN HUMPHRIES (1890 – 1918)

9th Infantry Battalion/12th Machine Gun Company

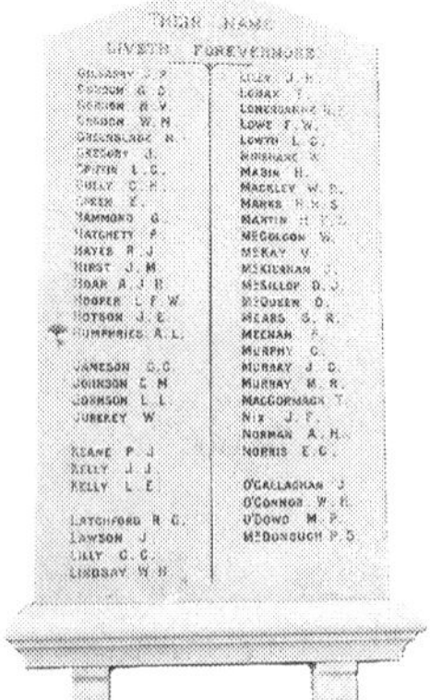

Albert Lincoln Humphries (Bert) was born in Ayr, Queensland, on the 19th of December 1890, the son of Edith Humphries and an unknown father. Edith Humphries married George Hill in 1893 when Bert Humphries was three years old. The Hill family lived in Townsville, and Bert was one of 11 children. He attended St Joseph's Catholic School and, once his school years were complete, became a motor trimmer.

On his 21st birthday, Bert Humphries married Stella Pearl Jenkins on the 19th of December 1911. They set up home in Sydney St, West End, Townsville.

Two years later, their only child was born in April 1913, a daughter whom they named Louisa Myrtle Pearle Humphries.

On the 14th of March 1917, Bert Humphries applied at Townsville to enlist in the Australian Imperial Force (AIF). He declared that he had no previous military experience, was 26 years four months old, married, and employed as a motor driver. His enlistment medical examination noted that he was 5'9" (175 cm) tall, weighed 145 lbs (66 kg), had a medium complexion, brown eyes, and dark hair. He identified as Church of England and listed his wife Stella as his next of kin, giving her address as Sydney Street, West End, Townsville.

Accepted into the AIF, Bert Humphries was allocated service number 759, assigned to the Infantry, and posted to the 9th Battalion, 19th Reinforcements. This was later amended to Machine Gun Company 12, Reinforcement 14. From enlistment until November 1917, Private Bert Humphries underwent recruit training in Australia and machine gun training in Seymour, Victoria.

Tragically, Bert's wife Stella died in July 1917, and their daughter passed to the care of Bert's mother-in-law, Mrs. Minnie Jenkins of Sydney Street, West End, Townsville. Guardianship of the child, Louisa, was granted to Mr. Henry Nottingham of Francis St, West End, Townsville.

On the 21st of November 1917, Private Bert Humphries' unit embarked for England via Egypt from Melbourne, Victoria, as part of Convoy No. 35. They sailed aboard HMAT A71 *Nestor*, which boarded 1,965 passengers. The entire convoy of six ships boarded 7,681 military personnel and 1,025 civilians on the voyage. Following brief stops in Egypt, the Reinforcements disembarked in Southampton, England, at the end of January 1918. From there, they were transported to the 1st Training Battalion at Sutton Veny near the town of Warminster in Wiltshire. In England, Private Bert Humphries' service number was changed to 759B.

Private Bert Humphries fell ill during his time in England and was admitted to hospital suffering from bronchitis in April 1918. It was not until the middle of May 1918 that he marched into 1st Australian Divisional Base Depot at Étaples, France. Two weeks later, he joined the 9th Battalion in the field in France on the 1st of June 1918.

In mid-April, the 1st Australian Division, including the 9th Battalion, was urgently redeployed from Flanders to northern France to contain a German breakthrough toward Hazebrouck, a key town on the Calais-Paris rail line. Its capture would disrupt Allied communications. The 9th Battalion played a crucial role in holding the line, fighting fiercely and successfully maintaining their positions despite being outnumbered.

It was during this operation that Private Bert Humphries was Killed in Action on the 21st of July 1918. He was 27 years seven months old and had served in the AIF for 16 months at the time of his death. He has no known grave, and his death was notified in the 423rd list of Australian Casualties released on the 17th of August 1918.

Albert Lincoln Humphries served on the Western Front in WW1. His name is inscribed on The Australian War Memorial Roll of Honour (panel 56), The Australian National Memorial, Villers-Bretonneux, France, The City of Townsville WW1 Honour Roll, The Townsville Cenotaph, and the Townsville No. 37 Magnetic Lodge P.A.F.S.O.A. Honor Board.

759/759B Private Albert Lincoln Humphries was awarded the British War Medal (no:59209) and the Victory Medal (no:58128). A memorial scroll and a memorial plaque were passed to his infant daughter in 1921 and 1922, respectively.

The following notice was published in the *Townsville Daily Bulletin* (Qld) on 7th August 1918:

HUMPHRIES. — In sad but loving memory of our dear nephew and cousin,

Private Albert Lincoln (Bert) Humphries, killed in action, July 21st, 1918.

> Somewhere in France, they have laid him,
> A soldier so brave and true;
> He died for his King and country,
> 'Neath the flag of the red, white and blue.

(Inserted by his sorrowing aunts, uncle and cousins.)

2273 PRIVATE CLAUDE A. M. ANDERSON (1898 – 1918)

41st Infantry Battalion

Claude Alexandra McClarence Anderson was born on the 2nd of July 1898 in Townsville, Queensland, the son of Alexander and Harriet Anderson (née Telfer). He was the oldest of three children. Alexander Anderson was a carpenter, and the family lived in a property on 3rd St Railway Estate, Townsville.

Claude Anderson grew up in Townsville and attended Ross Island State School. He went on to follow his father's example and became a carpenter. He also served for four years in the Citizen Forces and was still an active member when he applied on the 24th of July 1916, in Townsville, to join the Australian Imperial Force (AIF). Claude Anderson declared that he was single, 18 years and one month old, and was a qualified carpenter. His enlistment medical examination noted that he was 5'6½" (167 cm) tall, weighed 8st 12 lbs (56 kg), had a fair complexion, grey eyes, and light brown hair. He identified as Presbyterian, and he nominated his mother as his next of kin, giving her address as 3rd St Railway Estate.

Accepted into the AIF, Claude Anderson was allocated service number 2273, assigned to the Infantry, and posted to the 41st Battalion, 4th Reinforcements. The 41st Battalion was on its way to the Western Front when Claude Anderson enlisted. Following three months of initial training in Australia, the 4th Reinforcements embarked for England from Brisbane on the 21st of October 1916. They sailed as part of Convoy 25 aboard HMAT A36 *Boonah*, which boarded 1,235 military personnel. The entire convoy of 12 ships transported a total of 12,513 personnel and 1,800 horses on the voyage. They disembarked at Plymouth in January 1917.

Six months of training followed in England at the 11th Depot Battalion in Durrington. The Western Front was the next destination for Private Claude Anderson and his unit, and they landed there in June 1917, in time to participate in the Battle of Broodseinde and Passchendaele in September and

October. This battle saw the 41st Battalion leap-frogging the 42nd Battalion to continue the assault north of the Ypres–Roulers railway. It linked up with the 40th Battalion near the present-day Tyne Cot Cemetery. The 41st Battalion was then used to press home further attacks until relieved by Canadian forces who went on to capture Passchendaele.

In early 1918, the battalion was resting and refitting in northern France, preparing to return to the Belgian battlefields near Ypres when the German spring offensive started. The 41st Battalion was hastily ordered south from their rest camp at Quesques to stop the German advance. At Doullens, the 41st occupied defensive positions. From there, the Battalion was sent forward to defend beside the Bray–Corbie Road, near Amiens, remaining in the frontline against constant German attack for 39 days before being relieved. It also took part in fighting around Morlancourt at this time. The 41st Battalion then took its turn in manning the frontline and participated in the Battle of Villers-Bretonneux in April 1918, the Battle of Hamel in July 1918, and the Allied Hundred Days Offensive, which ultimately brought about an end to the war.

On the 29th of July 1918, two weeks after the Battle of Hamel, Private Claude Anderson received multiple wounds from shell fire. He was evacuated to the 5th Casualty Clearing Station; however, he succumbed and Died of Wounds on the 30th of July 1918. He was 22 years old and had served in the AIF for two years and six days. He had been a member of the 41st Battalion for 13 months. Private Claude Anderson was laid to rest in the Crouy British Cemetery, Crouy-sur-Somme, and his death was notified in the 425th list of Australian Casualties released on the 21st of August 1918.

Claude Alexandra McClarence Anderson's name is inscribed on The Australian War Memorial Roll of Honour (panel 133), The Ross Island State School Honour Role, The City of Townsville WW1 Honour Roll (incorrectly engraved as "G" instead of "C" Anderson) and The Townsville Cenotaph. Anderson Street, Railway Estate in Townsville, is named in Claude Anderson's memory.

2273 Private Claude Alexandra McClarence Anderson served in WW1 on the Western Front. He was awarded the British War Medal (no:50281) and the

Victory Medal (no:49723). These medals, as well as a memorial scroll and a memorial plaque, were passed to his father.

The following was published in the Daily *Townsville Daily Bulletin* on the 25th of December 1918:

TO MOTHER DEAR.

The following poem written in the trenches by Pte Alexander Claude Anderson just before Christmas 1917 was received by his parents Mr and Mrs A. Anderson, Third-street Railway Estate. The young soldier was Killed in Action in July last.

Remember me on Christmas Day!
I know in my old home the lamps are lighted:
And friends I love so well are gathered there.
But one is missing from our family —
I fancy I can see the vacant chair.
And when you gaze on the dear ones round you,
From the bullet's hiss and warfare's sorrow free.
Just think of your poor boy 'mid strife and danger
And in your humble prayers remember me,
Pray that God, who in time of trouble
None ever sought to find the comforter in vain,
May bring me through these awful scenes of carnage,
Back to the home I'd love to see again.
But should it be for me to fall in battle:
I'd like you all to think of me.
As having done my part for King and country
And the dear old flag of freedom.

2049 PRIVATE ERNEST JAMES CHANDLER (1889 – 1918)

41st Infantry Battalion

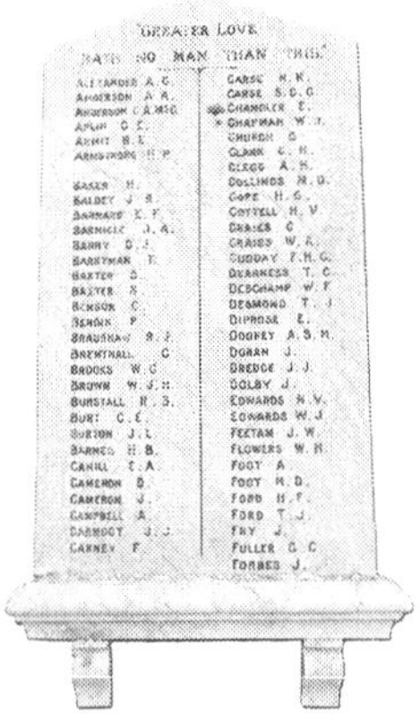

Ernest James Chandler (Mick) was born the fourth of seven children on the 9th of September 1889 at Townsville, Queensland. He was the son of James and Helen Chandler (née Ford). The Chandler family lived in a property on Stanley St in Townsville, and Ernest Chandler attended the Townsville State School.

After his school years were completed, Ernest Chandler qualified as a motor mechanic and worked as a Motor Driver in Brisbane. On the 8th of May 1916, he applied at Townsville to enlist in the Australian Imperial Force (AIF). He declared that he was 26 years, six months old, single, qualified as a motor car driver, and that he had no previous military experience. His medical examination noted that he was 5'3" (160 cm) tall, weighed 106 lbs (48 kg), had a medium complexion, grey eyes, and dark brown hair. He identified as Church of England and nominated his mother as his next of kin, giving her address as Stanley St, Townsville.

Accepted into the AIF, Ernest Chandler was allocated service number 2049, assigned to the Infantry and posted to the 41st Battalion, 3rd Reinforcement. The 41st Battalion sailed from Sydney, bound for England, in the same month that Ernest Chandler enlisted. On the 7th of September 1916, after four months of basic training in Australia, Private Ernest Chandler embarked for England with the rest of the contingent of the 3rd Reinforcements. They embarked from Brisbane, Queensland, as part of Convoy No. 24 and sailed aboard HMAT A46 *Clan MacGillivray*, which boarded 924 troops in Brisbane and a further 159 in Fremantle. The entire convoy of 13 ships transported a total of 12,600 troops on the voyage.

Eight weeks later, Private Ernest Chandler and his fellow reinforcements disembarked at Plymouth in England and were transported to Hurdcott Camp, near Fovant, Wiltshire. They stayed at Hurdcott for advanced training for the next five weeks. On the 20th of December 1916, the 3rd Reinforcements sailed from Folkestone for France. After more training at Étaples, they marched into the 41st Battalion in France on the 26th of January 1917.

Private Ernest Chandler suffered the bitter and harsh winter of 1916-17. The 41st Battalion rotated in and out of the front line, and when they weren't fighting, they spent their time training and performing labouring tasks in the rear areas. As the year progressed, Private Ernest Chandler fought with the Battalion at Messines on June 7th and at Broodseinde on the 4th of October 1917. He was wounded in this action and was evacuated with concussion from a shell wound on the 4th of October 1917. He re-joined the 41st Battalion five months later in March 1918 following recovery and some leave.

For the following five months, the 41st Battalion was stationed in Belgium, alternating between serving in the rear areas and on the front line. When the German Army launched its final major offensive in March 1918, the battalion was quickly deployed to France to help hinder the advance towards the crucial railway junction of Amiens. In the Allies' counter-offensive launched on the 8th of August 1918, the 41st Battalion actively participated in both the initial attack and the subsequent long advance. It was during this advance that Private Ernest Chandler was Killed in Action at Mericourt on the 12th of August 1918. He was 28 years, 11 months old at the time of his death.

Private Ernest Chandler had served with the AIF for two years, three months. He had been a member of the 41st Battalion for 18 months and was laid to rest at the Heath Cemetery, Harbonnieres, France.

Ernest James Chandler's name is inscribed on The Australian War Memorial Roll of Honour (panel 133), The Townsville West State School Memorial Gate, The City of Townsville WW1 Honour Roll, and The Townsville Cenotaph. Chandler Street, Garbutt, Townsville is named in his memory.

2049 Private Ernest James Chandler was awarded the British War Medal (no:46889) and the Victory Medal (no:46345). A memorial scroll and a memorial plaque were passed to his mother in 1921 and 1922, respectively.

The following tribute was published in *The Brisbane Courier* (Qld.) on the 30th of August 1918:

CHANDLER. —Killed in action, France, Pte Ernest James Chandler (Mick), 41st Battalion, 12th August 1918.

Inserted by his loving brother, T. F. Chandler, and sister-in-law.

7618 PRIVATE JOHN JAMES DREDGE (1894 – 1918)

42nd Infantry Battalion

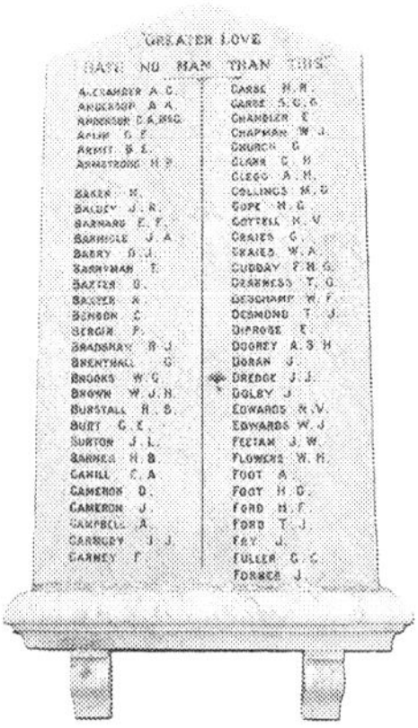

John Dredge was born in August 1894 at West Ham in England. He was the only son of seven children born to John and Rhubina Dredge (née Farrell). John Dredge Jr. moved to Australia sometime after 1907. Settling in Townsville, he found work as a cook, and on the 13th of May 1915, he married Ellen Jane Williams (Nellie). John and Nellie Dredge had two children during their marriage, and in 1917, they lived in a property on Mitchell Street, Townsville.

On the 28th of March 1917, John Dredge applied to enlist in the Australian Imperial Force (AIF) at Townsville. He declared that he was 22 years, seven months old, married with two children, qualified as a cook, and had no previous military experience.

His enlistment medical examination noted that he was 5'4½" (164 cm) tall, weighed 142 lbs (64 kg), had a fair complexion, blue eyes, and fair hair. He identified as Church of England and nominated his wife as his next of kin, giving her address as Mitchell St, Townsville. This was later changed to c/- Mrs W. Kepper, Wills St Townsville.

Accepted into the AIF, John Dredge was allocated service number 7618, assigned to the Infantry, and posted to the 15th Battalion, 25th Reinforcements. The 15th was fighting on the Western Front in France when John Dredge enlisted. April to June 1917 was spent completing Basic Training in Australia. On the 14th of June 1917, Private John Dredge and his fellow contingent of Reinforcements embarked from Sydney, New South Wales, as part of Convoy No. 32 bound for England. They sailed aboard HMAT A20

Hororata, which carried 1,744 troops. The entire convoy of eight ships transported a total of 6,406 troops on the voyage.

Ten weeks later, in September 1917, the contingent disembarked at Liverpool and were transported to the 4th Training Battalion and then the 12th Training Battalion, both located at Codford in Wiltshire. Three months of further training followed and then overseas to France on the 1st of January 1918. While undergoing training, Private John Dredge was re-allocated to the 42nd Battalion, and he marched into his unit, deployed near Locre in Belgium, on the 12th of January 1918.

For the next two months, the 42nd Battalion conducted operations in Belgium, alternating between the front line and rear areas. In March 1918, the battalion was quickly dispatched to France to help impede the German Army's final major offensive and played a part in stopping their advance towards the important railway hub of Amiens. The 42nd Battalion participated in the Allied offensive in August, which included the initial attack near Amiens and the subsequent advance aimed at breaking through the Hindenburg Line.

During this time, in August 1918, the Battalion was deployed close to Proyart, a village in the Somme Valley in France. The operation launched between the 10th and 12th of August, as part of the Battle of Amiens, was named after this village. It was during this battle that Private John Dredge was Killed in Action near Proyart on the 12th of August 1918. On that day, the 42nd Battalion War Diary records that: *"Acting on information received from Brigade regarding enemy retirement on our immediate front. Patrols were at once sent out to get in touch with the retreating force and determine their location."*

There are no recorded details of Private John Dredge's death, and he has no known grave. He was 24 years old at the time of his death. He had been a member of the AIF for 16 months and a member of the 42nd Battalion for seven months. His death was notified in the 430th list of Australian Casualties released on the 25th of September 1918.

Private John Dredge served on the Western Front in WW1. His name is inscribed on The Australian War Memorial Roll of Honour (panel 135), The Australian National Memorial, Villers-Bretonneux, France, The 42nd Infantry Battalion AIF Roll of Honour, The City of Townsville WW1 Honour Roll, and The Townsville Cenotaph. Dredge Street, Stuart, Townsville, is named in his memory.

7618 Private John James Dredge was awarded the British War Medal (no: 57442) and the Victory Medal (no: 56443). These medals were passed to his widow, as well as a memorial plaque and a memorial scroll, which were also passed to her in 1921. Ellen Dredge remarried Samuel Wynn in 1921.

The following tribute was published in the *Townsville Daily Bulletin* on 24th September 1918:

DREDGE. —Killed in Action on the 12th of August 1918, Pte John James Dredge, aged 24 years, dearly beloved husband of Nellie Dredge, and eldest son of Mrs. J. J. Dredge of Forest Gate, West Ham, England.

118 PRIVATE WILLIAM JOHN THOMAS ROWE (1896 – 1918)

15th Infantry Battalion

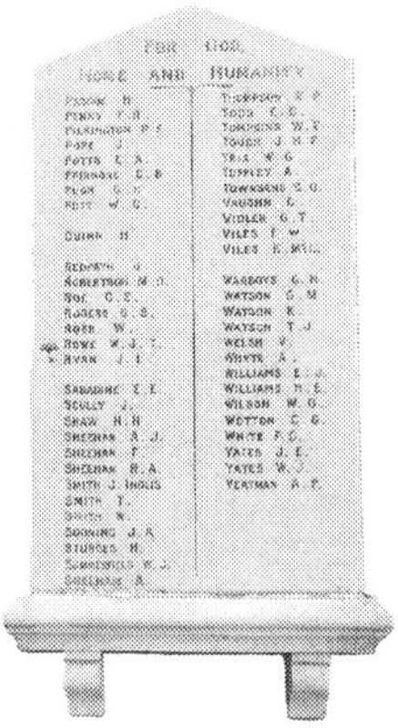

William John Thomas Rowe was born on the 14th of May 1896 in Bowen, the second son of William and Eliza Rowe (née Thomas). There were four children in the family, with the two youngest being girls, neither of whom survived.

Somewhere between 1896 and 1898, the Rowe family left Bowen and moved to Charters Towers. It was here that Eliza Rowe died on the 3rd of September 1899, the same year as the birth of her last child (named Eliza). Tragically, the young Eliza passed away the same year, when she was less than one year old. William Rowe was three years old when his mother died.

There is no known record of William Rowe's childhood or that of his surviving sibling, his brother Edward. However, in 1915 Edward Rowe lived in Aitkenvale, Townsville, and worked as a labourer.

On the 6th of December 1915, William Rowe applied in Townsville to enlist in the Australian Imperial Force (AIF). He declared that he was 19 years and six months old, single, a labourer, and that he had no previous military experience. His medical examination noted that he was 5'4" (163 cm) tall, weighed 8st 7lbs (54 kg), had a dark complexion, brown eyes, and dark brown curly hair. He identified as Church of England and nominated his brother as his next of kin, giving his address simply as Aitkenvale, Townsville.

Accepted into the AIF, William Rowe was allocated service number 118, assigned to the Infantry and posted to 'A' Squadron of the 14th Light Horse Regiment (LHR). The 14th LHR was raised in March 1916 as part of the AIF at Enoggera, Brisbane.

Six months of training followed enlistment for William Rowe until the 13th of May 1916, when the 14th LHR departed from Sydney bound for England. They sailed as part of Convoy No. 21 aboard HMAT A72 *Beltana*, which boarded 1,409 troops. The 18-ship convoy transported a total of 25,287 troops on the voyage.

Once in England, it was intended that the 14th LHR would be brought up to full strength to serve as the 3rd Division's light horse regiment. However, before it could be brought up to full strength, the establishment was reduced to only one squadron per division and as such only 'A' Squadron was formed. Trooper William Rowe disembarked with 'A' Squadron in July 1916, and after a couple of weeks training in England, was struck down with mumps and spent the next three weeks in the hospital at Fargo.

Soon after Trooper Rowe's discharge from the hospital, the divisional establishments of the Australian Army were changed again, this time removing mounted troops from the order of battle altogether. As a result, it was decided to disband the 14th Light Horse Regiment. Trooper Rowe was redesignated as Private Rowe and transferred as a reinforcement to the 15th Infantry Battalion, which was in the field in France.

At the end of September 1916, Private Rowe and the rest of the reinforcements were shipped to Étaples in France and marched into the 4th Australian Divisional Base Depot for another 10 days of training. That completed, they were taken on the strength of the 15th Battalion on the 16th of October 1916. On that date, the 15th Battalion was in the Zillebeke trenches, part of the Ypres salient in Belgium. Private Rowe was appointed as a driver in December 1916. A driver was a dangerous job, entailing the task of driving horse-drawn wagons along roads subject to accurate and frequent German bombardment.

In 1916-17, Europe experienced its harshest winter in 40 years. The soldiers of the 15th Battalion suffered greatly from illness and the cold as they alternated between defensive duties and labouring before moving to Gueudecourt. Driver William Rowe was among those who suffered, and he was hospitalized three times.

On the 21st of January 1917, Driver Rowe was court-martialled for the offence of absenting himself without leave for 10 days in January. It was during this absence that he married Nellie Newton, an English girl from Clayport. He was found guilty and forfeited a total of 87 days' pay. After the court-martial, Driver Rowe reverted to an infantryman, at his own request, and joined the 15th Battalion in the front line.

Three months passed until May 1917, when Private Rowe was hospitalized again, this time with influenza, which was sweeping through the AIF. He was so ill that he had to be evacuated back to a hospital in England. He was hospitalized for three months, and when released in late August 1917, he was granted two weeks' furlough. Private Rowe took an extra couple of days off and was charged with being absent. He was found guilty and awarded seven days' confinement to barracks, along with forfeiting four days' pay. It was to be another three months before Private Rowe rejoined the 15th Battalion in the field on the 8th of January 1918.

In March and April 1918, the battalion helped stop the German spring offensive. The battalion participated in the great Allied offensive of 1918, fighting near Amiens on the 8th of August 1918. This advance by British and Empire troops marked the greatest success in a single day on the Western Front.

On the 13th of August 1918, the 15th Battalion moved to a new position near Harbonnieres. At Bayonvillers, a German artillery bombardment killed nine and wounded two. One of those casualties was Private William John Thomas

Rowe, who was Killed in Action on the 13th of August 1918. He was 22 years old and had been a member of the AIF for two years and eight months.

Private William Rowe was initially buried in the Bayonvillers German Cemetery near Corbie but his body was later reinterred and laid to rest at Heath Cemetery (Plot IV, Row C, Grave No. 2), Harbonnieres, France.

William John Thomas Rowe's name is inscribed on The Australian War Memorial Roll of Honour (panel 77), The City of Townsville WW1 Honour Roll, and The Townsville Cenotaph.

118 Private William John Thomas Rowe was awarded the British War Medal (no:42987) and the Victory Medal (no:42521). A memorial scroll and a memorial plaque were passed to his wife Nellie in 1922.

401 PRIVATE ALEXANDER CAMPBELL (1887 – 1918)

1st Machine Gun Battalion

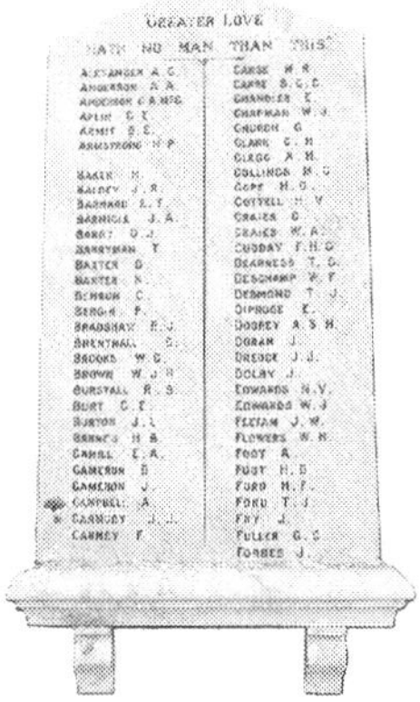

Alexander Campbell (Alex) was born on the 19th of September 1886 in Glasgow, Scotland. He was the oldest of two sons born to Donald and Bella Campbell (née McDiarmid). The Campbell family arrived in Queensland on the 5th of May 1887 when Alexander Campbell was an eight-month-old infant. The family settled in Townsville and lived in a property on Armstrong St, Hermit Park. Donald Campbell was a Justice of the Peace and a commercial traveller.

Alex Campbell grew up in Townsville and attended Townsville West State School. After his school years were completed, he worked as an accountant. He married Isabella Hill, who was a dressmaker, in Townsville, in 1915.

On the 17th of April 1916, Alex Campbell left Isabella in the care of his parents and applied to join the Australian Imperial Force (AIF). He declared that he was 29 years and eight months old, with no previous military experience.

He nominated his wife as his next of kin and gave her address as his parents' address in Armstrong St. Alex Campbell's medical assessment recorded that he was 5'7½" (171 cm) tall, weighed 10st 8lbs (67kg), had a dark complexion, blue eyes, and dark brown hair. He identified as Presbyterian.

Accepted into the AIF, Alex Campbell was allocated service number 401, assigned to the Machine Gun Corps, and was initially posted to the 13th Machine Gun Company. In March 1918, the 13th Machine Gun Company became part of the 1st Machine Gun Battalion. The battalion consisted of 64 medium machine guns and took part in the final stages of the war, seeing action during the Allied defensive operations during the German spring offensive and then the Allied Hundred Days Offensive.

After preliminary training, Private Alex Campbell embarked from Melbourne bound for England on the 19th of September 1916, as part of Convoy No.24. He sailed aboard HMAT A73 *Commonwealth*, which boarded 516 troops in Melbourne and a further 477 in Adelaide. The entire convoy of 13 ships transported a total of 12,600 troops on the voyage. From November 1916 until February 1917, Private Campbell trained in England before joining his unit (now the 21st Machine Gun Company) in France. He was promoted to Lance Corporal in March 1917; however, he was reduced to the rank of Private in May 1918 because of disciplinary action.

In September 1917, Private Campbell sustained a shrapnel wound to his right shoulder and was evacuated back to England. It took five months of treatment before he was fit enough to re-join his unit in France, which he did in February 1918. A month later in March 1918, Private Campbell was gassed and hospitalised again, re-joining his unit in April 1918. By this time, the battalion was heavily engaged in fighting to counter the German Spring Offensive.

After the German offensive was turned back, the battalion took part in the Allied Hundred Days Offensive. It was during this battle that Private Alex Campbell was Killed in Action by a shell burst at Mericourt on the 22nd of August 1918. He was buried in a grave with two other soldiers at Mourcourt Military Cemetery. His body was later reinterred and laid to rest at Heath Cemetery, Harbonnieres, France. At the time of his death Alex Campbell was 31 years old and had served with the AIF for two years and four months. He had been a member of the 1st Machine Gun Battalion for 18 months and his death was notified in the 433rd list of Australian Casualties released on the 13th of October 1918.

Alexander Campbell's name is inscribed on The Australian War Memorial Roll of Honour (panel 175), The City of Townsville WW1 Honour Roll, Townsville West State School Memorial Gate, and The Townsville Cenotaph. Campbell Street, Railway Estate, Townsville is named in his memory.

401 Private Alexander Campbell was awarded the British War Medal (no: 24771) and the Victory Medal (no: 24604). A memorial scroll and a memorial plaque were passed to his widow in 1922.

Alex had a brother, Robert, who enlisted on 18th September 1915 and served in WW1 with the 15th Field Ambulance. He survived and returned to Australia on 21st July 1918.

The following notice was published in the *Townsville Daily Bulletin* on the 10th of September 1918:

CAMPBELL — Killed in Action, 22nd August, Pte Alexander Campbell, the dearly beloved husband of Isabella S. Campbell, and eldest son of Mr and Mrs D. H. Campbell, also brother of Pte Robert Campbell, now returning to Australia.

Deeply Mourned.

881 PRIVATE ERNEST ARTHUR POTTS (1893 – 1918)

3rd Pioneer Battalion

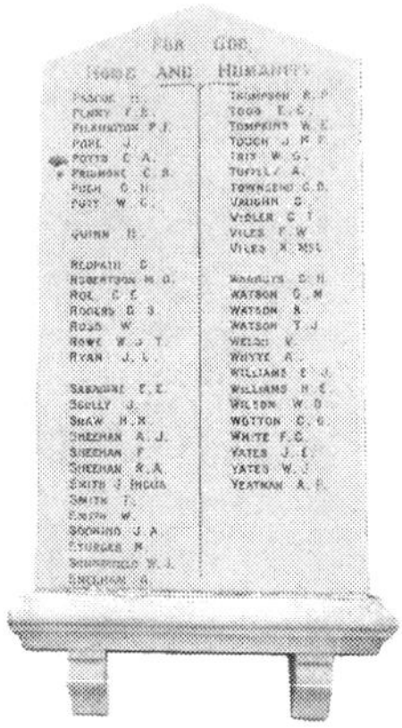

Ernest Arthur Potts was born on the 22nd of March 1893 in Townsville, the first-born child of David and Ada Potts (née Stokes). He had two brothers and three sisters. David Potts worked as a fireman and as a labourer, and the family lived in a property on Ackers St, Hermit Park, in Townsville.

As a child, Ernest Potts attended Townsville West State School, and after graduation, he worked as a shipping clerk with the Adelaide Steam Ship Company. On the 30th of September 1915, Ernest Potts left his job and applied at Townsville to enlist in the Australian Imperial Force (AIF). He declared that he was 22 years, six months old, single, and that he had two and a half years of previous military service with the Militia. Ernest Potts' medical examination noted that he was 5'9½" (177 cm) tall, weighed 135 lbs (61 kg), had a dark complexion, blue eyes, and black hair. He identified as Church of England and listed his father as his next of kin, giving his address as Ackers St, Hermit Park, Townsville.

Accepted into the AIF, Ernest Potts was allocated service number 881, assigned to the Engineers, and posted to the 3rd Australian Pioneer Battalion. Formed in Victoria in March 1916, the battalion consisted of volunteers from Victoria, Queensland, South Australia, and Western Australia. Pioneers, trained as infantry, performed light combat engineering tasks, enhancing troop protection and mobility, and could also fight alongside regular infantry when needed.

Nine months of training in Australia followed enlistment for Private Potts until his unit embarked from Melbourne bound for England on the 5th of June 1916 as part of Convoy 21. They sailed aboard HMAT A62 *Wandilla*, which carried 1,347 troops on the voyage. The 18-ship convoy transported 25,287 troops on the voyage. Fifty days later, Private Ernest Potts and his unit disembarked in Plymouth, England, on the 26th of July 1916.

After four months of training in England, Private Potts crossed to France on the 24th of November 1916. He was appointed and promoted to the position of Company Quartermaster Sergeant (CQMS) and joined the 3rd Pioneers near Nieppe, France. The Battalion was engaged in engineering works and sustaining casualties while building roads and other infrastructure.

In February 1917, after a brief hospital stay with mumps, CQMS Potts was appointed temporary Company Sergeant Major (CSM). He held this role until the end of June 1917, when he reverted to Sergeant at his own request.

At the end of July 1917, Sergeant Potts attended bomb training, and at the conclusion of the course, he was sent to England and attached to the Pioneer Training Battalion. Sergeant Potts returned to the 3rd Pioneer Battalion in mid-October 1917 and in December reverted to the rank of Private at his own request. He was detached to the 1st ANZAC Corps School located at Aveluy in France. A period of leave in England followed before Private Potts re-joined the 3rd Pioneer Battalion on the 4th of April 1918.

In late August 1918, the 3rd Pioneers were deployed at Etinehem on the Somme, near Bray and Mericourt. They were ordered to relieve the 35th Infantry Battalion in the Line on the night of the 20/21st of August 1918. The Unit War Diary records that, *"This was the first offensive action in which the Battalion had been engaged since its arrival in France, - as Infantry."* The diary goes on to record, *"but as we were the most Southerly Unit, and consequently on the flank, enemy shelling was heavy, and casualties rather heavy. Patrols endeavored to push into BRAY but owing to heavy enemy machine gun fire, had to withdraw."*

Private Ernest Potts was one of those casualties when, in front of Bray, he was wounded in the abdomen and hip by machine gun fire. He was evacuated to the 10th Field Ambulance but Died of Wounds at 2:30 p.m. on the 22nd of August 1918, aged 25 years and five months. He had served in the AIF for three years.

During his time in the 3rd Pioneer Battalion, he had held various ranks: Private, Sergeant, Warrant Officer, and was a Private at the time of his death. His death was listed in the 433rd Australian Casualty list, released on the 15th of October 1918. He was laid to rest at Beacon Cemetery, Sailly-Laurette, France (plot V, Row A, Grave No. 1).

Ernest Arthur Potts' name is inscribed on the Australian War Memorial Roll of Honour (panel 173), The Townsville West State School Memorial Gate, The Townsville Honor Roll, and the Townsville Cenotaph. Potts Street, Belgian Gardens, Townsville, is named in his memory.

881 Private Ernest Arthur Potts was awarded the British War Medal (no: 42190) and the Victory Medal (no: 41743). A memorial scroll and a memorial plaque were passed to his father in 1922.

6394 PRIVATE GEORGE THOMAS VIDLER (1890 – 1918)

26th Infantry Battalion

George Thomas Vidler was born on the 24th of May 1890 in Townsville. He was the youngest son and the third of five children born to John and Anne Vidler (née Morris). He had two brothers and two sisters. John Vidler was a dairyman, and the family lived at Cluden, which was in the Shire of Thuringowa and south of Townsville.

George Vidler grew up in Cluden and attended the Townsville West State School when it was a wooden structure, opened in 1887. After his school years were finished, George Vidler followed his father into the dairy business.

In October 1915, George Vidler's older brother, William, enlisted in the Australian Imperial Force (AIF). One year later, on the 16th of October 1916, George Vidler applied at Townsville to join the AIF. He declared that he was 26 years, four months old, single, employed as a dairyman, and had no previous military experience. His enlistment medical assessment noted that he was 5' 10½" (179 cm) tall, weighed 10st 10 lbs (68 kg), had a fair complexion, green eyes, and brown hair. George Vidler identified as Methodist and nominated his mother as his next of kin, giving her address as Cluden via Townsville.

Accepted into the AIF, George Vidler was allocated service number 6394, assigned to the Infantry, and posted to the 26th Battalion, 18th Reinforcements. The 26th was in the Somme Valley when George Vidler joined the AIF. Two months of training in Australia followed enlistment for Private George Vidler and the contingent of reinforcements until they embarked from Sydney on the 23rd of December 1916 as part of Convoy No. 16, bound for England. They boarded HMAT A64 *Demosthenes*, which carried 1,598 troops. The entire convoy of 11 ships transported 10,859 troops on the voyage.

Private George Vidler and his fellow reinforcements disembarked at Plymouth in March 1917 and, from there, were transported to Rollestone on the Salisbury Plain in Wiltshire, England, for further training. Three months later, Private Vidler crossed to France and, following a further two weeks' training at Le Havre, was taken on strength of the 26th Battalion on the 9th of July 1917. The 26th was deployed in the vicinity of Bapaume and was engaged in a period of training.

Later in 1917, the focus of the AIF's operations switched to Belgium. There, Private George Vidler fought in the battle of Menin Road on the 20th of September and participated in the capture of Broodseinde Ridge on the 4th of October.

In March 1918, Private Vidler was granted two weeks leave in England. He re-joined the 26th at the end of March 1918 to join the fight against the German Spring Offensive. In April 1918, the 26th Battalion was moved from the Messines sector to the Somme, where it fought alongside other Australian units. The 26th Battalion performed defensive duties in locations, such as Baizieux, Camon, and Ribemont, resulting in over 100 casualties before the German offensive was stopped.

During June and July, the battalion conducted a series of "peaceful penetration" operations, seizing small portions of the German front line during the calm before the last Allied offensive of the war. The first operation occurred on the 10th of June near Morlancourt, and another was conducted on the 17th of July 1918 around Monument Wood, near Villers-Bretonneux. In August, the Allies launched their Hundred Days Offensive, which ultimately brought an end to the war. On the opening day of the offensive, the 26th Battalion led the

7th Brigade's assault near Villers-Bretonneux. Following a stint in reserve, they advanced on the left flank of the brigade in late August during an offensive at Biaches, which saw the Allies push towards the Somme River.

On the 29th of August 1918, the battalion launched an attack near Barlieux, France, with the aim of reaching the Somme Canal. It was during this operation that Private George Vidler was Killed in Action by shellfire. He was 28 years old and had served with the AIF for one year, ten months. His death was notified in the 435th list of Australian casualties released on the 9th of October 1918. Private George Vidler's body was recovered, and he was laid to rest in the Assevillers New British Cemetery (Plot VII, Row B, Grave No. 8), France.

George Thomas Vidler's name is inscribed on The Australian War Memorial Roll of Honour (panel 109), The City of Townsville WW1 Honour Roll, and The Townsville Cenotaph. Vidler Street, Rosslea, Townsville, was originally named in his memory, but the street has been renamed Craftsman Street. Vidler Street, Stuart, Townsville, is named after the Vidler family. After the war, George Vidler's mother wrote on the Roll of Honour Particulars Form: *"His O.C. says that he was a willing horse, always to be found when wanted and one of the hardest worked. One of his comrade runners says that he was the coolest man in action that he ever saw and that he was the life and cheer of their section."*

6394 Private George Thomas Vidler served on the Western Front in The Great War. He was awarded the British War Medal (no: 55036) and the Victory Medal (no: 54197). George's father died a month after George was killed, and therefore the medals were passed to his widowed mother. A memorial scroll and a memorial plaque were also passed to his mother in 1922.

George's brother, Lieutenant William Vidler, survived the war and returned to Australia on the 18th of July 1919.

The following notice was published in the *Townsville Daily Bulletin* on the 28th of September 1918:

VIDLER. — On the 29th August, 1918, in France, George T. Vidler, the youngest and dearly loved son of Mr. and Mrs. J. Vidler, also brother of J. S. Vidler, Cadet W. H. O. Vidler, on Active Service, Mrs. F. Farenden, and Miss Vidler, of Cluden, nephew of Mrs. R. M. Urquhart, Hughenden, Mr.

and Mrs. O. Morris, Charters Towers, Mrs. J. Morris, and Mrs. H. Morris, Townsville, Mr. and Mrs. J. D. Stevens, Glasgow, Scotland.

2111 PRIVATE HERBERT HENRY SOMERSET MARKS (1897 – 1918)

41st Infantry Battalion

MARKS, H. H. S., Pte., 41st Bn

Herbert Henry Somerset Marks was born on the 29th of August 1897 in Townsville, Queensland. He was the oldest child of Herbert and Dora Marks (née Somerset). He had three brothers and two sisters. Herbert Marks snr. was well known in the Townsville community and was the Secretary of the Chamber of Commerce and Secretary of the Townsville Fire Brigade Board. He was also the captain in command of the No.2 Queensland Company, Garrison Artillery.

Herbert Marks jnr. attended Mundingburra School and Townsville Grammar School, where he passed the Sydney Junior examination. Winning a bursary at Gatton College, he attended for one year of his three-year term before leaving to volunteer. Herbert Marks jnr. was the first cadet to be registered under the new defence scheme in Townsville and possibly in the Commonwealth. Early on the morning the scheme came into operation, he was registered by his father, who was then the Area Officer at Townsville. Before that, Herbert Marks jnr. had been attached to the school cadets under the older system.

On the 20th of May 1916, Herbert Marks applied in Townsville to join the Australian Imperial Force (AIF). He declared that he was 18 years and nine months old, single, a station hand, and that he had previous military service with the Senior Cadets. Herbert Marks' medical examination noted that he was 5'8" (173 cm) tall, weighed 9st 3 lbs (58.5 kg), had a fair complexion, blue eyes, and light brown hair. He identified as Church of England, and he nominated his mother as his next of kin, giving her address as Bayswater Rd, Hyde Park, Townsville.

Accepted into the AIF, Herbert Marks was allocated service number 2111, assigned to the Infantry, and posted to the 41st Battalion, 3rd Reinforcements. The 41st Battalion, raised in Brisbane in February 1916, was sailing to England aboard HMAT *Demosthenes* when Herbert Marks joined. Following his enlistment, Herbert Marks underwent four months of initial training. Upon completion, Private Marks and his fellow Reinforcements embarked from Brisbane on the 7th of September 1916 as part of Convoy 24. They sailed to England aboard HMAT A46 *Clan MacGillivray* , which carried 925 troops from Brisbane and another 159 from Fremantle. The convoy, consisting of 13 ships, transported a total of 12,600 troops on the voyage.

The contingent of Reinforcements disembarked at Plymouth in November 1916 and were transported to Hurdcott Camp, near Fovant, Wiltshire, for three months of further training. They crossed to France on the 4th of February 1917 and marched into the 3rd Australian Division Base Depot at Étaples for another four days of training. Private Herbert Marks joined the 41st Battalion, which was deployed near Armentières, on the 9th of February 1917. This was the same day that the 41st Battalion relieved the 43rd Battalion on the front line. For the first six months of 1917, Private Marks alternated duty between the front line in Armentières and the Lys River in Belgium.

In June, during the Battle of Messines, the 41st Battalion held the entire 11th Brigade front as the war's first significant offensive unfolded. This battle began with the detonation of 19 underground explosive charges, known as the Messines mines, which the Allies had placed beneath the German lines. The simultaneous explosion obliterated the German defences and created massive craters, enabling the Allies to advance. During this action, Private Herbert Marks was wounded by gunshot wounds to his leg and arm and was evacuated to England for treatment. After recovering and being granted leave, he rejoined the 41st Battalion near Ypres in Belgium in October 1917.

In early 1918, the battalion was resting in northern France, preparing to return to Ypres, when the German spring offensive began. The 41st Battalion was rushed from Quesques to halt the German advance. At Doullens, they took defensive positions, later moving to defend the Bray–Corbie Road near Amiens, where they held the line under constant attack for 39 days before relief. They also fought near Morlancourt. Private Herbert Marks fought in these battles and, when the Allies attacked on 8 August 1918, was wounded again on the 16th of August. Despite his wounds he stayed on duty.

On the 1st of September 1918, the 41st Battalion fought an assault parallel to the Baupaume–Perrone Road near Haut Allaimes, France. It was during this fighting that Private Herbert Marks was Killed in Action. He was initially buried in the field but his body was later reinterred and laid to rest in the Peronne Communal Cemetery Extension (Plot IV, Row J, Grave No. 18), France.

Private Herbert Marks had turned 21 years old three days before his death and had served with the AIF for two years and three months. His death was notified in the 432nd Casualty List issued on the 6th of October 1918.

2111 Private Herbert Henry Somerset Marks was awarded the British War Medal (No: 46859) and the Victory Medal (No: 46315). A memorial scroll and a memorial plaque were passed to his father in 1921 and 1922, respectively.

The following tribute was published in The *Townsville Daily Bulletin* (Qld.) on the 16th of September 1918:

Son of Herbert and Dora MARKS, "Dalmuir", Stokes Street, Townsville, Queensland.

The news of the death of their son Private Herbert Henry Somerset Marks was received on Saturday by Captain and Mrs. H. B. Marks. The young soldier, who had just reached his twenty-first birthday on the last day of August, was killed on 1st September. He enlisted at Townsville on 20th May 1916 and joined the 41st Battalion. He received his education at Mundingburra School and at the Townsville Grammar School, where he passed the Sydney Junior. Winning a bursary at Gatton College, he attended there for one year out of his term of three, making a very promising record, and then left to volunteer. Private Marks was the first cadet to be registered under the new defence scheme in Townsville and probably in the Commonwealth. Early on the morning when the scheme came into operation, he was registered by his father, who was then Area Officer at Townsville. Previous to that he had been attached to the school cadets under the old system. There were many expressions of sorrow from friends in Townsville at the news of his death.

607 PRIVATE HERBERT VINCENT COTTELL (1897 – 1918)

41st Infantry Battalion

Herbert Vincent Cottell was born in Mackay on the 10th of October 1897. He was the fourth of six children born to Herbert and Hannah Cottell (née Sullivan). Herbert Snr. was a master mariner, and after moving to Townsville, the family lived in a property on Ahearn St, Hermit Park, Townsville.

Herbert Cottell Jnr. grew up in Townsville and attended the West End State School and served in the Cadets. When his school years were completed, he became a storekeeper. On the day after his 18th birthday, 11th of October 1915, Herbert Cottell applied to enlist in the Australian Imperial Force (AIF) at Townsville. He declared that he was 18 years old, single, a storekeeper with two years' service in the cadets. His medical examination noted that he was 5'6½" (169 cm) tall, weighed 9st (57kg), had a fair complexion, grey eyes, and dark brown hair. He identified as Roman Catholic and nominated his father as his next of kin, giving his address as Ahearn St, Townsville.

Accepted into the AIF, Herbert Cottell was allocated service number 607, assigned to the Infantry, and posted as an original member to "C" Company of the 41st Battalion. The 41st Battalion was raised in February 1916 with recruits from Brisbane, northern Queensland, and the northern rivers district of NSW. After six months of training in Australia, Private Herbert Cottell embarked with the 41st Battalion from Sydney on the 18th of May 1916. They were bound for England and sailed with Convoy No. 21 aboard HMAT A64 *Demosthenes*, which carried 1,610 troops. The entire convoy of 18 ships transported a total of 25,287 troops on the voyage. The 41st disembarked at Plymouth, England, on the 20th of July 1916.

From Plymouth, the battalion moved with the 3rd Division to training grounds on Salisbury Plain at Larkhill, where they underwent training until the 24th of November 1916. Private Herbert Cottell and his fellow soldiers of the 41st Battalion crossed to France and the Western Front, arriving on the 25th of November 1916. On Christmas Eve, the 41st took their position in the front line, for the very first time, at the Mushroom Salient, just east of La Chapelle-d'Armentières. Four soldiers were killed by artillery fire on the first night. During the harsh winter of 1916-17, they rotated between serving in the front line, engaging in training, and labouring tasks in the rear area.

Private Herbert Cottell was hospitalised with pneumonia and bronchitis in December 1916. He re-joined the 41st Battalion in mid-February 1917. Throughout the first half of 1917, Private Cottell served on the front line in and around Armentières in France and across the Lys River in Belgium. In June, at the Battle of Messines, the 41st Battalion held the entire 11th Brigade frontage until the mines were detonated and the attack began. Following this, the battalion engaged in a series of smaller "Bite and Hold" attacks aimed at capitalizing on the gains made during the Battle of Messines, including the assault on the 'Windmill' position of the Germans near Warneton, Belgium.

The 41st Battalion had to assault a longer than expected distance and maintain their gains. The flooded, muddy terrain made it nearly impossible to dig trenches, and the battalion had to make do with a series of linked-up, waterlogged shell craters. Despite this challenging situation, they repulsed several counterattacks over the next 18 days. Following this action, the battalion was withdrawn from the front line and sent to northern France for rest and further training to replenish its strength.

In September and October 1917, the battalion participated in the Battle of Broodseinde and Passchendaele. Private Herbert Cottell was wounded by gunshot in action on the 4th of October 1917. He was evacuated back to England and did not re-join the 41st Battalion in France until February 1918. He arrived in time for the launch of the German Army's last great offensive in March 1918. The 41st was rushed south to France and played a role in blunting the drive towards the vital railway junction of Amiens.

The Allies launched their own offensive on the 8th of August 1918, and the 41st Battalion played an active role both in the initial attack and the long advance that followed throughout August and into September 1918. It was during this phase of operations that Private Herbert Cottell sustained gunshot wounds to the head on the 25th of August 1918 in the vicinity of Étinehem. Evacuated to hospital, he Died of Wounds on the 4th of September 1918. He was 20 years, 10 months old and had served with the AIF for two years, 10 months. He was laid to rest in the St. Sever Cemetery Extension, Rouen, France.

Herbert Cottell's name is inscribed on The Australian War Memorial Roll of Honour (panel 133), The 41st Battalion Roll of Honour, The West End State School Honour Roll, The City of Townsville WW1 Honour Roll, and The Townsville Cenotaph. Cottell Street, Hyde Park, Townsville, is named in his memory.

607 Private Herbert Vincent Cottell was awarded the British War Medal (no:39793) and the Victory Medal (no:39409). A memorial scroll and a memorial plaque were passed to his father in 1922.

The following notices were published in the *Townsville Daily Bulletin* (Qld.) on the 17th of September 1918:

Captain and Mrs. H. P. Cottell, who were advised that their son, Pte Herbert Vincent Cottell had been seriously wounded, have now received the sad news that he had succumbed to his wounds, and passed away in France on September 4th. Great sympathy is felt for the bereaved family. Pte Cottell had been on service for three years and would have been 21 years of age in a few days' time.

IN MEMORIAM

COTTELL. —In loving memory of Pte Herbert Vincent Cottell, who departed this life in France on Wednesday, 4th September 1918, age 21 years, dearly beloved son of Hannah and Captain H.V. Cottell, also brother of Mrs C.G. Gilbert and Mrs F.J. Canning. R.I.P.

570 WARRANT OFFICER CLASS 2 LAWRENCE GLADSTONE LOWTH (1893 – 1918)

41st Infantry Battalion

Lawrence (Larry) Gladstone Lowth and his twin brother Stephen were born on the 21st of May 1893 in Townsville, Queensland, to Thomas and Margaret (Maggie) Lowth (née Smith). The twin brothers were the seventh and eighth children born. Larry had seven brothers and four sisters.

Thomas Lowth had been a police officer before retiring from the force in Townsville. He then entered the hotel industry and managed hotels in Townsville before he and Maggie built Lowth's Hotel in 1897 on the corner of Flinders and Stanley Streets, Townsville. In later years, Thomas Lowth became an Alderman on the Townsville City Council.

Larry Lowth attended the Townsville Grammar School and afterwards went to St Joseph's College, Hunters Hill, Sydney. After his school years were finished, he commenced a business career with Alfred Shaw Ltd. He was also a keen cricketer, footballer, and a fine athlete.

On the 18th of October 1915, Larry Lowth applied at Townsville to join the Australian Imperial Force (AIF). He declared that he was 22 years, five months old, single, and employed as a hardware salesman without previous military service. His medical examination noted that he was 5'9½" (177 cm) tall, weighed 10st 4 lbs (65 kg), had a fair complexion, blue eyes, and light brown hair. He identified as Roman Catholic and nominated his father as his next of kin.

Accepted into the AIF, Larry Lowth was allocated service number 570, assigned to the Infantry, and posted as an original member to "C" Company of the 41st Battalion. The battalion was raised at Enoggera, near Brisbane, in December 1915.

In July 1915, Larry Lowth's twin brother Steven, who was employed as a pharmacy assistant, had enlisted and was allocated to the Medical Corps and posted to the 7th Field Ambulance.

Six months of training in Australia was completed by the newly minted 41st Battalion. During this time, Private Larry Lowth was promoted to Corporal on the 3rd of March 1916. On the 18th of May 1916, the 41st Battalion embarked from Sydney bound for England. They sailed with Convoy No. 21 aboard HMAT A64 *Demosthenes*, which carried 1,610 troops. The entire convoy of 18 ships transported a total of 25,287 troops on the voyage. The ship docked at Plymouth, England, on the 20th of July 1916.

In England, Corporal Larry Lowth attended Musketry School and then undertook further training at Lark Hill. The 42nd Battalion crossed to France in November 1916. Further training at Étaples, France, followed before the Battalion deployed to the front. The winter of 1916/17 was horrendous, and the 41st Battalion spent much of it in the front line, the remainder being spent alternating between training and labouring in the rear areas. Corporal Lowth was promoted to Sergeant on the 7th of February 1917.

On the 23rd of February 1917, Sergeant Larry Lowth's younger brother Richard (Dick) Lowth enlisted and was allocated to the Artillery. Three Lowth brothers were now fighting in World War I.

Compared to other AIF battalions, the 41st Battalion's role in the battles in Belgium during 1917 was relatively straightforward. It provided support at Messines on the 7th of June and met its objectives at Broodseinde on the 4th of October with minimal difficulty. Furthermore, it avoided the severe losses experienced at Passchendaele.

Towards the end of June 1917, the 11th Brigade received orders to establish a new front-line position west of Warneton, exposed in full view of the German forces. During this period, work continued relentlessly, even under the relentless barrage of heavy shellfire. This phase came to be known within the battalion as "the 18 days."

As August 1917 commenced, the 41st was deployed to hold ground captured by two sister battalions in a feigned attack on the 31st of July. Sergeant

Larry Lowth and his fellow soldiers had to endure incessant rain, waterlogged trenches, and unrelenting artillery shelling. Many of the battalion's platoons saw their numbers dwindle from 35 men to less than 10.

In October 1917, about 10 days prior to Passchendaele, Sergeant Lowth was selected to return to England to work as part of the training staff at the 11th Training Battalion located at Dorrington, *Shropshire*. A month later, he was posted to the permanent cadre of the 9th Training Battalion, located at Fovant, Wiltshire. Sergeant Lowth completed his time as an instructor and, after leave in England, re-joined the 41st Battalion in the field in Belgium on the 8th of February 1918.

When the German Army launched its last great offensive in March 1918, the 41st was rushed south to France and played a role in blunting the drive towards the vital railway junction of Amiens. It was during these actions that Sergeant Lowth was gassed on the 4th of June 1918. He was evacuated back to the medical facilities in Rouen, France, where he was treated and recovered. He re-joined the 41st again in the field in the second week of July 1917.

On the 8th of August 1918, Sergeant Lowth was promoted to Temporary Warrant Officer Class Two and appointed as the Company Sergeant Major (CSM). On the same day, the Allies launched their own offensive, and the 41st played an active role both in the initial attack and the long advance that followed throughout August and into September 1918.

It was during this advance that CSM Larry Lowth was Killed in Action on the 7th of September 1918 at Mont St Quentin, France. He was buried in the field but his body was later reinterred and laid to rest at the Hem Farm Military Cemetery (Plot II, Row H, Grave No. 20), Hem-Monacu, France. CSM Larry Lowth was 25 years old and had served with the AIF for two years and 10 months.

Lawrence Gladstone Lowth's name is inscribed on The Australian War Memorial Roll of Honour (panel 134), the 41st Battalion Roll of Honour, the Townsville Grammar School Roll of Honour, The City of Townsville WW1 Honour Roll, and The Townsville Cenotaph. Lowth Street, Rosslea, Townsville is named in his memory.

Warrant Officer Class Two Lawrence Gladstone Lowth was awarded the British War Medal (no:39867) and the Victory Medal (no:39483). A memorial scroll and a memorial plaque were passed to his father in 1922.

Warrant Officer Larry Lowth's two brothers returned safely from the war, 6576 Private Stephen Lowth in January 1919 and 972 Gunner Richard Lowth in September 1919.

The following tribute was published in The *Brisbane Courier (Qld.)* on Monday the 16th of September 1918:

On Saturday evening Alderman and Mrs T Lowth received the sad news that Sergeant Lawrence Gladstone ('Larry') had been Killed in Action on 1st September. The deceased, who was in the 41st Battalion was 25 years of age. He enlisted at Townsville in October 1915, one month after his twin brother had joined up, the pair being greatly attached to each other. 'Larry' was a native of Townsville, and attended the Grammar School, and afterwards went to the College at Hunter's Hill, Sydney. He commenced his business career with Alfred Shaw Ltd., and afterwards joined Smellie and Co.'s, and had just started traveling for the firm when he enlisted for service. He was a keen cricketer and footballer and a fine athlete. On Sunday many messages and expressions of sympathy were received by the sorrowing parents.

7734 LANCE CORPORAL MURDOCH ROSS MURRAY (1895 – 1918)

15th Infantry Battalion

Murdoch Ross Murray was born in Townsville on the 6th of June 1895. He was the youngest of two sons, born to George and Margaret Murray (née Macgregor). George Murray was a contractor and a previous Mayor of Townsville in 1908, 1911, and 1916. The family lived in Alexander St, Stanton Hill.

Murdock Murray attended the Central State School, Townsville, and then qualified as a Carpenter. In March 1915, his older brother, James Murray, had enlisted in the AIF and had been Killed in Action in August 1916 at Pozières.

On the 30th of April 1917, Murdock Murray applied to enlist in the Australian Imperial Force (AIF) at Townsville. He was 21 years, 10 months old, employed as a carpenter and single. He declared that he had previous military experience in the Garrison Artillery. His medical examination noted that he was 5'11" (180cm) tall, weighed 159lbs (72kg), had a fair complexion, blue eyes, and brown hair. He identified as Presbyterian and nominated his father as his next of kin giving his address as Stanton Hill, Townsville.

Accepted into the AIF, Murdock Murray was allocated service number 7734, assigned to the Infantry, and posted to the 9th Battalion, 20th Reinforcements. He spent the next six months in Australia, training for overseas service. On the 31st of October 1917, the contingent of Reinforcements, including Private Murdock Murray, embarked from Sydney bound for England. They sailed as part of Convoy No. 35 aboard HMAT A14 *Euripides*, which carried 2,063 troops. The six-ship convoy transported a total of 6,701 troops, 980 naval personnel, and 1,025 civilians on the voyage.

Disembarking in Devenport, England, on Boxing Day 1917, the Reinforcements were soon moved to the 1st Training Battalion at Sutton Veny. Private Murdock Murray endured the freezing winter in England and in March 1918 was reallocated as a reinforcement for the 15th Battalion and crossed to France. He underwent further training in France and joined the 15th Battalion on the 17th of April 1918.

The 15th was located near the ruined village of Hébuterne in France, where the 4th Brigade Australians and the New Zealanders had checked the German advance early in April 1918. The war diary for the 15th has the following entry for the 17th of April 1918; *"Working on support line of our Sector. 200 reinforcements arrived – Good type of men."* Private Murdock Murray and his fellow reinforcements increased the 15th Battalion's manning to a strength of 57 officers and 955 other ranks. At the end of April 1918, the Battalion moved up to the support lines near Villers-Bretonneux.

Once deployed, the 15th Battalion assumed a support role, focusing on constructing defences. On the 22nd of May 1918, they relocated to Freschencourt and stayed there until the end of the month. In early June Private Murray marched with the 15th Battalion to Hamelet near Corbie. While deployed in this location, the Battalion was engaged in multiple patrol actions. The 15th was engaged in the great allied offensive of 1918, fighting near Amiens on the 8th of August 1918. This advance by Australian and Allied forces was the greatest success in a single day on the Western Front, one that German General Erich Ludendorff described as *"the black day of the German Army in this war."*

Private Murdock Murray was promoted to Lance Corporal on the 6th of September 1918. On the 18th of September, the 15th Battalion was in the Jeancourt area and moved into an attack at 0520hr (5:20 a.m). The attack was successful, and 400 Germans were taken prisoner. Seven Australians were killed, and tragically one of these was Lance Corporal Murdoch Murray, who was Killed in Action on the 18th of September 1918. He was 23 years, three months old and had been a member of the AIF for one year and four months.

Lance Corporal Murdoch Murray was initially buried in the field but his body was later reinterred and laid to rest at Jeancourt Communal Cemetery Extension (Plot I, Row A, Grave No. 20), France. His death was notified in the 442nd list of Australian Casualties released on the 18th of November 1918.

Murdock Ross Murray's name is inscribed on The Australian War Memorial Roll of Honour (panel 77), The City of Townsville WW1 Honour Roll, and The Townsville Cenotaph. Murray Street, Nelly Bay is named in the memory of Murdock and his brother James.

7734 Lance Corporal Murdock Ross Murray was awarded the British War Medal (no: 58558) and the Victory Medal (no: 57558). A memorial scroll and a memorial plaque were passed to his father in 1921 and 1922, respectively.

James' brother, 2Lt James Cassie Murray also served in the AIF. He was Killed in Action with the 7th MG Coy on 5th August 1916.

The following article was published in the *Cairns Post*, on 8th October 1918:

A DOUBLE LOSS. TOWNSVILLE, Oct .7, 1918. Mr. George Murray, contractor, received word on Saturday that his second son, Murdoch Ross Murray, was killed in France on September 18. He went to the front some twelve months ago, after his elder brother, Lieut. Murray, had paid the supreme sacrifice. Mr. Murray has now lost both his sons at the front.

The following article was published in the *Townsville Daily Bulletin,* on 9th May 1923:

TOWNSVILLE DEATH. MR GEORGE MURRAY DROWNED IN ROSS CREEK. (From Our Own Correspondent) Townsville, May 7. General regret was evidenced this morning when it was known that an old and respected citizen, Mr. George Murray, had been drowned. Deceased, who was about 60 years of age, was a contractor and builder. He was last seen after 11 o'clock last night in Stanley-street, approaching the direction of Ross Creek. This morning his body was found on the bank near the Causeway half a mile further up-stream, deposited there by the falling tide. He was fully dressed, including a tight-fitting cap. It was stated that a person dressed similarly was seen in this locality at midnight. There were no marks of violence on the body. Deceased took an active part in civic affairs and occupied the Mayoral Chair in 1908, 1911, and 1916. His only sons, Murdoch and James, sacrificed their lives in the war, which distressed him very greatly.

3548 SERGEANT ERNEST DIPROSE (1895 – 1918)

2nd Light Horse Field Ambulance

Ernest (Ernie) Diprose was born on the 2nd of April 1895 in Hastings, England. He was the third of six children born to Frederick and Ellen Diprose (née Callan). He had three brothers and two sisters. The Diprose family (mistakenly listed as Deprose) arrived in Australia in February 1899. It is not clear when the family moved to Townsville; however, they lived in a property on Alexandra St. in 1903, and Frederick Diprose worked as a commercial traveller.

Ernie Diprose grew up in Townsville and attended the Townsville Central State School. After his school years were completed, he worked as a clerk. He was also a member of the Queensland Naval Reserve.

On the 6th of September 1915, Ernie Diprose applied at Townsville to join the Australian Imperial Force (AIF). He declared that he was single, 20 years and four months old, and working as a clerk. His enlistment medical examination noted that he was 5'4" (163 cm) tall, weighed 118 lbs (54 kg), had a fresh complexion, brown eyes, and dark hair. Ernie Diprose identified as Plymouth Brethren and nominated his father as his next of kin.

Accepted into the AIF, Ernest Diprose was allocated service number 3548, assigned to the Infantry, and posted to the 9th Battalion, 11th Reinforcements. The 9th Battalion was raised within weeks of the declaration of war in August 1914 and embarked just two months later. The 9th was fighting at Gallipoli when Ernie Diprose enlisted.

From April to October 1915, Private Ernie Diprose underwent basic training in Australia at Enoggera in Brisbane. On the 21st of October 1915, Private Diprose and his fellow reinforcements embarked from Brisbane bound for Egypt. They sailed as part of Convoy No. 13 aboard HMAT A48 *Seang Bee*, which carried 1,013 troops. The entire convoy of 11 ships transported a total of 11,731 troops and 400 horses on the voyage.

In Egypt, Private Diprose was hospitalized in January 1916, suffering from mumps. After treatment, he was discharged from the hospital and returned to duty with the 9th Battalion. He trained with the 9th until he was transferred to the Australian Army Medical Corps and posted to the 2nd Light Horse Field Ambulance (2nd LHFA) at Serapeum on the 26th of February 1916.

Private Ernie Diprose was a member of the 2nd LHFA during The Battle of Romani, which was the Australian and New Zealand Mounted Division's (ANZM Division) first major victory on the 3rd-4th of August 1916. In October 1916, Private Diprose was promoted to Corporal and attached to the medical staff at Australian and New Zealand Mounted Division Headquarters (HQ's) at Romani. Corporal Diprose continued to serve at HQ's and in September 1917, he was promoted to Acting Sergeant.

Nearly a year later, tragedy struck on the 29th of September 1918 when Sergeant Diprose was accidentally Killed on Active Service *"by the explosion of a detonator in the unexploded portion of an aerial bomb"*. He was 23 years and six months old and had served in the AIF for three years and one month at the time of his death. He was initially buried in the field but his body was later was later reinterred and laid to rest in the Damascus British War Cemetery (Row C, Grave No. 27), Syria. Sergeant Ernest Diprose's death was notified in the 441st list of Australian Casualties released on the 14th of November 1918.

Ernest Diprose's name is inscribed on The Australian War Memorial Roll of Honour (panel 182), The City of Townsville WW1 Honour Roll, and The Townsville Cenotaph. Diprose Street, Pimlico, Townsville, is named in his memory.

3548 Sergeant Ernest Diprose was awarded the 1914/15 Star (no: 23506), the British War Medal (no: 21607), and the Victory Medal (no: 21531). A memorial scroll and a memorial plaque were passed to his father in 1921 and 1922, respectively.

The following notice was published in the *Townsville Daily Bulletin* (Qld) on the 27th of September 1919:

DIPROSE. — In loving memory of our dear son and brother, Sergt. Ernest Diprose, who was accidentally killed at Amman, in Palestine, September 27th, 1918.

No heathen hope of glory won by death,
Upon the battlefield where he was slain;
Can touch our faith with this, its subtle breath
Or solace give, to ease our hearts of pain.
No, 'tis our Saviour's love. His sacrifice,
That can alone eternal favor win;
His precious blood, it is, that doth suffice,
To gain "the glory," and to cleanse from sin.

LIEUTENANT JOHN BETT LAWSON M.C. (1891 – 1918)

41st Infantry Battalion

John Bett Lawson was born in September 1891 at Edentown, Fife, Scotland. He was the son of William and Annie Lawson (née Thomson). John Lawson grew up in Scotland and attended the Dumikiln School, Kirkealdy, Fifeshire.

In about 1910, at age 19, John Lawson entered Australia, and in 1915, he lived at Oolbun near Townsville. On the 18th of October 1915, he applied at Townsville to enlist in the Australian Imperial Force (AIF). He declared that he was 24 years old, single, and employed as a clerk without previous military service. For reasons that are not clear, he also declared that he had been born in Shanghai, China. John Lawson's enlistment medical examination noted that he was tall at 6'1" (185 cm), weighed 162 lbs (74 kg), had a fresh complexion, brown eyes, and dark hair. He identified as Roman Catholic and nominated his mother as his next of kin, giving her address as 8 Union St., Stirling, Scotland.

Accepted into the AIF, John Lawson was allocated service number 218, assigned to the Infantry, and posted as an original member to "A" Company of the fledgling 41st Battalion. The 41st was raised at Bell's Paddock Camp in Brisbane in February 1916 with recruits from Brisbane, northern Queensland, and the northern rivers district of New South Wales. Eight months of basic training in Australia followed enlistment for Private John Lawson. Once equipped and trained, the Battalion embarked from Sydney as part of Convoy No. 21 on the 18th of May 1916, bound for England. They sailed aboard HMAT A64 *Demosthenes*, which carried 1,610 troops. The entire convoy of 18 ships transported 25,287 troops and 78 other passengers on the voyage.

The 41st Battalion disembarked at Plymouth, England, on the 20th of July 1916. John Lawson had been promoted to Company Quartermaster Sergeant (CQMS) by this time. Following an additional four months of further training in England, the 41st Battalion crossed to France and arrived on the 25th of November 1916. The harsh winter of 1916-17 proved to be a gruelling experience for the battalion, as they spent a significant portion of it in the front line, with the remaining time being split between training and labouring in the rear areas.

On the 1st of March 1917, CQMS John Lawson was promoted to Warrant Officer Class 2 (WO2) and appointed as the Company Sergeant Major (CSM). In 1917, the 41st Battalion had a supporting role at Messines on the 7th of June 1917. Not long after this engagement, WO2 John Lawson was selected for Officer Training and was sent to England in August 1917 to complete a course at No. 5 Officer Cadet Battalion, Cambridge. He successfully completed the training and was commissioned as a 2nd Lieutenant (2Lt) in December 1917. A period of further training in England followed, and 2Lt Lawson re-joined the 41st Battalion in France on Christmas Eve 1917.

In March 1918, during the German Army's last significant offensive, the 41st Battalion was swiftly deployed to southern France to block the enemy advance towards the vital railway junction of Amiens. During the Allied offensive on the 8th of August 1918, the 41st Battalion actively participated in both the initial assault and the sustained advance extending from August into September of the same year. 2nd Lieutenant John Lawson was awarded the Military Cross (M.C.) for his actions on the 8th of August 1918. The official citation reads: *"For conspicuous gallantry in action. He led his platoon most ably, and during the advance, he mounted a tank and entered a wood, mopping up a dugout and taking seventy prisoners. His determination to reach the objective was a splendid example to all."*

2Lt John Lawson received his second pip on the 1st of September 1918 when he was promoted to Lieutenant. Between the 29th of September and the 2nd of October 1918, the 41st Battalion took part in its final significant military operation of the war. The Battalion fought as a unit of the Australian-American joint mission that successfully breached the formidable fortifications of the Hindenburg Line along the St Quentin Canal. It was during this advance that Lieutenant Lawson was Killed in Action on the 29th of September 1918 by direct shell fire near the village of Bony, Aisne, France. He was buried in the field and today has no known grave. Lieutenant John Lawson was 27 years old and had served with the AIF for two years and 11 months at the time of his death. His death was notified in the 443rd Casualty List issued on 2nd November 1918.

John Bett Lawson's name is inscribed on The Australian War Memorial Roll of Honour (panel 134), The Australian National Memorial, Villers-Bretonneux, France, The 41st Battalion Roll of Honour, The City of Townsville WW1 Honour Roll, The Townsville Cenotaph, and The Kirkcaldy War Memorial, Scotland. Lawson Street, Mysterton, Townsville, is named in his memory.

Lieutenant John Bett Lawson was awarded the Military Cross, the British War Medal (no: 39863) and the Victory Medal (no: 39479). A memorial scroll and a memorial plaque were passed to his mother in 1922.

The following tribute was published in the *Brisbane Courier* (Qld.) on the 29th of September 1920:

LAWSON. —In loving memory of Lieut. John Bett Lawson, M.C., 41st Battalion, killed in action, 29th September 1918. Inserted by Ellie Wilson.

631 PRIVATE DANIEL JOSEPH BARRY (1893 – 1918)

2nd Machine Gun Battalion

Daniel Joseph Barry was born on the 4th of November 1893 at Gurrumbah (Herberton), Queensland. He was the second of seven sons born to David and Mary Barry (née O'Meara). David Barry was a miner and, in later years, became a Publican in Yarraman.

Daniel Barry grew up in Herberton and, when old enough, was sent to further his education at St Joseph's College, Nudgee, Brisbane. He gained qualifications as an accountant and by 1916 he was living in Townsville.

Daniel Barry was 22 years and eleven months old when he applied at Townsville on the 14th of October 1916 to enlist in the Australian Imperial Force (AIF). He declared that he had no previous military experience, was single, and qualified as a clerk. His enlistment medical examination noted that he was 5'8¼ " (173 cm) tall, weighed 10st 5lbs (66 kg), had a dark complexion, grey eyes, and black hair. He identified as Roman Catholic and nominated his mother as his next of kin, giving her address as Gurrumbah via Ironbark, QLD.

Accepted into the AIF, John Lawson was allocated service number 631 assigned to the Infantry and posted to the 7th Machine Gun Company, 11th Reinforcements, which was part of the 2nd Machine Gun Battalion. The 7th Machine Gun Company was part of the 7th Brigade, and the men were drawn from the Battalions of that Brigade, the 25th (Qld), 26th (Qld/Tas) 27th (SA) and 28th (WA) Battalions. The 7th Machine Gun Company was absorbed into the 2nd Machine Gun Battalion later in March 1918.

On the 10th of May 1917, after seven months of basic training in Australia, Private Daniel Barry embarked for England from Melbourne with the rest of the contingent of Reinforcements. They sailed as part of Convoy No. 24 aboard HMAT A46 *Clan MacGillivray*, which boarded 467 troops and 328 civilians on the voyage. The entire convoy of 13 ships transported a total of 7,802 troops, 1,878 sailors, and 2,077 civilians on the voyage.

Private Daniel Barry and his fellow reinforcements disembarked at Plymouth, England, on the 28th of July 1917 and were moved to the training establishments located on the Salisbury Plains. Firstly Rollestone, then Codford and, finally, the Machine Gun Training Depot in Grantham, England.

After the training in England was finished, the freshly trained machine gunners crossed to France in late April 1918 and joined the 2nd Machine Gun Battalion in early May. They marched into the unit during the third and final stage of the Australian effort in attacking the Hindenburg Line. The final stage of the attack on the Hindenburg Line involved successful assaults in September 1918, resulting in the capture of key towns such as Bellicourt and Bony.

It was during this phase of the operations that Private Daniel Joseph Barry was Killed in Action near Bellicourt on the 3rd of October 1918, just a few weeks from the end of the war. He was 24 years and 10 months old at the time of his death and had served in the AIF for two years and in the 2nd Machine Gun Battalion for five months. He has no known grave, and his death was notified in the 435th list of Australian Casualties released on the 25th of November 1918.

Daniel Barry's name is inscribed on The Australian War Memorial Roll of Honour (panel 175), The Australian National Memorial, Villers-Bretonneux, France, The City of Townsville WW1 Honour Roll, and The Townsville Cenotaph. Barry Street, Stuart, in Townsville is named in his memory.

631 Private Daniel Joseph Barry served in WW1 on the Western Front. He was awarded the British War Medal (no:57043) and the Victory Medal (no:56083). These medals, as well as a memorial scroll and a memorial plaque, were passed to his father.

The following tribute was published in the *Cairns Post* on 23rd October 1918:

GUNNER BARRY KILLED.

News has been received from Base Records by Mr. and Mrs. D. P. Barry, of Gordonvale (formerly of Garrumbah), of the death of their second son,

Gunner Daniel Joseph Barry, of the 7th Machine-Gun Company, who was Killed in Action on October 3rd. Gunner Barry, who was for many months on the staff of Mr. P. J. Doyle, Cairns, enlisted at Townsville in October 1916.

3416 PRIVATE ALFRED ALEXANDER ANDERSON (1897 - 1918)

41st Infantry Battalion

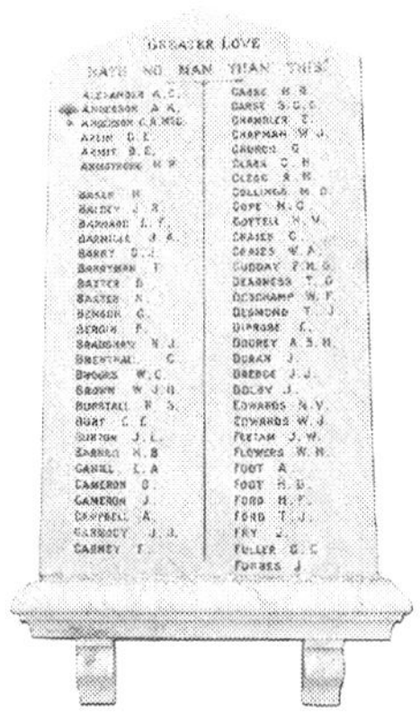

Alfred Alexander Anderson was born in 1897 on Thursday Island, Queensland, the only son of Rasmus and Sena Anderson (née Hansen). Rasmus Anderson was a carpenter, and after the family relocated from Thursday Island, they lived in a property on Warburton St, North Ward, Townsville. Alfred Anderson grew up in Townsville and attended West End State School. After his school years were completed, he gained a five-year apprenticeship as a carpenter with the Railways.

On the 1st of February 1917, Alfred Anderson applied, at Townsville, to enlist in the Australian Imperial Force (AIF). He declared that he was single, 20 years, nine months old, employed as a carpenter and had previously served for six months in the Citizen Military Forces in Townsville. His enlistment medical examination noted that he was 5'7½" (172 cm) tall, weighed 144 lbs (65kg), had a fair complexion, grey eyes and brown hair. He identified as Church of England and nominated his father as his next of kin, giving his address as Warburton St, North Ward.

Accepted into the AIF, Alfred Anderson was allocated service number 3440 and assigned to the Artillery but transferred to the Infantry following an appeal by the Army for volunteers due to a shortage of Infantry soldiers. He was then posted to the 41st Battalion, 8th Reinforcements. The 41st Battalion was fighting on the Western Front when Gunner Anderson transferred.

Alfred Anderson spent February to June 1917 completing Basic Training in Australia. On the 14th of June 1917, he and the rest of the contingent of

Reinforcements embarked from Sydney, New South Wales, as part of Convoy No 32 bound for England. They sailed aboard HMAT A20 *Hororata*, which carried 1,744 troops. The entire convoy of eight ships transported a total of 6,406 troops on the voyage. They disembarked in England on the 26th of August 1917 and were transported to Fovant on the Salisbury Plain for further training.

On the 1st of January 1918, the Reinforcements crossed to France, and seven days later they joined the 41st Battalion in the field. Belgium was the focus of the 41st Battalion 's activities when Private Anderson joined the unit. The Battalion alternated between serving in the rear regions and the front line.

In March 1918, during the German Army's last significant offensive, the 41st Battalion was quickly deployed to southern France to block the enemy advance towards the vital railway junction of Amiens. During the Allied offensive on the 8th of August 1918, the 41st Battalion actively participated in both the initial assault and the sustained advance extending from August into September of the same year.

Private Anderson was severely wounded (shell wound to the abdomen) in action on the 1st of September 1918 and admitted to the 9th Australian Field Ambulance and transferred to the 53rd Casualty Clearing Station on the same day. He was evacuated and admitted to the War Hospital England, and then transferred to the 1st Australian Auxiliary Hospital, Harefield.

It was here that Private Alfred Anderson died from primary diabetes secondary coma on the 5th of October 1918. He was 22 years, six months old at his time of death and had served in the AIF for one year and eight months. He had been a member of the 41st Battalion for nine months.

His death was notified in the 437th list of Australian Casualties released on the 29th of October 1918. Private Alfred Anderson was buried at Harefield Parish Churchyard; however, his final resting place was recorded as Cowley (St. Lawrence) Churchyard (Australian Section, Grave No. 70), Middlesex, England.

The following was noted in Private Alfred Anderson's military record:

"REMARKS OF INTEREST TO NEXT OF KIN:

Deceased was accorded a Military funeral. Wreath was given by Mr. Billyard Leake of Harefield. Firing Party, and bugler from Headquarters A.I.F. London were in attendance. Captain J. Murray (MC) was in charge, and a small number of the hospital staff attended. Two friends also attended. No 15497 Dvr W.T. Evans and one other. Chaplain Reverend T. Terry officiated."

Alfred Alexander Anderson's name is inscribed on The Australian War Memorial Roll of Honour (Panel 133), The City of Townsville WW1 Honour Roll, The Townsville Cenotaph, and the Townsville Railway Station Roll of Honour.

3416 Private Alfred Alexander Anderson served in WW1 on the Western Front. He was awarded the British War Medal (no: 57569) and the Victory Medal (no: 56562). These medals, as well as a memorial scroll and a memorial plaque, were passed to his father.

6284 PRIVATE ROBERT JAMES BRADSHAW (1895 – 1918)

19th Infantry Battalion

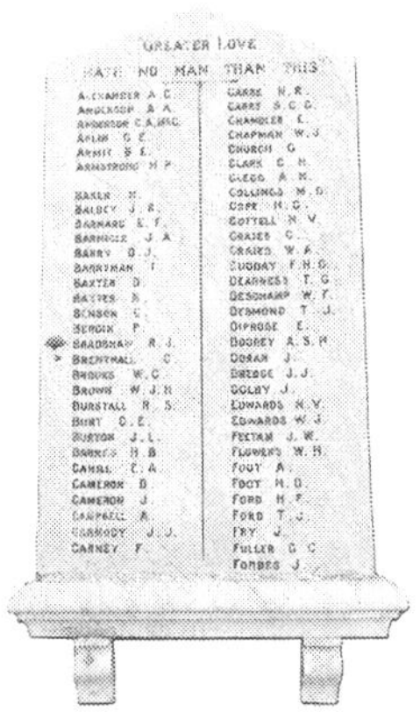

Robert James Bradshaw was born at Glebe, Sydney, New South Wales in February 1895. He was the second of three boys born to James and Mary Jane Bradshaw (née Bedford). James Bradshaw was a carter, and the family lived in Rockdale, NSW, in 1901. Robert Bradshaw grew up in Sydney and attended St. Benedict's School. After his school years were completed, he gained employment as a horse driver. Tragedy struck in 1906 when Mary Jane Bradshaw died. Robert Bradshaw was 11 years old when his mother passed away.

On the 31st of October 1916, Robert Bradshaw applied at the Show Ground Camp in Sydney to join the Australian Imperial Force (AIF). He declared that he was 21 years, eight months old, single, qualified as a horse driver, and that he had no previous military experience. His enlistment medical examination noted that he was 5'7¼" (171 cm) tall, weighed 142 lbs (64.4 kg), had a fair complexion, brown eyes, and brown hair. He identified as Roman Catholic, and he listed a friend, Mrs. Mary McCann, as his next of kin, giving her address as Otford, NSW. At the time of Robert Bradshaw's enlistment, his father was living in Little Street, Belgian Gardens, Townsville.

Accepted into the AIF, Robert Bradshaw was allocated service number 6284, assigned to the Infantry, and posted to the 19th Battalion, 18th Reinforcements. The 19th Battalion was raised at Liverpool in NSW in March 1915 as part of the 5th Brigade and was fighting on the Western Front when Robert Bradshaw enlisted.

After a short initial training of less than two weeks, Private Robert Bradshaw and the rest of the reinforcements embarked for England on the 11th of November 1916, as part of Convoy No. 27. They sailed aboard the A29 *Suevic*, which carried 1,593 troops. The entire convoy of 13 ships carried 15,278 troops and 311 other passengers on the voyage. Private Bradshaw contracted measles on the voyage and was offloaded at Cape Town and admitted to the hospital on the 14th of December 1916. He stayed in Cape Town until the 13th of January 1917 when he boarded HMAT *Berrima* to continue the voyage to England. After disembarking in Devonport, England, on the 16th of February 1917, Private Bradshaw was moved to the 5th Training Battalion at Rollestone on Salisbury Plain in Wiltshire. In April 1917, Private Bradshaw was charged with the offence of failing to salute an officer. He was found guilty and confined to barracks for three days.

Training completed; Private Bradshaw crossed to France to join his unit on the 3rd of May 1917. After a few days of further training at Étaples in France, he marched into the 19th Battalion on the 9th of May 1917 at Fricourt, Somme. The 19th was in the process of reorganizing after having taken part in the Second Battle of Bullecourt on the 3rd – 4th of May 1917. The Battalion took part in two more major battles before the year was out, Menin Road (20th – 22nd September) and Poelcappelle (9th – 10th October) in Belgium.

It was in this second action at Poelcappelle that Private Robert Bradshaw was Wounded in Action by gunshot wounds to both thighs and the left leg. He was evacuated back to London and eventually admitted to the 3rd Southern General Hospital, Oxford, for treatment. After recovering from his wounds,

Private Bradshaw was granted furlough in England from the 10th – 24th December 1917. He decided to extend his furlough for a further four days by going absent without leave. When he reported back to the 4th Command Depot at Hurdcott on the 28th of December, he was charged. Private Robert Bradshaw was found guilty and awarded three days of Field Punishment No. 2, which consisted of heavy labouring duties while, possibly, being restrained in handcuffs or fetters. He also forfeited eight days' pay.

On the 1st of February 1918, Private Bradshaw crossed back to France, and on the 7th of February 1918, he re-joined the 19th Battalion at Senighem in Northern France, where they were undergoing training. He was only back for a short time before he was admitted to the hospital suffering from scabies. He was released and returned to the battalion on the 16th of March 1918.

On the 23rd of July 1918, the 19th Battalion was deployed at Villers-Bretonneux and was in the process of relieving the 20th Battalion in the front line. It was during this activity that Private Robert Bradshaw was wounded a second time, this time by gas. He was evacuated to the 2nd Stationery Hospital at Abbeville, Somme. He recovered about six weeks after he was admitted to the hospital and was discharged back to his battalion on the 6th of September 1918. The 19th Battalion was located at Frise, about 30 kms from Amiens, when Private Bradshaw rejoined the unit.

In the spring of 1918, the last great German offensive was launched, and it was the 19th Battalion that played a key role in halting its progress. After the failure of this final effort by the enemy, the 19th Battalion fought in a sequence of battles that gradually pushed the German Army towards their ultimate surrender. Key encounters included the Battle of Amiens on the 8th of August 1918, the assault on Mont St Quentin on the 31st of August 1918, and the successful forcing of the Beaurevoir Line around Montbrehain on the 3rd of October 1918.

It was during this last action that Private Robert Bradshaw was gassed and wounded in the face on the 3rd of October 1918. He was evacuated to the 50th Casualty Clearing Station on the 4th of October but Died of Wounds (gas poisoning) on the 5th of October 1918. He was 23 years, eight months old and had served with the AIF for one month short of two years. He served with the 19th Battalion for 17 months. Private Robert Bradshaw is laid to rest at the Tincourt New British Cemetery, France. His death was notified in the 447th list of Australian casualties released on the 28th of November 1918.

Robert Bradshaw's name is inscribed on the Australian War Memorial Roll of Honour (panel 88), the City of Townsville WW1 Honour Roll, and the Townsville Cenotaph.

6284 Private Robert James Bradshaw was awarded the British War Medal (no: 52764) and the Victory Medal (no: 52003). These medals, as well as a memorial scroll and a memorial plaque, were passed to his father.

The following tributes were published in the *Daily Telegraph* (Sydney) on 24th October 1918:

BRADSHAW. — Died of wounds, in France, October 5th, 1918, Pte Robert James Bradshaw, after 3 years' active service, 19th Battalion.

My friend is sleeping his last long sleep,

And his grave I may never see:

But some gentle hand in that distant land

May scatter some flowers for me.

Inserted by his sorrowing friend, Gladys M'Cann.

BRADSHAW. — In sad and loving memory of Pte R. J. Bradshaw, 19th Battalion, died of wounds, in France, October 5, 1918.

> We pictured our friend returning,
> And longed to clasp his hand:
> But God has postponed the meeting,
> It will be in a better land.

Inserted by his loving friends, Mrs. M'Cann and Ned.

1705 PRIVATE EDGAR GREEN (1891 – 1918)

15th Infantry Battalion

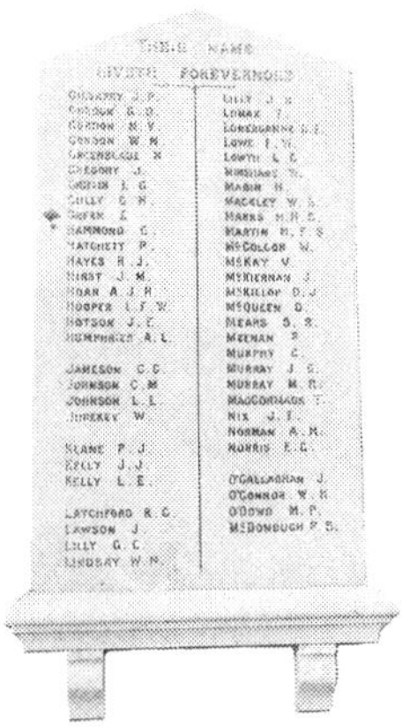

Edgar Green was born in February 1891 in Clapham, London. He was the youngest son of Tom and Matilda Green (née Jennings). He grew up in England and attended the Emmanuel School, Wandsworth, London. After his school years were finished, Edgar Green went on to study as a clerk. In 1912, he emigrated to Australia, settled in Brisbane, and then Townsville. He lived at a property on Ninth Avenue, South Townsville, and worked as a stenographer clerk.

On the 15th of January 1915, Edgar Green applied at Townsville to enlist in the Australian Imperial Force (AIF). He declared that he was single, 23 years and 11 months old, qualified as a Stenographer, and that he had no previous military experience. Edgar Green's medical assessment noted that he was 5'6" (167 cm) tall, weighed 8st 5lbs (53kg), had a dark complexion, brown eyes, and dark hair. He identified as Church of England. Edgar Green's parents were both deceased when he enlisted, and he nominated his brother, Tom Green, who was living in Sydney, as his next of kin.

Accepted into the AIF, Edgar Green was allocated service number 1705, assigned to the Infantry, and posted to the 15th Battalion, 4th Reinforcements. The 15th was at sea, on the way to Egypt, when Edgar Green enlisted. He was promoted to Corporal on enlistment; however, he reverted to Private when he eventually joined his unit in the field in June 1915. After three months of basic training, Private Edgar Green and his fellow reinforcements embarked from Brisbane, bound for Egypt, on the 8th of April 1915. They joined Convoy No.4 and sailed aboard HMAT A15 *Star of England*, which carried 507 troops and 560 horses. The entire convoy of five ships carried a total of 3,289 troops and 2,124 horses on the voyage.

The reinforcements disembarked in Egypt in May 1915 and joined the 15th Battalion on the Gallipoli Peninsula on the 2nd of June 1915. Between May and August 1915, the 15th Battalion was extensively engaged in establishing and defending the ANZAC beachhead. In August 1915, the 4th Brigade launched an assault on Hill 971, which was captured by the Australians at a considerable price. However, the arrival of Turkish reinforcements compelled the Australians to abandon their position and withdraw.

It was after this action that Private Edgar Green was reported missing on the 16th of August 1915. Two weeks later, on the 30th of August 1915, a cable was received from the United States Ambassador in Constantinople (now Istanbul). The Ambassador reported that Private Edgar Green was a Prisoner of War in hospital in Constantinople, suffering from a gunshot wound to the left leg. Private Green was reported as captured in the 78th list of Australian Casualties released on the 17th of September 1915.

Information continued to be passed by the Red Crescent about Private Green. In February 1917, Private Green was transferred from Bilemedik to Afion Karahissar Camp, which was the principal concentration camp of Australian prisoners of war in Turkey. In September 1917, word was received that Private Green was again transferred, this time from Afion Karahissar Camp to Yozgad prison camp, which was in the Vilayet of Angora in central Anatolia in Asia Minor. Prisoners faced a serious influenza epidemic in Yozgad in 1918. Private Edgar Green was one of the casualties of the outbreak when he contracted the Spanish Influenza and Died of Disease on the 10th of October 1918. He was 27 years and nine months old and had been a member of the AIF for three years and eight months. He had been a POW for three years and two months.

Private Edgar Green was buried at Yozgad; however, in 1927, his body was re-interred *"with every measure of care and reverence"* in Baghdad (North Gate) War Cemetery (Plot XXI, Row G, Grave No. 2), Iraq. Edgar Green's name is inscribed on The Australian War Memorial Roll of Honour (panel 76), The City of Townsville WW1 Honour Roll, and The Townsville Cenotaph.

1705 Private Edgar Green served in WW1 at Gallipoli. He was awarded the 1914/15 Star (no:15724), the British War Medal (no:9987), and the Victory Medal (no:9944). A memorial scroll and a memorial plaque were passed to his brother, Tom, in 1922.

The following notice was published in the *Brisbane Courier* on 10th October 1919:

GREEN. —In memory of the late Corporal Edgar Green, late 15th Battalion, A.I.F., who died in Yozgad, Turkey, after three years and two months prisoner of war.

Better love hath no man than to lay down his life for his country.

Inserted by his sincere friends, Mr. and Mrs. A. D. Brown and family, of Sandgate, Queensland.

LIEUTENANT WALTER HENRY FLOWERS (1884 - 1918)

49th Infantry Battalion

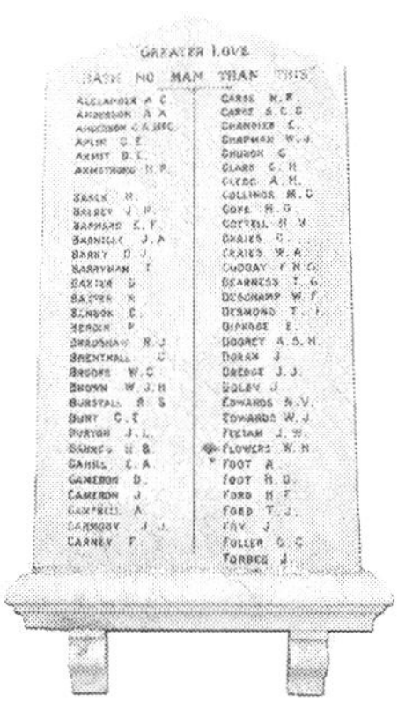

Walter Henry Flowers was born on the 7th of January 1884 in Townsville, Queensland. He was the only child of William and Mercy Flowers (née Richardson). William Flowers worked as a labourer, and Mercy Flowers served as a nurse. The Flowers family lived in a property on Norman St in Townsville. William Flowers Snr. passed away in England in 1920.

Growing up in Townsville, Walter Flowers Jnr. attended Townsville West State School, and after completing his school years, worked as a grocer. On the 11th of February 1909, he married Edith Elsie Ruth Mundy, and they made their home on Margaret Street in West End.

On the 20th of September 1915, at the age of 32, Walter Flowers applied to enlist in the Australian Imperial Force (AIF) in Townsville. He declared that he was married, employed in mercantile pursuits, and that he had prior military service as a member of the Australian Garrison Artillery. He also declared his rank as a Lieutenant in the 2nd Infantry, Kennedy Regiment. Walter Flowers' medical examination noted that he was 5'4½" (164 cm) tall, weighed 152lbs (69kg), with a fair complexion, blue eyes, and light brown hair. He identified as Methodist and designated his wife Elsie as his next of kin, with her address as c/- Mrs. K. Wall, "Warriston," Petrie Terrace, Red Hill, Brisbane.

Walter Flowers' previous service and rank led to his appointment as a 2nd Lieutenant (2Lt) with the 49th Battalion, 3rd Reinforcements in April 1916. The 49th Battalion was established in Egypt on the 27th of February 1916, following the expansion of the AIF. Its composition included Gallipoli veterans from the 9th Battalion, making up approximately half of its members, and new recruits from Australia forming the remaining half. Given that the 9th Battalion was primarily composed of soldiers from Queensland, the majority of the 49th Battalion were recruited from the same state. In June 1916, the 49th Battalion departed from Alexandria in Egypt and embarked for the Western Front to join the British Expeditionary Force.

Following Walter Flowers' enlistment, eight months of initial training was completed in Australia until the 1st of May 1916, when he set sail with the rest of the Reinforcements from Brisbane, bound for Egypt. They joined Convoy No. 22 and sailed aboard HMAT A46 *Clan MacGillivray,* carrying 713 troops from Brisbane and an additional 326 from Sydney. The entire convoy of 13 ships transported a total of 8,403 troops during the voyage. They disembarked in Egypt on the 13th of June 1916 and engaged in two months of training until the 5th of August 1916 when 2Lt Flowers and the rest of the reinforcements sailed from Alexandria, heading for England to join the British Expeditionary Force.

Upon arriving in England, 2Lt Walter Flowers was assigned to the 13th Training Battalion at Rollestone for pre-deployment training before transferring to the Reinforcement Camp at Étaples, France, on the 16th of October 1916. Tragedy struck on the 24th of October 1916 when 2Lt Walter Flowers' wife, Edith, passed away in Brisbane. Just five days after this sad event, 2Lt Flowers joined the 49th Battalion in the field.

Before 2Lt Flowers joined the battalion, the Battle of Mouquet Farm in August and early September had had a severe impact on the 49th, resulting in the loss of 14 officers and 417 other ranks who were either killed or wounded. Consequently, the Battalion was withdrawn from participating in significant offensives for the remainder of the year. Instead, they rotated through the front lines near Ypres and later returned to the Somme. During this period, their activities included patrolling, conducting raids, engaging in training, and performing manual labour in the rear areas, interspersed with periods of rest.

On the 13th of January 1917, 2Lt Flowers was detached to a School of Instruction, rejoining the 49th Battalion on the 9th of February 1917. Just five days after his return, he fell ill and was evacuated to the 45th Casualty Clearing Station. From there, he was transferred by Ambulance Train No. 11 and admitted to the 9th General Hospital at Étretat, near Le Havre, on the 19th of February 1917, suffering from tachycardia. 2Lt Flowers was discharged to duty on the 19th of March 1917.

Following his release from the hospital, 2Lt Flowers was posted to 4th Australian Divisional Base Depot at Étaples, where he remained until May 1917 before rejoining the 49th Battalion. The 49th Battalion participated in the Battle of Messines on the 9th of June 1917 and the Battle of Polygon Wood on the 26th of September 1917. 2Lt Flowers was promoted to Lieutenant in June 1917 after the Battle of Messines.

Another winter of trench routine followed. With Russia's internal disintegration in October 1917, expectations grew of a major German offensive on the Western Front in early 1918. Towards the end of March 1918, this offensive began in France, prompting the 4th Division to shift its forces to defend and hold positions around Dernancourt on the River Ancre. On the 5th of April 1918, Lieutenant Walter Flowers saw significant action as the 49th Battalion played a crucial role in repelling a large German attack and later launched a critical counter-attack that afternoon. The German threat persisted until late April, when Australian troops from the 13th and 15th Brigades successfully executed a counter-attack, driving the enemy out of Villers-Bretonneux. During this assault, the 49th Battalion remained in reserve.

In July 1918, Lieutenant Walter Flowers was detached to Merricourt as the Town Major. While on leave in October 1918, he visited his aunt in Peterborough, England. Unfortunately, he fell ill with influenza and bronchitis during his leave and was admitted to the Peterborough Infirmary, but tragically did not recover. He Died of Disease on the 25th of October 1918. He was laid to rest with full military honours at the Fletton Cemetery on the 30th of October 1918 (Plot number NC. 423). Lieutenant Walter Flowers was 35 years old and had been a member of the AIF for three years and one month. His death was reported in the 448th/449th list of Australian Casualties released on the 30th of November 1918.

Walter Flowers' name is inscribed on The Australian War Memorial Roll of Honour (panel 148), The Townsville West State School Memorial Gate of Honour, The City of Townsville WW1 Honour Roll, and The Townsville Cenotaph. Flowers Street, Railway Estate, Townsville, is named in his memory. Lieutenant Walter Henry Flowers served in WW1 on the Western Front. He was awarded the British War Medal (no:37680) and the Victory Medal (no:37375). A memorial scroll and a memorial plaque were presented to his mother in 1922.

The following notice was published in the *Townsville Daily Bulletin* (Qld) on the 29th of October 1918:

DEATH

FLOWERS. — Lieutenant Walter Henry on Active Service, on 25th October, 1918, at his aunt's residence, Fletton, Peterboro, England. Aged 35 years.

Deeply regretted.

The following article was published in the *Townsville Daily Bulletin* (Qld) on the 14th of January 1919:

AUSTRALIAN OFFICER'S FUNERAL

Lieut. Walter Henry Flowers, Australian Imperial Force, a nephew of Mrs. Groves, Fletton Avenue, Peterborough, who died in Peterborough infirmary from pneumonia following influenza, was buried with full military honours in old Fletton cemetery on Wednesday afternoon. Lieut. Flowers (says a Fletton paper) was a native of Queensland, Australia. He was stationed at Northampton, and when on leave was in the habit of visiting his aunt Mrs. Groves. A week last Wednesday he went to see her, intending to leave the following day, but he had a slight cold which developed into influenza, and consequently he was unable to travel. On the Thursday following, he was removed to Peterborough Infirmary where he unfortunately passed away the next morning. He was 35 years of age. The deceased had been in France two years and seen much active service during that period, and quite recently had been awarded military honours, including the Military Medal. The funeral service, the first part of which was held in the Cemetery Chapel, was conducted by the Rev. R.C. Frew, military chaplain. Many sympathizers were present. The chief mourners were Mr. E. Flowers (uncle), Mrs. Webb (Aunt), Mrs. Washbourne (Aunt), Mrs. Clarke (Cousin), Mrs. Johnson (Cousin), Mrs. Tee, and Mr. and Mrs. R. R. Coggan. Among those in attendance were Capt. G. D. Coleman (O. C. troops at Peterborough), Lieut. J. M. Woodthorpe, Lieut A. W. Keeble, Lieut. J. T. Dickinson (commanding Fletton Platoon of the Hunts, Volunteers), Corporal Edmondson (representing Australian Headquarters), C.S.M. Vittery (musketry instructor to the 1st Hunts. Volunteers Regt.). The firing party was provided by the Fletton section of the Hunts Volunteers, under Sergt. Major Gibson R.D.C., and the following N.C.Os. were present: Sergt. Medlock, Sergt. Briddell, Corporals Hawkins, Rowells, and Petts. Bugler Munson of the R.D.C. sounded the "Last Post." The coffin bore the inscription "Lieut. Henry Walter Flowers, 9th Battalion, A.I.F. 1918, aged 34 years." The floral tribute bore cards as follows: "With dearest love from Auntie Groves and Auntie Washbourne," "With our fond love to our dear nephew, from Aunt Eliza and Uncle Will." In tender memory of our dear Wal from Merci, Walter, and Mona," "With Nancy's love," "In loving memory of a dear one from Rose and Bert," "With deepest sympathy from his brother officers," "In remembrance from Mr. and Mrs. Coggan," "In loving memory of dear Walter from Mr. and Mrs. Tee and family," 197 Lincoln-street Peterborough.

(N.B. There is no record of the award or recommendation for the award of a Military Medal for Lieutenant Walter Henry Flowers).

3069 SAPPER PATRICK JAMES KEANE (1899 - 1918)

AIF Railway Unit

Patrick James Keane was born on the 28th of November 1899 at Ayr, Queensland. He was the youngest child of Michael and Sarah Keane (née Kneeshaw) and had two brothers and one sister. In 1902, when Patrick was three years old, Michael Keane passed away in Bowen, Queensland. Subsequently, Sarah Keane relocated the family to Townsville. In May 1910, she married Charles Wakeford, who worked as a storeman. The family lived in a property on Surrey St in Hyde Park, and Patrick Keane attended Mundingburra State School. Once his school years were completed, Patrick Keane worked as a Telegraphist with the Railways.

On the 28th of May 1918, Patrick Keane applied in Townsville to enlist in the Australian Imperial Force (AIF). His two brothers had already enlisted. He stated that he was 18 years and six months old, single, and had previous service with the 2nd Infantry, a unit of the Citizen Forces. His enlistment medical examination noted that he was 5'10" (178 cm) tall, weighed 144 lbs (65 kg), had a fresh complexion, brown eyes, and brown hair. He identified as Roman Catholic and he nominated his mother, Mrs. Sarah Wakeford, as his next of kin, giving her address as Surrey St, Hyde Park, Townsville.

Accepted into the AIF, Patrick Keane was allocated service number 3069, assigned to the Engineers, and posted to the AIF Railway Companies, May 1918 Reinforcements. In 1915, the Railway Operating Division (ROD) was formed under the Royal Engineers to oversee railway operations in various war zones. Staffed primarily by seasoned railway workers, the ROD managed both standard and narrow-gauge railways. Its affiliated companies were crucial in transporting troops and essential supplies to and from the front lines.

After completing six weeks of basic training in Australia, Sapper Patrick Keane and his fellow Reinforcements departed from Sydney, bound for England, on the 17th of July 1918. They joined Convoy No. 41 and sailed aboard HMAT

A30 *Borda*, which carried 1,419 troops. The entire convoy, consisting of eight ships, transported a total of 3,728 troops, 157 naval personnel, and 180 civilians during the voyage. They disembarked in London at the end of September 1918 and marched into the Australian Railway Operating Division Depot at Longmoor in Hampshire.

Sapper Keane fell ill in October 1918 and was admitted to Hospital in Aldershot on the 14th of October. Tragically, he succumbed to influenza and pneumonia on the 27th of October 1918, just two weeks before the war's end, at Frensham Hill Hospital. He was one month short of his nineteenth birthday when he passed away and had been a member of the AIF for five months. He was laid to rest at Brookwood Military Cemetery (Plot IV, Row F, Grave No. 20), Surrey, England, with full military honours. Sapper Patrick Keane's death was reported in the combined 448th and 449th list of Australian Casualties released on the 30th of November 1918.

Patrick Keane's name is inscribed on the Australian War Memorial Roll of Honour (panel 26), The Townsville Railway Station Roll of Honour, The Townsville West State School Memorial Gate, and the Townsville Cenotaph. Sapper Patrick Keane was the last of the soldiers whose names are inscribed on the Townsville Cenotaph to die and one of the youngest. Keane Street in Currajong, Townsville, is named in his memory and in recognition of his two brother's service in WW1.

3069 Sapper Patrick James Keane served in WW1 and was awarded the British War Medal (no:59776). This medal, along with a memorial scroll and a memorial plaque, was presented to his widowed mother in 1922.

Tributes to Patrick Keane were published in The *Townsville Daily Bulletin* on the 27th of October 1919:

KEANE. — In loving memory of our dear son and brother, Private Patrick James Keane, Railway Operation Division, who died on active service on October 27th, 1918. (Inserted by his father, mother, sisters, and brothers.)

KEANE. — In loving memory of Private Patrick Keane, who died whilst on active service on the 27th of October 1918. (Inserted by his loving comrades, T. and G. Richmond.)

PART 5 – AFTER THE WAR

World War I left an enduring legacy on Townsville and North Queensland, shaping its people, culture, and spirit. Despite being thousands of miles from the front lines, the war had a profound impact on this small city, altering its social landscape, commemorative practices, and community structures.

The Great Sacrifice and Loss

Townsville's commitment to World War I was significant, echoing the widespread dedication across Australia. With a population of 17,000, Townsville contributed substantially to the war effort as young men enlisted to defend the British Empire. Half of all eligible Australian men enlisted, and many cities, including Townsville, faced a disproportionate toll. Of the 1,786 men from Townsville who volunteered, 172 never came home—representing almost 10% of the town's enlisted soldiers.

For the community, this meant profound loss and mourning that continued long after victory celebrations. Families of those who didn't return faced an enduring period of grief and upheaval. For many who returned, reintegration was challenging. Some soldiers, physically or psychologically wounded, struggled to find work or faced isolation due to their disabilities, creating a lasting impact on the social fabric of the town.

Political Divides and the Conscription Debate

Conscription was among the most divisive issues during the war, impacting Townsville as it did the rest of Australia. Two referendums were held, in 1916 and 1917, to decide if compulsory military service should be introduced to sustain the war effort. Both were defeated, highlighting a deep societal rift. Townsville voted "No" both times, reflecting a mix of views: some saw conscription as a betrayal of freedom, while others believed it necessary for the Empire's defence. This divide affected relationships within the city, as families, friends, and neighbours found themselves on opposing sides.

Honouring the Fallen: Townsville's Cenotaph and Post-War Remembrance

After the war, towns across Australia erected monuments to honour those who served and died. In 1924, Townsville unveiled its Soldiers Memorial, later known as the Cenotaph—a clock tower designed by local masons Melrose and Fenwick to commemorate World War I soldiers. Funded by public donations, it quickly became a central site for memorial events, including Anzac Day.

In 1932, the area surrounding the memorial was named Anzac Memorial Park, and in 1934, a memorial arch was added at the entrance, marking the park as a sacred space for remembrance. Over time, the Cenotaph grew to represent all Australians who served in subsequent conflicts. Encircling the memorial are 27 small plinths with bronze plaques, each a quiet tribute to those who served.

Social and Economic Challenges for Townsville

The end of World War I brought a period of social and economic hardship for Townsville. Many veterans struggled to resume their pre-war lives, and the city faced the challenge of providing adequate support for those with lasting physical and mental injuries. Organizations like the Returned Soldiers and Sailors Imperial League of Australia (RSL) emerged to support returning servicemen, advocating for their needs and creating social networks to aid in reintegration.

One notable initiative was the Scartwater Trust, a pastoral project established by A.H.W. Cunningham and local graziers to support ex-servicemen and their families. Using land along the Suttor and Belyando Rivers, the trust developed a cattle operation, with profits directed toward scholarships for veterans' children and community causes. By 1927, it was financially self-sustaining, providing annual contributions to the Townsville branch of the RSL. This initiative demonstrated North Queensland's pragmatic approach to supporting veterans, focusing on sustainable, community-led solutions over traditional memorials.

The Legacy organisation was established in Australia in 1923 to support the families of soldiers who died or were injured during World War I. Founded by ex-servicemen, it aimed to provide financial aid, education, and care for widows and children, offering a compassionate network for those profoundly affected by war. In Townsville, a Legacy group operated from 1947 under the guidance of the parent body in Brisbane. In 1948 the local Group become a fully-accredited Legacy organisation which still operates today. The basic objective of. Legacy is to care for widows, sons and daughters of comrades who have passed on.

The Enduring Legacy of World War I in Townsville

World War I's legacy in Townsville is deeply embedded. Memorials like the Cenotaph serve as constant reminders of the sacrifices made by a generation of Australians. Each year, Anzac Day and Remembrance Day gatherings in Townsville reflect the toll the war took on the community and the resilience that emerged in its wake. The story of the Kennedy Regiment is remembered as a symbol of Townsville's commitment to the Empire's defence.

On a broader level, World War I altered Townsville's identity, aligning it with a growing sense of national unity and shared sacrifice. The physical and emotional costs of the war fostered a culture of remembrance and resilience. Townsville learned to honour its fallen while supporting those who returned, forever changed.

Today, the echoes of World War I are woven into the city's landscape, from Anzac Memorial Park to street names that pay homage to the fallen. Though the war ended more than a century ago, its impact on Townsville remains a vital part of its history and identity.

> They shall grow not old, as we that are left grow old:
> Age shall not weary them, nor the years condemn.
> At the going down of the sun and in the morning
> We will remember them.

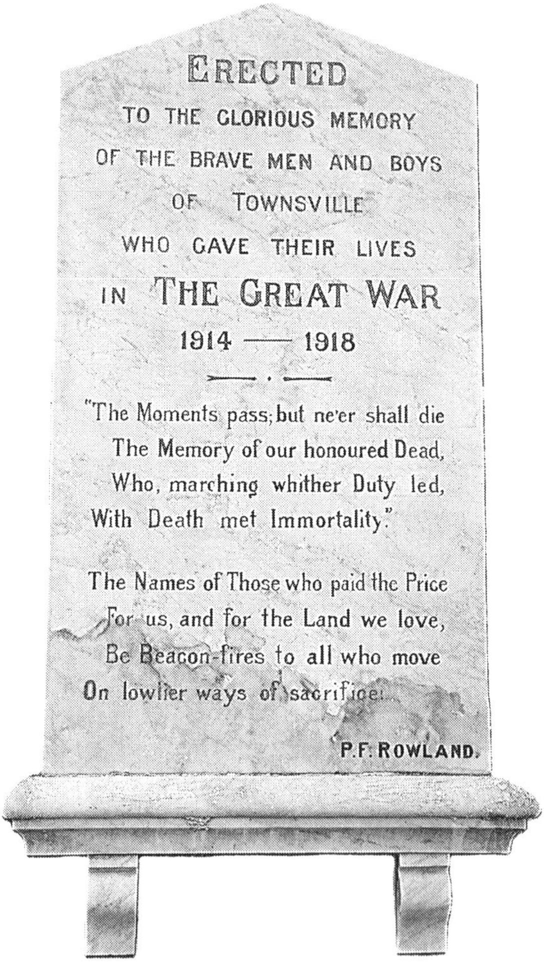
ERECTED
TO THE GLORIOUS MEMORY
OF THE BRAVE MEN AND BOYS
OF TOWNSVILLE
WHO GAVE THEIR LIVES
IN THE GREAT WAR
1914 — 1918
"The Moments pass; but ne'er shall die
The Memory of our honoured Dead,
Who, marching whither Duty led,
With Death met Immortality."
The Names of Those who paid the Price
For us, and for the Land we love,
Be Beacon-fires to all who move
On lowlier ways of sacrifice.
P.F. ROWLAND.

ABOUT THE AUTHOR

LIEUTENANT COLONEL GARY RICHARD EWART (RETD.)

Gary Ewart joined the Australian Regular Army on 27 July 1971. After completing recruit training, he was assigned as a Private Soldier to the Royal Australian Army Medical Corps. Rising steadily through the ranks to Warrant Officer Class 2, he was selected for long-term schooling and commissioned as a Lieutenant on 25 August 1986.

As a commissioned officer, Lieutenant Colonel Ewart undertook a variety of roles across multiple units and headquarters, including operational service in East Timor. He also served as the Commanding Officer of a Combat Services Support Battalion. After an exemplary career, Lieutenant Colonel Ewart retired from the Defence Force in 2018.

BIBLIOGRAPHY

OFFICIAL RECORDS

Australian War Memorial, Canberra

AWM 4: Australian Imperial Force unit war diaries, 1914-18 War

AWM 38: Official History, 1914-18 War: Records of C E W Bean Official Historian

National Archives of Australia, Canberra

Series B2455: Attestation Paper of Persons Enlisted for Service Abroad

BOOKS AND BOOK CHAPTERS

Bolton G., A Thousand Miles Away, A history of North Queensland to 1920 ANU Press between 1965–1991

Carlyon L., Gallipoli, Pan Macmillan, Sydney, 2001

- The Great War, Pan Macmillan, Sydney, 2006

Emery M., They rode into History: The Story of the 8th Light Horse Regiment Australian Imperial Force 1914-1919, Slouch Hat Publications, 2009

Grey J., A Military History of Australia, 3rd edition, Cambridge University Press, 2008.

Roberts C., The Landing at ANZAC 1915 (Australian Army Campaigns Series Book 12), Big Sky Publishing 2013

Grainger J., The Battle for Palestine 1917, The Boydell Press 2006

Edgar P., Counter Attack Villers-Bretonneux (Australian Army Campaigns Series Book 27), Big Sky Publishing 2020

Lee R., The Battle of Menin Road 1917 (Australian Army Campaigns Series Book 20), Big Sky Publishing 2018

Passlow J., The Battle of Polygon Wood 1917 (Australian Army Campaigns Series Book 19), Big Sky Publishing 2018

Deayton C., The Battle of Messines 1917 (Australian Army Campaigns Series Book 18), Big Sky Publishing 2017

Coombes D., The Battles of Bullecourt 2017 (Australian Army Campaigns Series Book 19), Big Sky Publishing 2017

Westerman W., The Battle of Broodseinde Ridge 1917 (Australian Army Campaigns Series Book 21), Big Sky Publishing 2017

Mathew J., Highways and Byways, The Origin of Townsville Street Names Townsville Library Service 2008

Wildman O., Queenslanders who fought in the Great War, Besley and Pike Ltd, 1919

Tregarthen G., Sea Transport of the AIF, Naval Transport Board, c 1919

Larard C. & Newnham M., The Townsville Cenotaph World War 1 Soldiers Hastings Printers / Allclear Print and Signs 2017

ONLINE RESOURCES

Ancestry.com.au

Australian War Memorial: www.awm.gov.au

National Archives of Australia: www.naa.gov.au

Trove (digitised newspaper reports and images), National Library of

Australia: www. trove.nla.gov.au

NEWSPAPERS

Argus (Melbourne, Vic.)

Australasian (Melbourne, Vic.)

Brisbane Courier (Qld.)

Bundaberg Mail (Qld.)

Burnett Advertiser (Qld.)

Cairns Post (Qld.)

Catholic Press (Sydney, NSW)

Charters Towers Evening Telegraph (Qld.)

Croydon Mining News (Qld.)

Daily Mercury, Mackay (Qld.)

Daily Telegraph (Sydney, NSW)

Lithgow Mercury (NSW)

Northern Herald (Cairns, Qld.)

Northern Miner (Charters Towers Qld.)

Queensland Times (Ipswich Qld.)

Queenslander (Brisbane, Qld.)

Sun (Sydney)

Sydney Morning Herald (NSW)

Tamworth Daily Observer (NSW)

Telegraph, (Brisbane, Qld.)

Townsville Daily Bulletin (Qld.)

INDEX

Made in United States
Orlando, FL
05 November 2026

90229882R00284